T0139823

Studies in Systems, Decision and Control

Volume 94

Series editor

Janusz Kacprzyk, Polish Academy of Sciences, Warsaw, Poland
e-mail: kacprzyk@ibspan.waw.pl

About this Series

The series "Studies in Systems, Decision and Control" (SSDC) covers both new developments and advances, as well as the state of the art, in the various areas of broadly perceived systems, decision making and control- quickly, up to date and with a high quality. The intent is to cover the theory, applications, and perspectives on the state of the art and future developments relevant to systems, decision making, control, complex processes and related areas, as embedded in the fields of engineering, computer science, physics, economics, social and life sciences, as well as the paradigms and methodologies behind them. The series contains monographs, textbooks, lecture notes and edited volumes in systems, decision making and control spanning the areas of Cyber-Physical Systems, Autonomous Systems, Sensor Networks, Control Systems, Energy Systems, Automotive Systems, Biological Systems, Vehicular Networking and Connected Vehicles, Aerospace Systems, Automation, Manufacturing, Smart Grids, Nonlinear Systems, Power Systems, Robotics, Social Systems, Economic Systems and other. Of particular value to both the contributors and the readership are the short publication timeframe and the world-wide distribution and exposure which enable both a wide and rapid dissemination of research output.

More information about this series at http://www.springer.com/series/13304

Alejandro Peña-Ayala
Editor

Learning Analytics: Fundaments, Applications, and Trends

A View of the Current State of the Art
to Enhance e-Learning

 Springer

Editor
Alejandro Peña-Ayala
WOLNM & Escuela Superior de Ingeniería,
  Mecánica y Eléctrica, Zacatenco
Instituto Politécnico Nacional
Mexico City, DF
Mexico

ISSN 2198-4182   ISSN 2198-4190 (electronic)
Studies in Systems, Decision and Control
ISBN 978-3-319-85029-0   ISBN 978-3-319-52977-6 (eBook)
DOI 10.1007/978-3-319-52977-6

This Springer imprint is published by Springer Nature
The registered company is Springer International Publishing AG
The registered company address is: Gewerbestrasse 11, 6330 Cham, Switzerland

Preface

Learning is the result of student's inner and outer actions with her own thoughts as well as his tutor, companions, and learning environment. In consequence, natural inquiries arise to consider the efficacy, efficiency, and the quality of the learning resources, processes, and outcomes, as well as other subtle issues such as learners' attitude, engagement, performance, behavior, attrition, cheating, and collaboration.

In addition, tutors' support, companions' influence and interaction, system scaffolding, content utility, interface friendship, and other items recreate a learning setting that bias learner's achievements. The evaluation of such factors claims specialized on–off-line data gathering procedures, huge databases, accurate models, reliable methods and techniques to examine information, powerful visualization tools, and qualified criteria to interpret knowledge and discover findings.

In this context, learning analytics (LA) arises as an emergent discipline that pursues improvement in teaching and learning by a critical evaluation of raw data and the generation of patterns that characterize learner habits, predict learner responses, and provide timely feedback. What is more, LA supports decision-making, tailors readable content, facilitates realistic assessments, and provides personal supervision of learner's progress. The goal is to scale the real-time exploitation of LA by learner, academics, and educational computer-based systems to enhance learners' accomplishments at course and individual tier.

This book shapes a glance of recent research, studies, and applications of LA in the field of education as a way to trace a conceptual and practical view of the LA field to recreate a state of the art and a vision of future trends to encourage forthcoming labor. Therefore, this book builds on recent activities in LA and presents works that report recent advances, carry out innovative explorations, and establish foundations for further research. According to the nature of the contributions accepted for this volume, the following four topics are presented in this book:

- *Reviews* highlight specific topics of interest that describe in detail a particular line of the LA arena, as well as sketch a broad landscape to define the nature, grounds, and applications of the emergent field.

- *Approaches* contribute with a particular paradigm and instrument to deal with specific issues concerned with the personalization of learning support and the exploitation of huge log data repositories.
- *Conceptual* introduces a particular topic (e.g., a priori knowledge), provides arguments to ground the concept, and explains how to apply it in LA settings with the purpose to enhance certain functionalities.
- *Applications* explains how to use and exploit LA perspectives, techniques, and approaches in order to achieve a given goal concerned to the analysis of assessment repositories and the scheduling on cloud servers.

This volume is the product of the research recently achieved by authors, who are engaged to promote their views, methodologies, results, and findings to the community of practitioners, pedagogues, psychologists, computer scientists, academics, and students interested in the emergent domain of LA!

As a result of the workflow that encompasses the submission of proposals and their respective evaluation, as well as the edition of the complete manuscript with the corresponding revision, tuning, and decision according to the Springer quality principles, nine works were approved, edited as chapters and organized according to the following sequence:

Chapter 1: Surveys LA works that apply some techniques to deal with particular issues in higher education settings. Moreover, the sample of works is organized into clusters according to the stakeholder traits to identify trends.

Chapter 2: By means of a review of related works, introduces teaching and learning analytics as a synergy between both teaching analytics and LA to transfer LA underlying elements to academics for improving teaching practice.

Chapter 3: Sketches a landscape of LA to define the nature, roots, and related domains of the field. Such scenery identifies related domains, learning paradigms, underlying elements and legal concerns, as well as approaches.

Chapter 4: Focuses on computer-based adaptive assessment and the way it can be optimized by means of LA approaches. The goal is to provide formative adaptive tests to learners by instructing the next actions to be fulfilled by user.

Chapter 5: Encourages the provision of personalized feedback and support to learners by means of a student relationship engagement system that follows a data-driven strategy that considers a holistic and human-centric view of data.

Chapter 6: Tackles the challenge of extracting meaningful information and discovering valuable knowledge from huge data sets collected from massive open online courses by using LA dashboards that facilitate the interpretations.

Chapter 7: Claims for addressing LA approaches according to theory-driven strategy that considers a priori knowledge. Thus, a two-level framework is proposed that defines LA as a meta-level process to guide five components.

Chapter 8: Aims at knowledge discovery in big data by the application of educational data mining and visualization tools. As result, clusters of students' profiles are organized and interpreted to measure the quality of education.

Chapter 9: Takes advantage of LA methods for inspiring the design of job scheduling on cloud servers. Such an approach defines that the cloud broker acts as a teacher, while local schedulers of cloud sites play as students.

I express my gratitude to authors, reviewers, the Springer editorial team, and the editors Dr. Thomas Ditzinger and Prof. Janusz Kacprzyk for their valuable collaboration to develop this work.

I also acknowledge the support given by the Consejo Nacional de Ciencia y Tecnología (CONACYT) and the Instituto Politécnico Nacional (IPN), both are Mexican Government institutions, through the grants: CONACYT–SNI-36453, CONACYT 264215, IPN-Sabbatical Leave: DG:2015–118–1–196 and CPE/PIAS/ 1357–15, IPN–SIP/DI/DOPI/EDI–888/16; IPN-COFAA-SIBE-ID: 9020/2015– 2016, IPN–SIP–20160899.

Last but not least, I acknowledge the strength given by my Father, Brother Jesus, and Helper, as part of the research projects of World Outreach Light to the Nations Ministries (WOLNM).

Mexico City, Mexico   Alejandro Peña-Ayala
November 2016

Contents

Contributors

Kathryn Bartimote-Aufflick Quality and Analytics Group, The University of Sydney, Sydney, NSW, Australia

Yolaine Bourda Université Paris-Sud, Orsay, France

Adam J. Bridgeman Faculty of Science, The University of Sydney, Sydney, NSW, Australia

Éric Bruillard ENS Cachan—Bât. Cournot, Cachan, France

Leonor Adriana Cárdenas-Robledo Escuela Superior de Ingeniería Mecánica y Eléctrica Unidad Zacatenco, Instituto Politécnico Nacional, Gustavo A. Madero, Mexico City, Mexico

Mahsa Chitsaz UNSW Australia, Kensington, NSW, Australia

Andrew Clayphan UNSW Australia, Kensington, NSW, Australia

Martin Ebner Educational Technology, Graz University of Technology, Graz, Austria

Mohammad Samadi Gharajeh Islamic Azad University, Tabriz, Iran

Tommi Kärkkäinen Department of Mathematical Information Technology, University of Jyväskylä, Jyväskylä, Finland

Mohammad Khalil Educational Technology, Graz University of Technology, Graz, Austria

Philipp Leitner Educational Technology, Graz University of Technology, Graz, Austria

Danny Yen-Ting Liu Faculty of Science, The University of Sydney, Sydney, NSW, Australia

Abelardo Pardo Faculty of Engineering and Information Technology, The University of Sydney, Sydney, NSW, Australia

Alejandro Peña-Ayala WOLNM: Artificial Intelligence on Education Lab, Iztapalapa, Mexico City, Mexico; Escuela Superior de Ingeniería Mecánica y Eléctrica Unidad Zacatenco, Instituto Politécnico Nacional, Gustavo A. Madero, Mexico City, Mexico

Fabrice Popineau Université Paris-Sud, Orsay, France

Mirka Saarela Department of Mathematical Information Technology, University of Jyväskylä, Jyväskylä, Finland

Demetrios G. Sampson Department of Digital Systems, University of Piraeus, Piraeus, Greece; School of Education, Curtin University, Bentley, Perth, WA, Australia

Stylianos Sergis Department of Digital Systems, University of Piraeus, Piraeus, Greece

Jean Simon Université de la Reunion, Saint-Denis, France

Humberto Sossa Centro de Investigación en Computación, Instituto Politécnico Nacional, Gustavo A. Madero, Mexico City, Mexico

Jill-Jênn Vie Université Paris-Sud, Orsay, France

Lorenzo Vigentini School of Education, UNSW Australia, Kensington, NSW, Australia; UNSW Australia, Kensington, NSW, Australia

Xia Zhang UNSW Australia, Kensington, NSW, Australia

Chapter 1
Learning Analytics in Higher Education—A Literature Review

Philipp Leitner, Mohammad Khalil and Martin Ebner

Abstract This chapter looks into examining research studies of the last five years and presents the state of the art of Learning Analytics (LA) in the Higher Education (HE) arena. Therefore, we used mixed-method analysis and searched through three popular libraries, including the Learning Analytics and Knowledge (LAK) conference, the SpringerLink, and the Web of Science (WOS) databases. We deeply examined a total of 101 papers during our study. Thereby, we are able to present an overview of the different techniques used by the studies and their associated projects. To gain insights into the trend direction of the different projects, we clustered the publications into their stakeholders. Finally, we tackled the limitations of those studies and discussed the most promising future lines and challenges. We believe the results of this review may assist universities to launch their own LA projects or improve existing ones.

Keywords Learning analytics · Higher education · Stakeholders · Literature review

P. Leitner (✉) · M. Khalil (✉) · M. Ebner
Educational Technology, Graz University of Technology,
Münzgrabenstraße 35A/I, 8010 Graz, Austria
e-mail: philipp.leitner@tugraz.at

M. Khalil
e-mail: mohammad.khalil@tugraz.at

M. Ebner
e-mail: martin.ebner@tugraz.at

© Springer International Publishing AG 2017
A. Peña-Ayala (ed.), *Learning Analytics: Fundaments, Applications,
and Trends*, Studies in Systems, Decision and Control 94,
DOI 10.1007/978-3-319-52977-6_1

1

Abbreviations

AA	Academic analytics
ACM	Association for computing machinery
EDM	Educational data mining
HE	Higher education
ITS	Intelligent tutoring system
LA	Learning analytics
LAK	Learning analytics and knowledge
LMS	Learning management system
MOOC	Massive open online course
NMC	New media consortium
PLE	Personal learning environment
RQ	Research question
SNA	Social network analysis
VLE	Virtual learning environment
WOS	Web of science

1.1   Introduction

The aim of LA is to evaluate user's behavior in the context of teaching and learning, further to analyze and interpret it to gain new insights and to provide the stakeholders with new models for improving teaching, learning, effective organization, and decision making (Siemens and Long 2011). A key fact is the return of the resulting knowledge to the teachers and students to optimize their teaching and learning behavior, to promote the development of skills in the area, and to better understand education as well as the connected fields, e.g. university business and marketing. Available resources can be used more efficiently to provide better support and individual care to develop potentials.

In the area of HE, LA has proven to be helpful to colleges and universities in strategic areas such as resource allocation, student success, and finance. These institutions are collecting more and more data than ever before, to maximize strategic outcomes. Based on key questions data is analyzed and predictions are made to gain insights and set actions. Many examples of successful analytics and frameworks use are available across a diverse range of institutions (Bichsel 2012). Ethical and legal issues of collecting and processing students' data are seen as barriers by the HE institutions in LA (Sclater 2014).

In this chapter, we present a literature review to evaluate the progress of LA in HE since its early beginning in 2011. We conducted the search with the three popular libraries: the LAK conference, the SpringerLink, and the WOS databases.

We then refined the returned results and settled on including 101 relevant publications. This chapter mainly contributes by analyzing them and lists the used

LA methods, limitations and stakeholders. It is expected that this study will be a guide for academicians who would like to improve existing LA projects or assist universities to launch their own.

The next section gives a short introduction on the topic of LA and describes LA in HE in detail. The subsequent sections are concerned with our research design, methodology and execution of the review. The outcomes of the research questions and the literature survey are presented in the third section. The penultimate section discusses the findings and shows the conclusion of our survey. A glance of future trends are presented in the last section.

1.2 A Profile of Learning Analytics and Learning Analytics in Higher Education

In this section we present a profile of LA in general and describe the analysis process. Further, we give emphasis to LA in HE, discuss challenges and identify the involved stakeholders.

1.2.1 Learning Analytics

Since its first mention in the New Media Consortium (NMC) Horizon Report 2012 (Johnson et al. 2012), LA has gained an increasing relevance. LA is defined as "the measurement, collection, analysis and reporting of data about learners and their contexts for purposes of understanding and optimizing learning and the environments in which it occurs" (Elias 2011). Another definition states "the use of intelligent data, learner-produced data, and analysis models to discover information and social connection, and to predict and advise on learning" (Siemens 2010).

The NMC Horizon Report 2013 identified LA as one of the most important trends in technology-enhanced learning and teaching (Johnson et al. 2013). Therefore, it is not surprising, that LA is the subject of many scientific papers. The research and improvement of LA involves doing the development, the use and integration of new processes and tools to improve the performance of teaching and learning of individual students and of teachers. LA focuses specifically on the process of learning (Siemens and Long 2011).

Due to its connections with digital teaching and learning, LA is an interdisciplinary research field with connections to the field of teaching and learning research, computer science and statistics (Johnson et al. 2013). The available data is collected, analyzed and the gained insights are used to understand the behavior of the students to provide them additional support (Gašević et al. 2015).

A key concern of LA is the gathering and analyzation of data as well as the setting of appropriate interventions to improve the learners learning experience (Greller

Fig. 1.1 The five steps of the
analysis process

et al. 2014). These "actionable intelligence" from Educational Data Mining
(EDM) is supporting the teaching and learning and provides ideas for customization,
tutoring and intervention within the learning environment (Campbell et al. 2007).

According to Campbell and Oblinger (Campbell and Oblinger 2007), an analysis
process has five steps, shown in Fig. 1.1.

Capturing, data is captured and collected in real-time from different sources like
Virtual Learning Environments (VLE), Learning Management Systems (LMS),
Personal Learning Environment (PLE), web portals, forums, chat or rooms, and
combined with student information (Lauría et al. 2012; Tseng et al. 2016).

Reporting, the collected data is used to generate accurate models for identifying
and measuring the student's progress. Often visualization is used in LA dashboards
for a better understanding of the data (Muñoz-Merino et al. 2013; Leony et al.
2013).

Predicting, the data is used to identify predictors for student success, outcomes
and for identifying at-risk students. Further, it is used for decision-making about
courses and resource allocation which then is used by the decision-makers of the
institutions (Akhtar et al. 2015; Lonn et al. 2012).

Acting, the information gained from the data analyzation process is used to set
appropriate interventions in e.g. teaching or supporting students who are at risk of
failure or dropping out (Freitas et al. 2015; Palmer 2013).

Refining, the gathered information is used in a cyclical process for continuous
improvements of the used model in teaching and learning (Nam et al. 2014; Pistilli
et al. 2014).

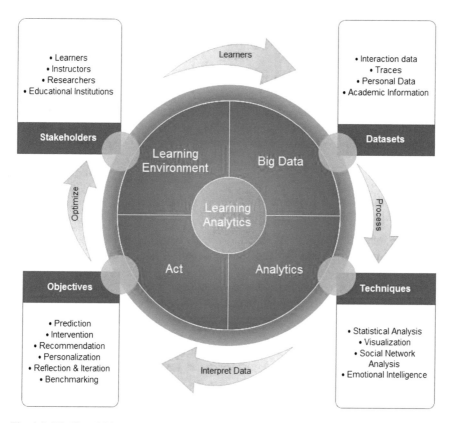

Fig. 1.2 Khalil and Ebner LA life cycle (Khalil and Ebner 2015)

Although research in the field of LA in recent years celebrates boom, LA is still in its infancy. Students, researchers and educational managers need to discuss ideas and opportunities on how to integrate these possibilities in their research and practice (Ferguson 2012).

In 2015, an LA approach which depicted a life cycle was introduced by Khalil and Ebner (2015), as shown in Fig. 1.2.

The cycle includes four main stages:

- Generation of data: this process starts from the learning environments where different stakeholders reside in MOOC, LMSs or any other VLEs.
- Data storage: learners leave a lot of traceable data behind them. Learners are not just consumers but also producers of data.
- Analysis: analytics methods seek to discover hidden patterns inside educational datasets. Analytics techniques are various. The authors defined them mainly into quantitative and qualitative analysis methods.

- Act: the analysis outcome should be interpreted to actions. In this stage, action is considered as prediction, intervention, recommendation, personalization and reflection.

At the end, the life cycle loop is closed by introducing the "optimization" process. Similar to Campbell and Oblinger (2007), they realized that there are similarities in phases between the available LA approaches in the literature. LA is an open loop of stages that should be closed at the end by optimizing learning environments and stakeholders (learners, tutors, decision makers…etc.).

1.2.2 Learning Analytics in Higher Education

HE looks forward to a future of uncertainty and change. In addition to the national and global as well as political and social changes, the competition on university level increases.

HE needs to increase financial and operational efficiency, expand local and global impact, establish new funding models during a changing economic climate and respond to the demands for greater accountability to ensure organizational success at all levels (van Barneveld et al. 2012). HE must overcome these external loads in an efficient and dynamic manner, but also understand the needs of the student body, who represents the contributor as well as the donor of this system (Shacklock 2016).

In addition to the strong competition, universities have to deal with the rapidly changing technologies that have arisen with the entry of the digital age. In the course of this, institutions collected enormous amounts of relevant data as a by-product. For instance, when students take an online course, use an Intelligent Tutoring System (ITS) (Arnold and Pistilli 2012; Bramucci and Gaston 2012; Fritz 2011; Santos et al. 2013) play educational games (Gibson and de Freitas 2016; Holman et al. 2013, 2015; Westera et al. 2013) or simply use an online learning platform (Casquero et al. 2014, 2016; Wu and Chen 2013; Ma et al. 2015; Santos et al. 2015; Softic et al. 2013).

In recent years, more universities use methods of LA in order to obtain findings on the academic progress of students, predict future behaviors and recognize potential problems in an early stage. Further, LA in the context of HE is an appropriate tool for reflecting the learning behavior of students and provide suitable assistance from teachers or tutors. This individual or group support offers new ways of teaching and provides a way to reflect on the learning behavior of the student.

Another motivation behind the use of LA in universities is to improve the inter-institutional cooperation, and the development of an agenda for the large community of students and teachers (Atif et al. 2013).

Table 1.1 Overview of the stakeholders (Romero and Ventura 2013)

Stakeholder	Objectives, benefits and perspectives
Learner	Support the learner with adaptive feedback, recommendations, response to his or her needs, for learning performance improvement
Educators	Understand students' learning process, reflect on teaching methods and performance, understand social, cognitive and behavioral aspects
Researchers	Use the right EDM technique which fits the problem, evaluation of learning effectiveness for different settings
Administrators	Evaluation of institutional resources and their educational offer

On an international level, the recruitment, management and retention of students have become as high level priorities for decision makers in institutions of HE. Especially improving the student retention starts and the understanding of the reason behind and/or prediction of the attrition has come in the focus of attention due to the financial losses, lower graduation rates, and inferior school reputation in the eyes of all stakeholders (Delen 2010; Palmer 2013).

Despite that LA focuses strongly on the learning process, the results still in the beneficial for all stakeholders. Romero and Ventura (2013) divided those involved stakeholders based on their objectives, benefits and perspectives in the four groups shown in Table 1.1.

1.3 Research Design, Methodology and Execution

This research aims at the elicitation of an overview on the advancement of the LA field in HE since it emerged in 2011. The proposed Research Questions (RQ) to answer are:

- **RQ1**: What are the research strands of the LA field in HE (between January 2011 and February 2016)?
- **RQ2**: What kind of limitations do the research papers and articles mention?
- **RQ3**: Who are the stakeholders and how could they be categorized?
- **RQ4**: What methods do they use in their papers?

1.3.1 Literature Review Procedure

In accordance to this objective, we performed a literature review following the procedure of Machi and McEvoy (2009). Figure 1.3 displays the six steps for a literature review used in this process.

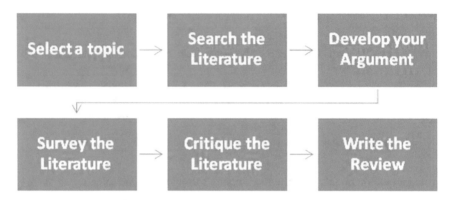

Fig. 1.3 The literature review: six steps to success (Machi and McEvoy 2009)

After we selected our topic, we identified data sources based on their relevance in the computing domain:

- The papers of the LAK conference published in the Association for Computing Machinery (ACM) Digital Library,
- The SpringerLink, and
- The Thomson Reuters WOS database.

and the following search parameters:

In the LAK papers, we didn't need to search for the "Learning Analytics" term because the whole conference covers the LA discipline. We searched the title, the abstract and the author keywords for "Higher Education" and/or "University".

In the SpringerLink database, we searched for the "Learning Analytics" term in conjunction with either "Higher Education" or "University" ("Learning Analytics AND (Higher Education OR University)").

In the WOS database, we searched for the topic "Learning Analytics" in conjunction with either "Higher Education" or "University" and in the research domain "science technology".

The defined inclusion criteria of the fetched papers from the libraries were set to be: (a) written in English, and (b) published between 2011 till the February 2016. We superficially assessed the quality of the reported studies, considering only articles that provided substantial information for LA in HE. Therefore, we excluded articles that did not meet the outlined inclusion principles.

The literature survey was conducted in February and March 2016. In the initial search, we found a total of 135 publications (LAK: 65, SpringerLink: 37, WOS: 33). During the first stage, the search results were analyzed based on their titles, author keywords and abstracts. After this stage, 101 papers remain for the literature survey. We fully read each publication and actively searched for their research questions, techniques, stakeholders, and limitations. Regular meetings between the authors were set on a weekly basis to discuss the results. Additionally, we added to our spreadsheet the Google Scholar[1] citation count as a measurement of article's impact.

In order to present our findings, we analyze each of the research questions separately. This section presents our findings.

1.3.2   Response to Research Question 1

In order to answer the RQ1, which corresponds to "What are the research strands of the LA field in HE (between January 2011 and February 2016)?", we tried to extract the main topics from the research questions of the publications.

We identified that many of the publications do not outline their research questions clearly. Many of the examined publications described use cases. This concerns in particular the older publications of 2011 and 2012, and is probably resulting from the young age of the scientific field of LA.

As a result, we did a brief text analysis on the fetched abstracts in order to examine the robust trends in the prominent field of LA and HE. We have collected all the article abstracts, processed them through the R software, and then refined the resulted corpus. In the final stages, we demonstrated the keywords and chose the Word cloud as a representation tool of the terms as shown in Fig. 1.4. The figure was graphically generated using one of the R library packages called "wordcloud".[2]

In order to ease reading the cloud, we adopted four levels of representation depicted in four colors. The obtained list of words that have been used were classified into singular phrases, bi-grams, tri-grams and quad-grams. The most cited singular words were "academic", "performance", "behavior" and "MOOCs". "Learning environment", "case study" and "online learning" were the most repeated bi-grams. The highest tri-grams used in the abstracts were "learning management systems", "Higher Education institutions" and "social network analysis". While quad-grams were only limited to "massive open online courses" which were merged at the final filtering stage with the "MOOCs" term.

The word cloud shows a glance about the general topics when LA is ascribed with HE. LA researchers focused on utilizing its techniques towards enhancing performance and students' behaviors. The popular adopted educational environment was Massive Open Online Course (MOOC) platforms. Furthermore, LA was also used to perform practices of interventions, observing dropout, videos, dashboards and engagement.

In Fig. 1.5 the collected articles are from the library data sources. Results show an obvious increase in the number of publications since 2011. For instance, there were 32 papers in 2015, incremented from 26 articles in 2014 and 17 articles in 2013. However, there were 5 articles only in 2011 and 12 articles in 2012. Because

---

[1]Online: http://scholar.google.com.

[2]Online: https://cran.r-project.org/web/packages/wordcloud/index.html.

Fig. 1.4 Word cloud of the prominent terms from the abstracts

Fig. 1.5 Collected articles distributed by source and year

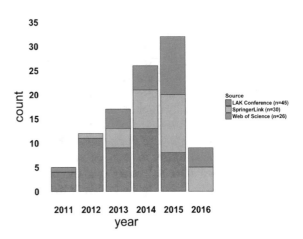

February 2016 was the date of collecting the publications in this study, the 2016 year was not indexed with many papers. On the other hand, the figure shows the apparent involvement of the journal articles from the SpringerLink and WOS libraries from 2013.

We cross-referenced the relevant publications with Google Scholar to derive their citation impact. Table 1.2 shows the 10 most cited publications.

Table 1.2 Citation impact of the publications

Paper title	Year of publication	No. of Google citations (Feb. 2016)
Course signal at Purdue: using learning analytics to increase student success (Arnold and Pistilli 2012)	2012	164
Social learning analytics: five approaches (Ferguson and Shum 2012)	2012	94
Classroom walls that talk: using online course activity data of successful students to raise self-awareness of underperforming peers (Fritz 2011)	2011	52
Goal-oriented visualizations of activity tracking: a case study with engineering students (Santos et al. 2012)	2012	46
Where is research on massive open online courses headed? A data analysis of the MOOC research initiative (Gasevic et al. 2014)	2014	46
Course correction: using analytics to predict course success (Barber and Sharkey 2012)	2012	36
Improving retention: predicting at-risk students by analyzing clicking behavior in a virtual learning environment (Wolff et al. 2013)	2013	34
Learning designs and learning analytics (Lockyer and Dawson 2011)	2011	33
The pulse of learning analytics understandings and expectations from the stakeholders (Drachsler and Greller 2012)	2012	30
Inferring higher level learning information from low level data for the Khan Academy platform (Muñoz-Merino et al. 2013)	2013	28

1.3.3 Response to Response to Research Question 2

We identified for RQ2, which corresponds to "What kind of limitations do the research papers and articles mention?", three different limitations, either clearly mentioned in articles or being tacitly within the context.

Limitations through time, some of the publications stated that continuous work is needed (Elbadrawy et al. 2015; Ifenthaler and Widanapathirana 2014; Koulocheri and Xenos 2013; Lonn et al. 2012; Palavitsinis et al. 2011; Sharkey 2011). Either a longitudinal study would be necessary to prove hypotheses or because of the shortage of the project (Fritz 2011; Nam et al. 2014; Ramírez-Correa and Fuentes-Vega 2015).

Limitations through the size, other publications talked about the need for more detailed data (Barber and Sharkey 2012; Best and MacGregor 2015; Rogers et al. 2014), the small group sizes (Junco and Clem 2015; Jo et al. 2015; Martin and

Whitmer 2016; Strang 2016), the unsure scalability, possible problems in wider context and the problem of the generalization of the approach or method (Prinsloo et al. 2015; Yasmin 2013).

Limitations through the culture, many of the publications mention that their approach might only work in their educational culture and is not applicable somewhere else (Arnold et al. 2014; Drachsler and Greller 2012; Grau-Valldosera and Minguillón 2014; Kung-Keat and Ng 2016). Additionally, the ethics differ strongly around the world, so cooperation projects between different universities in different countries needs different moderation as well as the use of data could be ethically questionable (Abdelnour-Nocera et al. 2015; Ferguson and Shum 2012; Lonn et al. 2013; Park et al. 2016).

Furthermore, ethical discussions about data ownership and privacy have recently arisen. Slade and Prinsloo (2013) pointed out that LA touches various research areas and therefore overlaps with ethical perspectives in areas of data ownership and privacy. Questions about who should own the collected and analyzed data were highly debated. As a result, the authors classified the overlapping categories in three parts:

- The location and interpretation of data,
- Informed consent, privacy and the de-identification of data, and
- The management, classification and storage of data.

These three elements generate an imbalance of power between the stakeholders which they addressed by proposing a list of 6 grounding principles and considerations: LA as moral practice, students as agents, student identity and performance are temporal dynamic constructs, student success is a complex and multidimensional phenomenon, transparency, HE cannot afford to not use data (Slade and Prinsloo 2013).

1.3.4 Response to Response to Research Question 3

In order to answer the RQ3, which corresponds to "Who are the stakeholders and how could they be categorized?", we determined the stakeholders from the publications and categorized them into three types. As a basis, we took the four stakeholders as mentioned in Sect. 1.2.2 and introduced in (Machi and McEvoy 2009). We merged the Researchers and Administrators from the original classification into one distinct group. Therefore, the institutional perspective [Academic Analytics (AA)] is separated from the learners' and teachers' one (LA).

Figure 1.6 depicts the defined LA stakeholders as a Venn-Diagram. The figure shows that there had been more research conducted concerning the Researchers/Administrators with overall 65 publications and 40 of them only concerning themselves, than in the field of Learners with a total of 53 publications and 21 single mentions. Also, it seems that Teachers are only a "side-product" of this field with only 20 mentions and only 7 dedicated to them alone.

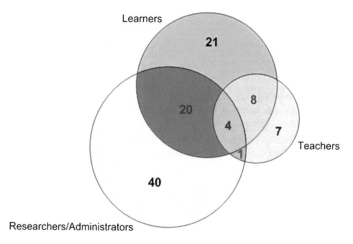

Fig. 1.6 Venn-diagram of stakeholders in the publications

Most of the combined articles addressed Researchers/Administrators together with Learners (20 publications). Only 8 articles can be found with an overlap between Learners and Teachers, which should be one of the most researched and discussed combinations within LA in HE. Nearly no work has been done by combining Researchers/Administrators with Teachers (in 1 publications) and only 4 paper combined all 3 stakeholders. This lack of research will be a matter of debate in the discussion section.

1.3.5 Response to Response to Research Question 4

By analyzing the selected studies to answer RQ4, which corresponds to "What techniques do they use in their papers?", we identified the techniques used in LA and HE publications. We took into account the methods presented by Romero and Ventura (2013), Khalil and Ebner (2016) and Linan and Perez (2015). We propose an overview of the used techniques of the different articles in Table 1.3.

The results of Fig. 1.7 show, that the research is focused mainly on prediction with a total of 36 citations. Outlier detection for pointing out at-risk or dropping out students with a citation count of 29. Distillation of data for human judgment in form of a visualization with a citation count of 33 than in all other parts including rarely used techniques like gamification or machine learning with a total amount of 102 counts.

Table 1.3 Overview of the used LA techniques of this study

Techniques	Key applications	Examples
Prediction	Predicting student performance and detecting student behaviors	AbuKhousa and Atif (2016), Cambruzzi et al. (2015), Harrison et al. (2015)
Clustering	Grouping similar materials or students based on their learning and interaction patterns	Aguiar et al. (2014), Asif et al. (2015), Scheffel et al. (2012)
Outlier detection	Detection of students with difficulties or irregular learning processes	Grau-Valldosera and Minguillón (2011), Manso-Vázquez and Llamas-Nistal (2015), Sinclari and Kalvala (2015)
Relationship mining	Identifying relationships in learner behavior patterns and diagnosing student difficulties	Kim et al. (2016), Pardo et al. (2015), Piety et al. (2014)
Social network analysis	Interpretation of the structure and relations in collaborative activities and interactions with communication tools	Hecking et al. (2014), Tervakari et al. (2013), Vozniuk et al. (2014)
Process mining	Reflecting student behavior in terms of its examination traces, consisting of a sequence of course, grade and timestamp	Menchaca et al. (2015), Vahdat et al. (2015), Wise (2014)
Text mining	Analyzing the contents of forums, chats, web pages and documents	Gasevic et al. (2014), Lotsari et al. (2014), Prinsloo et al. (2012)
Distillation of data for human judgment	Helping instructors to visualize and analyze the ongoing activities of the students and the use of information	Aguilar et al. (2014), Grann and Bushway (2014), Swenson (2014)
Discovery with models	Identification of relationships among student behaviors and characteristics or contextual variables. Integration of psychometric modelling frameworks into machine-learning models	Gibson et al. (2014), Kovanović et al. (2015), Lockyer and Dawson (2011)
Gamification	Include possibilities for playful learning to maintain motivation; e.g. integration of achievements, experience points or badges as indicators of success	Holman et al. (2013), Øhrstrøm et al. (2013), Westera et al. (2013)
Machine learning	Find hidden insights in data automatically (based on models who are exposed to new data and adapt itself independently)	Corrigan et al. (2015), McKay et al. (2012), Nespereira et al. (2016)
Statistic	Analysis and interpretation of quantitative data for decision making	Clow (2014), Khousa and Atif (2014), Simsek et al. (2015)

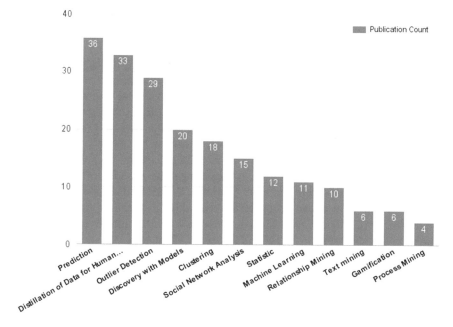

Fig. 1.7 The publication count of the used LA techniques

1.4 Discussion and Conclusion

In this chapter, we examined hundreds of pages to introduce a remarkable literature review of the LA field in the HE domain. We presented a state-of-the-art study of both domains based on analyzing articles from three major library references: the LAK conference, SpringerLink and WOS. The total number of relevant publications was equal to 101 articles in a period between 2011 and 2016.

In this literature review study, we followed the procedure of Machi and McEvoy (2009) in which we selected the topic, searched the literature to get the answers to the research questions, surveyed and critiqued the literature and finally introduced our review. Using this big dataset, we identified the research strands of the relevant publications. Most of the publications described use cases rather than comprehensive research—especially the prior publications, which is comprehensible because at the time, the universities had to figure out how to handle and harness the abilities offered by LA for their benefit.

To make a better holistic overview on the advancement of LA field in HE, we proposed four main RQs. These questions were related to the research strands of LA in HE, limitations, stakeholders and what techniques were used by LA experts in the HE domain, respectively.

The first research question was answered by generating a word cloud of a final corpus which was formed from all abstracts of the included papers. Results revealed that the usage of MOOCs, enhancing learning performance, students behavior, and benchmarking learning environments were strongly researched by LA experts in the domain of HE.

The paper with the title "Course signals at Purdue: using learning analytics to increase student success" by Arnold and Pistilli (2012), was the most cited article of our inclusion, which focused on a tool of prediction. Also, we identified that there was a clear increment of publications since 2011 till 2015, Further it was shown the apparent involvement of the journal articles from the SpringerLink and WOS libraries in 2013 and 2015 over the LAK conference publications.

The second research questions showed that limitations were mainly concerning the needed time to prepare data or getting the results, the size of the available dataset and examined group and ethical reasons. While the discussions of privacy and ownership have arisen dramatically after 2012, we found that the ethical constraints drive the limitations to the greatest extent of this literature review study similar to the arguments in Khalil and Ebner (2015, 2016b).

The analysis shows that there was clamor regarding who are the main stake-holders of LA and HE. As the leading stakeholders of LA should be learners and students (Khalil and Ebner 2015), we found that researchers play a major role of the loop between HE and LA. Figure 1.6 demonstrated the high use of researchers and administrators in carrying out decisions. The direct overlap between learners and teachers was not evidently identified in our study.

At the final stage, we tried to elaborate what were the most used techniques of LA in HE. This research question was answered based on solid articles that discussed the LA techniques. The scanning showed that prediction, distilling of data for human judgment, and outlier detection were the most used methods in the HE domain. General data mining methodologies from text mining to Social Network Analysis (SNA) were identified with high usage in the analyzed publications. On the other hand, we noticed that there are new techniques that seem to be used more frequently in the past two years such as serious gaming, which belongs to the gamification techniques.

1.5 Future Trends

In this section we tackle the future development in the field of LA in HE, which can be divided into short-term (1–2 years) and long term (3–5 years) trends.

1.5.1 Short-Term Trends

Over the next 1 to 2 years, universities must adjust to the social and economic factors, which postulated the change in the capabilities of the students (Johnson et al. 2016). The tuning of the areas analysis, consultation, examination of individual learning outcomes and the visualization of continuously-available, aggregated information in dashboards are gaining more and more importance. Students expect real-time feedback during learning with critical self-reflection on the learning progress and learning goal which strengthens their expertise in self-organization. If adequate quantities of data from students are available, they can be carried out for subsequently, predictive analytics (Johnson et al. 2016).

1.5.2 Long-Term Trends

The relevance of LA in HE will mint even more over the next 3–5 years. This trend is promoted by the strong interest of students for individual evaluations and care. To serve this market, dashboards and analysis applications that specifically address the needs of each customer will develop stronger. This approach offers many advantages: Accessing your own data in an appropriate form allows better self-reflection and a healthy rivalry among the fellow students. The teachers can survey a large amount of students and precisely recognize those who need their help. University and college dropouts can be better detected by appropriate analyzing and with targeted interventions they remain in the university system (Shacklock 2016).

To master the associated problems, the LA market will have to change. Currently, many different systems and analytical approaches are used. The fragmentation of the market will grow even further in the future, which makes the interuniversity comparison very difficult or even impossible. Therefore, the creation of standards is essential (Shacklock 2016). Furthermore, a change in the type of analysis is foreseeable. Most current and past data have been used to measure the success of students. Today, advances in predictive analytics (predictive analysis) are important. By using the analysis of existing data sets of many students, predictive models can be developed and warn thus students who are at risk not to meet their learning success (Shacklock 2016).

Acknowledgements This research project is co-funded by the European Commission Erasmus+ program, in the context of the project 562167-EPP-1-2015-1-BE-EPPKA3-PI-FORWARD.

References

Abdelnour-Nocera J, Oussena S, Burns C (2015) Human work interaction design of the smart university. In: Human work interaction design. Work analysis and interaction design methods for pervasive and smart workplaces. Springer International Publishing, pp 127–140
AbuKhousa E, Atif Y (2016) Virtual social spaces for practice and experience sharing. In: State-of-the-Art and Future Directions of Smart Learning. Springer, Singapore, pp 409–414

Aguiar E, Chawla NV, Brockman J, Ambrose GA, Goodrich V (2014) Engagement vs performance: using electronic portfolios to predict first semester engineering student retention. In: Proceedings of the fourth international conference on learning analytics and knowledge. ACM, pp 103–112

Aguilar S, Lonn S, Teasley SD (2014) Perceptions and use of an early warning system during a higher education transition program. In: Proceedings of the fourth international conference on learning analytics and knowledge. ACM, pp 113–117

Akhtar S, Warburton S, Xu W (2015) The use of an online learning and teaching system for monitoring computer aided design student participation and predicting student success. Int J Technol Des Edu, pp 1–20

Arnold KE, Pistilli MD (2012) Course signals at Purdue: using learning analytics to increase student success. In: Proceedings of the 2nd international conference on learning analytics and knowledge. ACM, pp 267–270

Arnold KE, Lonn S, Pistilli MD (2014) An exercise in institutional reflection: the learning analytics readiness instrument (LARI). In: Proceedings of the fourth international conference on learning penetrating the black box of time-on-task estimation and knowledge. ACM, pp 163–167

Asif R, Merceron A, Pathan MK (2015) Investigating performance of students: a longitudinal study. In: Proceedings of the fifth international conference on learning analytics and knowledge. ACM, pp 108–112

Atif A, Richards D, BilginA, Marrone M (2013) Learning analytics in higher education: a summary of tools and approaches. In: 30th Australasian Society for computers in learning in tertiary education conference, Sydney

Barber R, Sharkey M (2012) Course correction: using analytics to predict course success. In: Proceedings of the 2nd international conference on learning analytics and knowledge. ACM, pp 259–262

Best M, MacGregor D (2015) Transitioning design and technology education from physical classrooms to virtual spaces: implications for pre-service teacher education. Int J Technol Des Edu, pp 1–13

Bichsel J (2012) Analytics in higher education: benefits, barriers, progress, and recommendations. EDUCAUSE Center for Applied Research

Bramucci R, Gaston J (2012) Sherpa: increasing student success with a recommendation engine. In: Proceedings of the 2nd international conference on learning analytics and knowledge. ACM, pp 82–83

Cambruzzi WL, Rigo SJ, Barbosa JL (2015) Dropout prediction and reduction in distance education courses with the learning analytics multitrail approach. J UCS 21(1):23–47

Campbell JP, Oblinger DG (2007) Academic analytics, EDUCAUSE white paper. Retrieved 10 Feb 2016 from https://net.educause.edu/ir/library/pdf/PUB6101.pdf

Campbell JP, DeBlois PB, Oblinger DG (2007) Academic analytics: a new tool for a new era. EDUCAUSE Rev 42(4):40–57

Casquero O, Ovelar R, Romo J, Benito M (2014) Personal learningenvironments, highereducation and learninganalytics: a study of theeffects of servicemultiplexityonundergraduatestudents' personal networks/Entornos de aprendizaje personales, educación superior y analítica del aprendizaje: un estudio sobre los efectos de la multiplicidad de servicios en las redes personales de estudiantes universitarios. Cultura y Educación 26(4):696–738

Casquero O, Ovelar R, Romo J, Benito M, Alberdi M (2016) Students' personal networks in virtual and personal learning environments: a case study in higher education using learning analytics approach. Interact Learning Environ 24(1):49–67

Clow D (2014) Data wranglers: human interpreters to help close the feedback loop. In: Proceedings of the fourth international conference on learning analytics and knowledge. ACM, pp 49–53

Corrigan O, Smeaton AF, Glynn M, Smyth S (2015) Using educational analytics to improve test performance. In: Design for teaching and learning in a networked world. Springer International Publishing, pp 42–55

Delen D (2010) A comparative analysis of machine learning techniques for student retention management. Decis Support Syst 49(4):498–506

Drachsler H, Greller W (2012) The pulse of learning analytics understandings and expectations from the stakeholders. In: Proceedings of the 2nd international conference on learning analytics and knowledge. ACM, pp 120–129

Elbadrawy A, Studham RS, Karypis G (2015) Collaborative multi-regression models for predicting students' performance in course activities. In: Proceedings of the fifth international conference on learning analytics and knowledge. ACM, pp 103–107

Elias T (2011) Learning analytics: definitions, processes and potential

Ferguson R (2012) Learning analytics: drivers, developments and challenges. Int J Technol Enhanced Learning 4(5/6):304–317

Ferguson R, Shum SB (2012) Social learning analytics: five approaches. In: Proceedings of the 2nd international conference on learning analytics and knowledge. ACM, pp 23–33

Freitas S, Gibson D, Du Plessis C, Halloran P, Williams E, Ambrose M, Dunwell I, Arnab S (2015) Foundations of dynamic learning analytics: using university student data to increase retention. Br J Educational Technol 46(6):1175–1188

Fritz J (2011) Classroom walls that talk: using online course activity data of successful students to raise self-awareness of underperforming peers. Internet Higher Edu 14(2):89–97

Gasevic D, Kovanovic V, Joksimovic S, Siemens G (2014) Where is research on massive open online courses headed? A data analysis of the MOOC research initiative. Int Rev Res Open Distrib Learning, 15(5)

Gašević D, Dawson S, Siemens G (2015) Let's not forget: learning analytics are about learning. TechTrends 59(1):64–71

Gibson D, de Freitas S (2016) Exploratory analysis in learning analytics. Technol Knowl Learning 21(1):5–19

Gibson A, Kitto K, Willis J (2014) A cognitive processing framework for learning analytics. In: Proceedings of the fourth international conference on learning analytics and knowledge. ACM, pp 212–216

Grann J, Bushway D (2014) Competency map: visualizing student learning to promote student success. In: Proceedings of the fourth international conference on learning analytics and knowledge. ACM, pp 168–172

Grau-Valldosera J, Minguillón J (2011) Redefining dropping out in online higher education: a case study from the UOC. In: Proceedings of the 1st international conference on learning analytics and knowledge. ACM, pp 75–80

Grau-Valldosera J, Minguillón J (2014) Rethinking dropout in online higher education: The case of the UniversitatOberta de Catalunya. Int Rev Res Open Distrib Learning, 15(1)

Greller W, Ebner M, Schön M (2014) Learning analytics: from theory to practice–data support for learning and teaching. In: Computer assisted assessment. Research into e-assessment. Springer International Publishing, pp 79–87

Harrison S, Villano R, Lynch G, Chen G (2015) Likelihood analysis of student enrollment outcomes using learning environment variables: a case study approach. In: Proceedings of the fifth international conference on learning analytics and knowledge. ACM, pp 141–145

Hecking T, Ziebarth S, Hoppe HU (2014) Analysis of dynamic resource access patterns in a blended learning course. In: Proceedings of the fourth international conference on learning analytics and knowledge. ACM, pp 173–182

Holman C, Aguilar S, Fishman B (2013) GradeCraft: what can we learn from a game-inspired learning management system? In: Proceedings of the third international conference on learning analytics and knowledge. ACM, pp 260–264

Holman C, Aguilar SJ, Levick A, Stern J, Plummer B, Fishman B (2015) Planning for success: how students use a grade prediction tool to win their classes. In: Proceedings of the fifth international conference on learning analytics and knowledge. ACM, pp 260–264

Ifenthaler D, Widanapathirana C (2014) Development and validation of a learning analytics framework: two case studies using support vector machines. Technol Knowl Learning 19(1–2):221–240

Jo IH, Yu T, Lee H, Kim Y (2015) Relations between student online learning behavior and academic achievement in higher education: a learning analytics approach. In: Emerging issues in smart learning. Springer, Berlin, pp 275–287

Johnson L, Adams S, Cummins M (2012) The NMC horizon report: 2012 higher education edition. The New Media Consortium, Austin

Johnson L, Adams Becker S, Cummins M, Freeman A, Ifenthaler D, Vardaxis N (2013) Technology outlook for Australian tertiary education 2013–2018: an NMC horizon project regional analysis. New Media Consortium

Johnson L, Adams S, Cummins M, Estrada V, Freeman A, Hall C (2016) NMC horizon report: 2016 higher education edition. The New Media Consortium, Austin. http://cdn.nmc.org/media/2016-nmc-horizon-report-he-EN.pdf

Junco R, Clem C (2015) Predicting course outcomes with digital textbook usage data. Internet High Edu 27:54–63

Khalil M, Ebner M (2015) Learning analytics: principles and constraints. In: Proceedings of world conference on educational multimedia, hypermedia and telecommunications, pp 1326–1336

Khalil M, Ebner M (2016a) What is learning analytics about? A survey of different methods used in 2013–2015. In: Proceedings of smart learning conference, Dubai, UAE, 7–9 Mar. HBMSU Publishing House, Dubai, pp 294–304

Khalil M, Ebner M (2016b) De-identification in learning analytics. J Learning Anal 3(1), pp 129–138 http://dx.doi.org/10.18608/jla.2016.31.8

Khousa EA, Atif Y (2014) A learning analytics approach to career readiness development in higher education. In: International conference on web-based learning. Springer International Publishing, pp 133–141

Kim J, Jo IH, Park Y (2016) Effects of learning analytics dashboard: analyzing the relations among dashboard utilization, satisfaction, and learning achievement. Asia Pac Edu Rev 17(1):13–24

Koulocheri E, Xenos M (2013) Considering formal assessment in learning analytics within a PLE: the HOU2LEARN case. In: Proceedings of the third international conference on learning analytics and knowledge. ACM, pp 28–32

Kovanović V, Gašević D, Dawson S, Joksimović S, Baker RS, Hatala M (2015) Penetrating the black box of time-on-task estimation. In: Proceedings of the fifth international conference on learning analytics and knowledge. ACM, pp 184–193

Kung-Keat T, Ng J (2016) Confused, bored, excited? An emotion based approach to the design of online learning systems. In: 7th International conference on university learning and teaching (InCULT 2014) proceedings. Springer, Singapore, pp 221–233

Lauría EJ, Baron JD, Devireddy M, Sundararaju V, Jayaprakash SM (2012) Mining academic data to improve college student retention: an open source perspective. In: Proceedings of the 2nd international conference on learning analytics and knowledge. ACM, pp 139–142

Leony D, Muñoz-Merino PJ, Pardo A, Kloos CD (2013) Provision of awareness of learners' emotions through visualizations in a computer interaction-based environment. Expert Syst Appl 40(13):5093–5100

Liñán LC, Pérez ÁAJ (2015) Educational data mining and learning analytics: differences, similarities, and time evolution. Revista de Universidad y SociedaddelConocimiento 12(3): 98–112

Lockyer L, Dawson S (2011) Learning designs and learning analytics. In: Proceedings of the 1st international conference on learning analytics and knowledge. ACM, pp 153–156

Lonn S, Krumm AE, Waddington RJ, Teasley SD (2012) Bridging the gap from knowledge to action: Putting analytics in the hands of academic advisors. In: Proceedings of the 2nd international conference on learning analytics and knowledge. ACM, pp 184–18

Lonn S, Aguilar S, Teasley SD (2013) Issues, challenges, and lessons learned when scaling up a learning analytics intervention. In: Proceedings of the third international conference on learning analytics and knowledge. ACM, pp 235–239

Lotsari E, Verykios VS, Panagiotakopoulos C, Kalles D (2014) A learning analytics methodology for student profiling. In: Hellenic conference on artificial intelligence. Springer International Publishing, pp 300–312

Ma J, Han X, Yang J, Cheng J (2015) Examining the necessary condition for engagement in an online learning environment based on learning analytics approach: the role of the instructor. Internet High Edu 24:26–34

Machi LA, McEvoy BT (2009) The literature review: six steps to success. Corwin Sage, Thousand Oaks

Manso-Vázquez M, Llamas-Nistal M (2015) A monitoring system to ease self-regulated learning processes. IEEE RevistaIberoamericana de TecnologiasdelAprendizaje 10(2):52–59

Martin F, Whitmer JC (2016) Applying learning analytics to investigate timed release in online learning. Technol Knowl Learning 21(1):59–74

McKay T, Miller K, Tritz J (2012) What to do with actionable intelligence: E 2 coach as an intervention engine. In: Proceedings of the 2nd international conference on learning analytics and knowledge. ACM, pp 88–91

Menchaca I, Guenaga M, Solabarrieta J (2015) Project-based learning: methodology and assessment learning technologies and assessment criteria. In: Design for teaching and learning in a networked world. Springer International Publishing, pp 601–604

Muñoz-Merino PJ, Valiente JAR, Kloos CD (2013) Inferring higher level learning information from low level data for the Khan Academy platform. In: Proceedings of the third international conference on learning analytics and knowledge. ACM, pp 112–116

Nam S, Lonn S, Brown T, Davis CS, Koch D (2014) Customized course advising: investigating engineering student success with incoming profiles and patterns of concurrent course enrollment. In: Proceedings of the fourth international conference on learning analytics and knowledge. ACM, pp 16–25

Nespereira CG, Elhariri E, El-Bendary N, Vilas AF, Redondo RPD (2016) Machine learning based classification approach for predicting students performance in blended learning. In: The 1st International conference on advanced intelligent system and informatics (AISI2015), 28–30 Nov 2015, BeniSuef, Egypt. Springer International Publishing, pp 47–56

Øhrstrøm P, Sandborg-Petersen U, Thorvaldsen S, Ploug T (2013) Teaching logic through web-based and gamified quizzing of formal arguments. European conference on technology enhanced learning. Springer, Berlin, pp 410–423

Palavitsinis N, Protonotarios V, Manouselis N (2011) Applying analytics for a learning portal: the organic. Edunet case study. In: Proceedings of the 1st international conference on learning analytics and knowledge. ACM, pp 140–146

Palmer S (2013) Modelling engineering student academic performance using academic analytics. Int J Eng Educ 29(1):132–138

Pardo A, Mirriahi N, Dawson S, Zhao Y, Zhao A, Gašević D (2015) Identifying learning strategies associated with active use of video annotation software. In: Proceedings of the fifth international conference on learning analytics and knowledge. ACM, pp 255–259

Park Y, Yu JH, Jo IH (2016) Clustering blended learning courses by online behavior data: a case study in a Korean higher education institute. Internet High Educ 29:1–11

Piety PJ, Hickey DT, Bishop MJ (2014) Educational data sciences: framing emergent practices for analytics of learning, organizations, and systems. In: Proceedings of the fourth international conference on learning analytics and knowledge. ACM, pp 193–202

Pistilli MD, Willis III JE, Campbell JP (2014) Analytics through an institutional lens: definition, theory, design, and impact. In: Learning analytics. Springer New York, pp 79–102

Prinsloo P, Slade S, Galpin F (2012) Learning analytics: challenges, paradoxes and opportunities for mega open distance learning institutions. In: Proceedings of the 2nd international conference on learning analytics and knowledge. ACM, pp 130–133

Prinsloo P, Archer E, Barnes G, Chetty Y, Van Zyl D (2015) Big (ger) data as better data in open distance learning. Int Rev Res Open Distrib Learning, 16(1)

Ramírez-Correa P, Fuentes-Vega C (2015) Factors that affect the formation of networks for collaborative learning: an empirical study conducted at a Chilean university/Factores que afectanla formación de redes para el aprendizajecolaborativo: unestudioempíricoconducidoenunauniversidadchilena. Ingeniare: RevistaChilena de Ingenieria, 23(3), 341

Rogers T, Colvin C, Chiera B (2014) Modest analytics: using the index method to identify students at risk of failure. In: Proceedings of the fourth international conference on learning analytics and knowledge. ACM, pp 118–122

Romero C, Ventura S (2013) Data mining in education. Wiley Interdiscip Rev Data Min Knowl Discovery 3(1):12–27

Santos JL, Govaerts S, Verbert K, Duval E (2012) Goal-oriented visualizations of activity tracking: a case study with engineering students. In: Proceedings of the 2nd international conference on learning analytics and knowledge. ACM, pp 143–152

Santos JL, Verbert K, Govaerts S, Duval E (2013) Addressing learner issues with StepUp!: an evaluation. In: Proceedings of the third international conference on learning analytics and knowledge. ACM, pp 14–22

Santos JL, Verbert K, Klerkx J, Duval E, Charleer S, Ternier S (2015) Tracking data in open learning environments. J Univ Comput Sci 21(7):976–996

Scheffel M, Niemann K, Leony D, Pardo A, Schmitz HC, Wolpers M, Kloos CD (2012) Key action extraction for learning analytics. European conference on technology enhanced learning. Springer, Berlin, pp 320–333

Sclater N (2014) Code of practice "essential" for learning analytics. http://analytics.jiscinvolve. org/wp/2014/09/18/code-of-practice-essential-for-learning-analytics/

Shacklock X (2016) From bricks to clicks: the potential of data and analytics in higher education. The Higher Education Commission's (HEC) report

Sharkey M (2011) Academic analytics landscape at the University of Phoenix. In: Proceedings of the 1st international conference on learning analytics and knowledge. ACM, pp 122–126

Siemens G (2010) What are learning analytics. Retrieved 10 Feb 2016 from http://www. elearnspace.org/blog/2010/08/25/what-are-learning-analytics/

Siemens G, Long P (2011) Penetrating the fog: analytics in learning and education. EDUCAUSE Rev 46(5):30–40

Simsek D, Sándor Á, Shum SB, Ferguson R, De Liddo A, Whitelock D (2015) Correlations between automated rhetorical analysis and tutors' grades on student essays. In: Proceedings of the fifth international conference on learning analytics and knowledge. ACM, pp 355–359

Sinclair J, Kalvala S (2015) Engagement measures in massive open online courses. In: International workshop on learning technology for education in cloud. Springer International Publishing, pp 3–15

Slade S, Prinsloo P (2013) Learning analytics ethical issues and dilemmas. Am Behav Sci 57 (10):1510–1529

Softic S, Taraghi B, Ebner M, De Vocht L, Mannens E, Van de Walle R (2013) Monitoring learning activities in PLE using semantic modelling of learner behaviour. Human factors in computing and informatics. Springer, Berlin, pp 74–90

Strang KD (2016) Beyond engagement analytics: which online mixed-data factors predict student learning outcomes? Education and information technologies, pp 1–21

Swenson J (2014) Establishing an ethical literacy for learning analytics. In: Proceedings of the fourth international conference on learning analytics and knowledge. ACM, pp 246–250

Tervakari AM, Marttila J, Kailanto M, Huhtamäki J, Koro J, Silius K (2013) Developing learning analytics for TUT Circle. Open and social technologies for networked learning. Springer, Berlin, pp 101–110

Tseng SF, Tsao YW, Yu LC, Chan CL, Lai KR (2016) Who will pass? Analyzing learner behaviors in MOOCs. Res Pract Technol Enhanced Learning 11(1):1

Vahdat M, Oneto L, Anguita D, Funk M, Rauterberg M (2015) A learning analytics approach to correlate the academic achievements of students with interaction data from an educational simulator. In: Design for teaching and learning in a networked world. Springer International Publishing, pp 352–366

van Barneveld A, Arnold KE, Campbell JP (2012) Analytics in higher education: establishing a common language. EDUCAUSE Learning Initiative 1:1–11

Vozniuk A, Holzer A, Gillet D (2014) Peer assessment based on ratings in a social media course. In: Proceedings of the fourth international conference on learning analytics and knowledge. ACM, pp 133–137

Westera W, Nadolski R, Hummel H (2013) Learning analytics in serious gaming: uncovering the hidden treasury of game log files. In: international conference on games and learning alliance. Springer International Publishing, pp 41–52

Wise AF (2014) Designing pedagogical interventions to support student use of learning analytics. In: Proceedings of the fourth international conference on learning analytics and knowledge. ACM, pp 203–211

Wolff A, Zdrahal Z, Nikolov A, Pantucek M (2013) Improving retention: predicting at-risk students by analysing clicking behaviour in a virtual learning environment. In: Proceedings of the third international conference on learning analytics and knowledge. ACM, pp 145–149

Wu IC, Chen WS (2013) Evaluating the practices in the e-learning platform from the perspective of knowledge management. Open and social technologies for networked learning. Springer, Berlin, pp 81–90

Yasmin D (2013) Application of the classification tree model in predicting learner dropout behaviour in open and distance learning. Dis Educ 34(2):218–231

Chapter 2
Teaching and Learning Analytics to Support Teacher Inquiry: A Systematic Literature Review

Stylianos Sergis and Demetrios G. Sampson

Abstract *Teacher inquiry* is identified as a key global need for driving the continuous improvement of the teaching and learning conditions for learners. However, specific barriers (mainly related to teachers' data literacy competences), can defer teachers from engaging with inquiry to improve their teaching practice. To alleviate these barriers and support teacher inquiry, the concept of *Teaching and Learning Analytics (TLA)* has been proposed, as a complementing synergy between Teaching Analytics and Learning Analytics. Teaching and Learning Analytics aims to provide a framework in which the insights generated by Learning Analytics methods and tools can become meaningfully translated for driving teachers' inquiry to improve their teaching practice, captured through Teaching Analytics methods and tools. In this context, TLA have been identified as a research challenge with significant practical impact potential. This chapter contributes the first systematic literature review in the emerging research field of Teaching and Learning Analytics. The insights gained from the systematic literature review aim to (a) transparently outline the existing state-of-the-art following a structured analysis methodology, as well as (b) elicit insights and shortcomings which could inform future work in the Teaching and Learning Analytics research field.

Keywords Teaching analytics · Teacher inquiry · Learning analytics · Educational design · Teacher reflection

S. Sergis · D.G. Sampson (✉)
Department of Digital Systems, University of Piraeus,
Androutsou 150, 18532 Piraeus, Greece
e-mail: Demetrios.Sampson@curtin.edu.au

S. Sergis
e-mail: steliossergis@iti.gr

D.G. Sampson
School of Education, Curtin University, Building 501 Level 4,
Bentley Campus, Kent Street, Bentley, Perth, WA 6102, Australia

Abbreviations

ED   Educational design
RQ   Research question
SLR   Systematic literature review
SNA   Social network analysis
TLA   Teaching and learning analytics

2.1   Introduction

Data-driven teacher appraisal is among the key priorities of educational policies worldwide for continuously monitoring and improving the teaching and learning conditions offered to learners (OECD 2013). Data-driven teacher appraisal can be related either to (OECD 2009):

- *Meeting external accountability mandates*, which take a summative standpoint towards assessment of teachers' educational design and delivery practice or
- *Developing a continuous cycle of self-improvement*, which is guided by the teachers themselves and takes a formative standpoint towards improvement.

However, since the latter can also be considered as a pre-requisite for the former, explicit focus is being placed for supporting teachers to engage in self-evaluation and improvement of their practice (namely educational design and delivery), in a process commonly termed as *teacher inquiry* (Check and Schutt 2012).

Teacher inquiry refers to a continuous process of investigation, reflection and improvement of teaching practice, based on the collection, analysis and interpretation of diverse educational data (Avramides et al. 2015). However, despite the emerging global need for teachers to engage in inquiry, specific barriers can hinder its wide adoption. Examples of such barriers include teachers' low *data literacy* competences for collecting, analyzing and interpreting educational data (Marsh and Farrell 2014), the need for timely data collection and analysis (Kaufman et al. 2014) as well as the quality of educational data that can be manually collected (Mandinach 2012). To address these barriers, specific data Analytics strands have emerged, as follows:

- *Teaching Analytics*, which refers to the methods and digital tools to help teachers analyze and improve the educational designs prior to the delivery. Furthermore, more recent developments on Teaching Analytics also support analysis of how the teacher delivers the educational designs (e.g., Gauthier 2013; Prieto et al. 2016).[1]

---

[1]In this book chapter, we will consider this extended strand of Teaching Analytics research as part of the *proposed concept of Teaching and Learning Analytics (TLA)* and not as part of the Teaching Analytics strand.

- *Learning Analytics*, which refer to the methods and digital tools that allow the measurement, collection, analysis and reporting of data about learners and their contexts, for purposes of understanding and optimizing learning and the environments in which it occurs (SOLAR 2011).

However, each Analytics strand focuses on supporting specific inquiry tasks, namely Teaching Analytics mainly focus on capturing and analyzing the teacher actions during the educational design and delivery process, while Learning Analytics mainly focus on capturing and analyzing the learner actions, despite the explicit mention of "[educational] **context**" in their definition. Therefore, each digital Analytics strand can offer fragmented support to teachers towards reflecting on and improving their educational design and delivery. More specifically, Teaching Analytics do not account for the learners' actions and, therefore may have limited value for evaluating the impact of educational designs. On the other hand, Learning Analytics have not yet fully accounted for the aspect of context (namely educational design and delivery), which is a significant factor that can affect learners' performance and progress (e.g., Dyckhoff 2011; Toetenel and Rienties 2016). The latter limitation is also noticeable in a number of recent analyses of the Learning Analytics research field (e.g., Papamitsiou and Economides 2014; Sin and Muthu 2015; Nunn et al. 2016).

As a response to this need, a new Analytics strand has been proposed, which can be termed *Teaching and Learning Analytics (TLA)*. TLA is presented as a synergy between Teaching Analytics and Learning Analytics in order to holistically support the process of teacher inquiry. More specifically, TLA argues for the need for methods and tools that will allow teachers to analyze their educational design and delivery process and also utilize learners' educational data for evidence-based evaluation, reflection on and improvement of this process (McKenney and Mor 2015). This synergy has been considered as one of the key research challenges in the field of Technology-enhanced Education (Lockyer et al. 2013; Wasson et al. 2016).

In this context, this book chapter reports on the first systematic literature review (SLR) in the emerging research field of TLA. The contribution of this book chapter is that it analyzes the current state-of-the-art in the TLA research, using the concept of teacher inquiry as a backbone analysis framework, with the aim of providing transparent overview of overarching insights and shortcomings.

The remainder of the book chapter is structured as follows: Sect. 2.2 presents the background of this work related to the concept of teacher inquiry and Teaching and Learning Analytics. Section 2.3 presents the methodology followed in the systematic literature review process. Section 2.4 presents the results of the systematic literature review. Finally, Sect. 2.5 discusses the main findings and conclusions. The Appendix section contains the full analysis of the state-of-the-art research TLA works, following the analysis framework described in Sect. 2.3.

2.2 Background

This section will initially present the foundational concepts, namely *teacher inquiry* and *Teaching and Learning Analytics* in corresponding sections. Furthermore, it will outline the manner in which the two concepts are connected and how TLA can provide more holistic support to teachers for engaging in the full spectrum of tasks associated with teacher inquiry.

2.2.1 Teacher Inquiry

Teacher inquiry is defined as a sequence of actions in which "*teachers identify questions for investigation in their practice and then design a process for collecting evidence about student learning that informs their subsequent educational designs*" (Avramides et al. 2015). Essentially, teacher inquiry is a form of action research, in which teachers define specific questions regarding their educational design and delivery and collect evidence to answer these questions (Altrichter et al. 2008). Therefore, this process can guide reflection and improvement in a systematic and evidence-based manner (Dana and Yendol-Hoppey 2014).

Teacher inquiry generically follows a cycle of steps (Timperley et al. 2010; Hansen and Wasson 2016), which is outlined as follows (also depicted in Fig. 2.1):

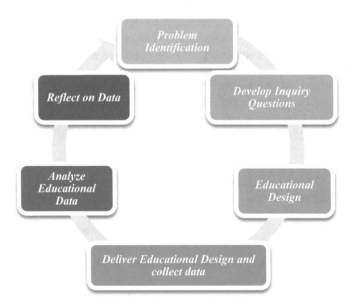

Fig. 2.1 Overview of the teacher inquiry cycle

- *Step 1: Problem Identification.* During this step, the teacher identifies a specific aspect of their educational design and/or delivery that they wish to investigate/evaluate in order to improve it.
- *Step 2: Develop Inquiry Questions.* During this step, the teacher defines the specific questions that they will investigate, related to evaluating or investigating aspects of their educational design and/or delivery. Furthermore, the teacher defines which *educational data* they will need to collect during delivery to answer the specific question they defined, as well as the method for collecting these data.
- *Step 3: Educational Design.* During this step, the teacher formulates the educational design which they will deliver in order to implement their inquiry.
- *Step 4: Deliver Educational Design and collect data.* During this step, the teacher delivers the educational design to the learners and collects the educational data using the collection method.
- *Step 5: Analyze educational data.* After the teacher has collected the educational data, they analyse them in order to elicit insights to answer the inquiry question they have defined.
- *Step 6: Reflect on data.* Finally, the analysed data are used by the teacher in order to answer the defined inquiry question and (if needed) revise the practice in which they conduct their educational design and/or delivery.

As aforementioned, teacher inquiry is gaining momentum globally as teachers are continuously expected to improve the teaching and learning conditions for their learners on an evidence-based manner (OECD 2013). However, despite this emerging push, specific barriers exist that hinder teachers perform each step of the inquiry cycle, including teachers' low *data literacy* competences to collect, analyze and interpret educational data (Marsh and Farrell 2014), untimely collection and analysis of educational data (Kaufman et al. 2014) and low quality of educational data that can be manually collected by the teacher (Mandinach 2012). To address these issues and facilitate teachers in performing the tasks of the inquiry cycle, a research synergy to exploit the potential of Teaching Analytics and Learning Analytics has been recently proposed, namely TLA. The following section describes the concept of TLA and how it can support the process of teacher inquiry.

2.2.2 Teaching and Learning Analytics

The emerging research strand of TLA refers to the methods and tools for supporting teachers engage in inquiry for reflecting on and improving their educational design and delivery. To do that, TLA aims to combine the individual capacity of Teaching Analytics and Learning Analytics in order to exploit:

- The potential of *Teaching Analytics* to analyze the educational designs in the constituent elements (e.g., learning and assessment activities and educational resources/tools) and the interrelations between these elements.
- The potential of *Learning Analytics* to measure, collect, analyse and report on learners' educational data and the learning context that they are generated, aiming to improve the learning conditions for individual learners or groups of learners (Papamitsiou and Economides 2014).

Essentially, TLA introduces a framework that aims to combine the focal points of the existing Analytics strands and re-purpose them towards addressing a new challenge, which is currently under-supported by each individual strand in isolation. More specifically, TLA argues that insights generated by Learning Analytics methods and tools can be mapped to the analyzed (through Teaching Analytics tools) elements of teaching practice that generated them, and therefore support teachers to reflect on and improve their educational design and delivery based on evidence (Greller et al. 2014; Greller and Drachsler 2012; Emin-Martinez et al. 2014; Bakharia et al. 2016). In this regard, TLA is appropriate to support the concept of teacher inquiry (Mor et al. 2015), as defined in the previous section, and it can be directly linked to all teacher inquiry cycle, as indicated in Table 2.1.

Therefore, in this book chapter, TLA will be defined as a framework to guide the process of teachers' reflection on their educational design and delivery, based on evidence from educational data related to both their learners, as well as their own.

As aforementioned, the research field of TLA is still relatively new but highly emerging and important (Wasson et al. 2016). Based on this fact, it is reasonable to argue for the need to have a systematic and critical overview of the current research state-of-the-art. This overview will provide insights on how the existing works have been aligned to the overarching challenge that TLA is aimed to address, namely

Table 2.1 Mapping between TLA and the steps of teacher inquiry cycle

Teacher inquiry cycle steps	How TLA can contribute
1. Problem identification	Teaching analytics can be used to capture and analyze the educational design and facilitate the teacher to:
2. Develop inquiry questions	• pinpoint the specific elements of their educational design that relate to the problem they have identified and
3. Educational design	• elaborate on their inquiry question by defining explicitly the educational design elements they will monitor and investigate in their inquiry
4. Deliver educational design and collect data	Learning analytics can be used to collect the learner/teacher educational data that have been defined to answer their inquiry question.
5. Analyze data	Learning analytics can be used to analyse and report on the collected data and facilitate sense-making
6. Reflect on data	The combined use of TLA can be used to answer the inquiry questions and support reflection on educational design and delivery

support the process of teacher inquiry. Furthermore, these insights could also outline shortcomings that future TLA research could aim to address.

In this context, the contribution of this book chapter is to perform a systematic literature review in the research field of TLA and provide the aforementioned insights following a systematic approach. Using the teacher inquiry cycle as a backbone framework, the SLR was structured and implemented based on a specific step-by-step methodology, which is described in detail in the following section.

2.3   Systematic Literature Review Methodology

The systematic literature review followed the widely accepted methodology of Kitchenham and Charters (2007). More specifically, the methodology included the definition of (a) the analysis framework of existing research works (depicting the research questions addressed), (b) the literature inclusion and exclusion criteria and (c) the literature search strategy adopted (Brereton et al. 2007). Each of these methodology steps are described in the following sections.

2.3.1   Research Questions: Research Work Analysis Framework

In order to provide a structured method to analyze the existing research works in Teaching and Learning Analytics, a set of research questions were defined. These research questions aimed to collect insights on how the current state-of-the-art in TLA supports the steps of the inquiry cycle, as they were outlined in Table 2.1.

The research questions (analysis framework) were defined as follows:

- *RQ1. What Teaching Analytics tasks were employed?* This Research Question was related to the steps of the inquiry cycle related to "Problem Identification", "Develop Inquiry Questions" and "Educational Design". It aimed to elicit the Teaching Analytics tasks that each research work adopted in terms of analyzing the educational design and, thus, supporting the teacher to clearly define inquiry questions based on the problems they had identified.
- *RQ2. Which educational data types were collected regarding the learner?* This Research Question was related to the step of the inquiry cycle "Develop Inquiry Questions" and aimed to identify the educational data types that each research work collected, related to learners.
- *RQ3. Which educational data types were collected regarding the teacher?* This Research Question was related to the step of the inquiry cycle "Develop Inquiry Questions" and aimed to identify the educational data types that each research work collected, related to teachers.

- *RQ4. What data analysis method was used to process the collected teacher/learner data?* This Research Question was related to the steps of the inquiry cycle "Deliver Educational Design and collect data" and "Analyze educational data". It aimed to identify the (Learning Analytics) methods that each research work exploited towards processing the learners' and teachers' educational data.
- *RQ5. Which was the focus of reflection?* This Research Questions was related to the step of the inquiry cycle "Reflect on Data". It aimed to identify which aspect of teachers' practice the TLA work provided reflective insights for.
- *RQ6. Were teachers provided with recommendations for supporting reflection?* This Research Question was related to the step of the inquiry cycle "Reflect on Data". It aimed to elicit whether the research work provided recommendations to support teachers' reflection and sense-making, or whether the teacher had to engage in ad hoc reflective insights based on their own reasoning.

2.3.2 Inclusion and Exclusion Criteria

In order to ensure that the identified research works were relevant to answer the Research Questions of this SLR, a set of inclusion and exclusion criteria was defined. The adopted inclusion and exclusion criteria are as follows:

- **Inclusion Criterion**:
  - Publications should describe original research work related to the use of TLA methods/tools for supporting the teacher to reflect on their teaching design and delivery.
  - No restriction was imposed on the date of publication of the publications.

- **Exclusion Criteria**:
  - Publications should not focus solely on the use of Teaching Analytics methods and tools that do not take into account the delivery of the educational design.
  - Publications should not focus solely on the use of Learning Analytics methods and tools to exclusively facilitate the teacher support individual learners' progress (but not reflection on their educational design and delivery).
  - Publications should not be included in the conference proceedings as posters (in case of conference publications).
  - Publications should be written in English.
  - Abstract-only publications were not considered.
  - Updated versions of the same publications were only considered once.

2.3.3 Literature Search Strategy

The literature search strategy was devised in order to identify and collect research works and use them to answer the proposed research questions. Following recommended practice in systematic literature reviews (Brereton et al. 2007), the search strategy adopted the following protocol in terms of keyword. The *keywords* for guiding the search were selected. In order to ensure that any relevant research papers would not be excluded at this point, general keywords were used, namely *"Teaching Analytics"*, *"Learning Analytics"*, *"Educational Analytics"*, *"Teacher Inquiry"*, *"Analytics"*. Additionally, the use of Boolean operators (OR, AND) among the general keywords was also performed in order to extend the search results. The keywords were appropriately selected in order to include the key concepts relevant to the focus of the SLR. By adopting general keywords, research works that were relevant to the SLR but did not explicitly use terms such as "teacher inquiry", were also captured. The timeframe in which this literature search was conducted was May–June 2016.

Regarding the *digital databases* used in the search, these included prestigious scientific journals and international conference proceedings relevant to the field of Teaching Analytics and Learning Analytics, as follows:

- Journal of Learning Analytics [http://learning-analytics.info/].
- Computers & Education [http://www.journals.elsevier.com/computers-and-education].
- British Journal of Educational Technology [http://onlinelibrary.wiley.com/journal/10.1111/(ISSN)1467-8535].
- Journal of Educational Technology & Society [http://www.ifets.info].
- IEEE Transactions on Learning Technologies [https://www.computer.org/web/tlt].
- Computers in Human Behavior [http://www.journals.elsevier.com/computers-in-human-behavior].
- Proceedings of the Learning Analytics and Knowledge (LAK) Conference (2012–2016) [https://solaresearch.org/events/lak/].
- IEEE Conference on Advanced Learning Technologies (ICALT) (2012–2015) [http://ieeexplore.ieee.org/].
- Furthermore, relevant research works directly cited in the initially identified publications from the above databases were also considered.

The research works selection process was conducted in two steps, as follows:

- *Step 1.* All research works retrieved using the literature search strategy were assessed based on the inclusion and exclusion criteria (defined in Sect. 2.3.2). At this step, each research work was initially assessed in terms of the title and abstract in order to identify and reject papers that were not relevant to the aims and Research Questions of this SLR.

- *Step 2.* All research works that were initially approved during Step 1, were more deeply analyzed based on the full text in order to ensure that they were relevant to the Research Questions.

After the aforementioned process was finalized, a pool of *54 research works* remained, which was used for addressing the defined Research Questions.

2.4 Systematic Literature Review Results

This section will present the results of the SLR for each of the Research Questions. The results for each of the RQ is outlined in a separate sub-section, presenting both a discussion of results as well as quantitative analyses of the collected data. Furthermore, a detailed table depicting the full quantitative results of the full SLR can be found in the Appendix section.

2.4.1 Results Related to the Teaching Analytics Tasks Employed (RQ1)

The RQ1 was aimed to elicit which Teaching Analytics tasks each research work adopted in order to support the first three steps of the inquiry cycle. The critical analysis of existing works highlighted a set of three overarching and recurring Teaching Analytics tasks, which are depicted in Table 2.2 and Fig. 2.2.

As Table 2.2 depicts, the analysis of research works led to the definition of the following Teaching Analytics tasks:

- *Analysis of ED in terms of elements* ($N = 52$, $x = 96.3\%$). This Teaching Analytics task related to the basic analysis of the educational design in terms of its constituent elements. More specifically, this task aims to create a structured representation of the educational design, where each element (i.e., each learning activity, assessment activity and/or educational resource/tool) is explicitly defined. The *main aim* of this task is to support the teacher transparently

Table 2.2 Teaching analytics tasks

#	Overarching teaching analytics task	Occurrence frequency ($N = 54$)	Percentage (%)
1	Analysis of educational design (ED) in terms of elements	52	96.3
2	Capturing the flow of learning and assessment activities	20	37.0
3	Analysis of learning and assessment activity types	13	24.1

The teaching analytics tasks were not mutually exclusive in each research work

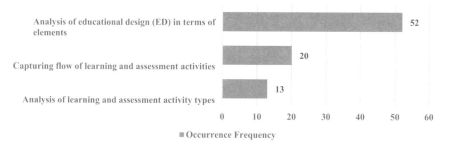

Fig. 2.2 Overview of the teaching analytics tasks

'decompose' their educational design and, therefore, be able to define and investigate inquiry questions on each comprising element (e.g., Romero et al. 2008; Hung et al. 2012).

- *Capturing the flow of learning and assessment activities* (N = 20, x = 37.0%). This Teaching Analytics task extended the previous task, by not only capturing the learning and assessment activities of the educational design but also defining the specific flow in which these should be delivered. The *main aim* of this task is to enable the teacher to compare between their designed flow of activities and the flow that their learners follow during delivery (e.g., Camilleri et al. 2013).
- *Analysis of learning and assessment activity types* (N = 13, x = 24.1%). This Teaching Analytics task aimed to include another layer of detail when analyzing the educational design, by classifying the learning and assessment activities in specific types (which were defined based on the focus of each work). To give an example, Rienties et al. (2015) classified learning activities in seven types (*productive, assimilative, assessment, communication, finding and handling information, experiential, interactive*). The *main aim* of this task is to allow teachers to define and answer inquiry questions related to how different learning activity types can impact their learners' performance (e.g., Gómez-Aguilar et al. 2015) or their own actions when delivering the educational design (e.g., Prieto et al. 2016).

2.4.2 Results Related to the Educational Data Types Collected Regarding the Learner (RQ2)

The RQ2 was aimed to elicit the learner educational data types that TLA research adopt in order to support the second step of the inquiry cycle ("Develop Inquiry Questions"). Table 2.3 and Fig. 2.3 depict the resulting set of seven overarching earner educational data types that were elicited from the critical analysis of the TLA research works. It is mentioned that 52 (out of the overall 54) TLA research works utilized learners' educational data.

Table 2.3 Learner educational data types

#	Learner educational data type	Occurrence frequency (N = 52)	Percentage (%)
1	Assessment scores	28	53.8
2	Engagement in learning activities	27	51.9
3	Engagement with educational resources or tools	24	46.2
4	Engagement in discussion activities	21	40.4
5	Customizable list of educational data	11	21.2
6	Demographics	2	3.8
7	Behavior	2	3.8
8	Physical setting	2	3.8

The educational data types were not mutually exclusive in each research work

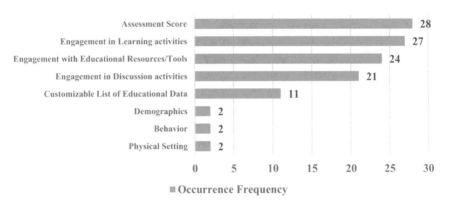

Fig. 2.3 Overview of the learner educational data types

As Table 2.3 depicts, the analysis of research works highlighted the following learner educational data types:

- *Assessment scores* (N = 28, x = 53.8%). This educational data type refers to formative or summative assessment performance of the learners during the delivery of the educational design. The *main aim* of this educational data type is to offer a benchmark for evaluating the impact on the learners' performance of specific educational design elements (e.g., specific learning activities) (Mirriahi and Dawson 2013; Hernández-García et al. 2015) or the teachers' actions during the delivery of the educational design (e.g., Howlin and Lynch 2014).
- *Engagement in learning activities* (N = 27, x = 51.9%). This educational data type refers to the level in which learners engaged with the learning activities, in terms of either time spent on the activities (e.g., Fernández-Gallego et al. 2013)

or frequency of attempts of the activities (e.g., El-Bishouty et al. 2015). The *main aim* of this educational data type was to identify the learning activities that were attributed with low learner engagement and therefore, might need revising. Moreover, learning activities which highly engaged learners could also be highlighted, in order to provide the teacher with a 'good-practice' blueprint.

- *Engagement with educational resources or tools* (N = 24, x = 46.2%). This educational data type refers to the level in which learners engaged with the educational resources and tools, in terms of either time spent on the educational resources/tools (e.g., Rienties et al. 2015) or frequency of access/use (e.g., Mirriahi and Dawson 2013). The *main aim* of this educational data type was to help the teacher pinpoint specific educational resources/tools that were not engaging to the learners and, therefore, might need to be revised or replaced.
- *Engagement in Discussion activities* (N = 21, x = 40.4%). This educational data type refers to the level of engagement of learners in the learning activities that explicitly included discussions between the learners or between the learners and the teacher (e.g., through a forum). The *main aim* of this educational data type was to provide evidence on which of these discussion activities were engaging (or not) to the learners and, inform teachers to possibly revise them (e.g., Ali et al. 2012).
- *Customizable List of Educational Data* (N = 11, x = 21,2%). This educational data 'type' aims to depict research works that either did not provide an exhaustive list of the learner educational data they collected (e.g., Mazza and Milani 2005) or allowed the teacher to define a custom array of educational data to be considered (from the presented set of seven overarching learner educational data types) (e.g., Kladich et al. 2013).
- *Demographics* (N = 2, x = 3.8%). This educational data type mainly refers to learners' past competences. The *main aim* of this educational data type was to allow teachers to reflect on their educational design/delivery (or specific elements), by also explicitly taking into account learners' prior competences (e.g., Dunbar et al. 2014).
- *Behavior* (N = 2, x = 3.8%). This educational data type mainly refers to learners' level of attendance during the delivery of the educational design. The *main aim* of this educational data type was to allow teachers to explicitly consider the level in which learners attended the delivery as an additional evaluation variable when they reflect on their educational design and delivery (e.g., Bos and Brand-Gruwei 2016).
- *Physical Setting* (N = 2, x = 3.8%). This educational data type was used in research works that aimed to study TLA in the context of informal settings. The *main aim* of this educational data type was to allow teachers to investigate whether their learners were following the designed flow of learning and assessment activities (in the physical space) and whether there were any deviations that could inform revisions in subsequent educational designs (e.g., Melero et al. 2015).

2.4.3   Results Related to the Educational Data Types Collected Regarding the Teacher (RQ3)

The RQ3 was aimed to elicit the teacher educational data types that TLA research adopt in order to support the second step of the inquiry cycle ("Develop Inquiry Questions"). As Table 2.4 and Fig. 2.4 depict, a set of three overarching teacher educational data types were elicited from a total of 15 research works that utilized such educational data.

As Table 2.4 depicts, the analysis of research works highlighted the following teacher educational data types:

- *Engagement in discussion activities* (N = 11, x = 73.3%). This educational data type refers to the frequency of teachers' participation in learning activities focused on discussion (e.g., through a forum). The *main aim* of this educational data type is to support teachers to reflect on the way they supported learners during these activities, in terms of feedback and scaffolding (e.g., Dawson 2010). Furthermore, this educational data type could also relate to analyzing the content of the interventions made by the teacher, in order to help them assess the 'quality' of feedback and scaffolding provided (e.g., van Leeuwen et al. 2015).
- *Engagement in learning activities* (N = 6, x = 40.0%). This educational data type refers to the level in which teachers participated in the learning activities, in terms of providing feedback and support to the learners as well as orchestrating the delivery of the learning activities (Prieto et al. 2011; Martinez-Maldonado et al. 2016). The *main aim* of this educational data type was to provide evidence to teachers on (a) whether they provided the level of feedback and support they had initially planned for or (b) whether they orchestrated the delivery of the

Table 2.4 Teacher educational data types

#	Teacher educational data type	Occurrence frequency (N = 15)	Percentage (%)
1	Engagement in discussion activities	11	73.3
2	Engagement in learning activities	6	40.0
3	Location/physical data	1	6.7

The educational data types were not mutually exclusive in each research work

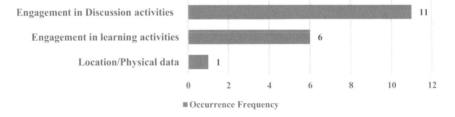

Fig. 2.4 Overview of the teacher educational data types

learning activities according to their initial design. In both cases, teachers gained access to evidence that could help them identify and improve potentials problematic aspects in their practice.

- *Location/Physical Data* (N = 1, x = 6.7%). This educational data type refers to teachers' physical data (e.g., eye-tracking) and physical setting data (e.g., position and point-of-view). The *main aim* of this educational data is to collect highly granulated evidence on the specific physical actions teachers made during the delivery of their educational design, without the need for manual data collection and analysis (e.g., Prieto et al. 2016).

2.4.4   Results Related to the Data Analysis Method Used to Process the Collected Teacher and Learner Educational Data (RQ4)

The RQ4 was aimed to elicit the analysis methods that existing TLA research works employ in order to support the "Deliver Educational Design and collect data" and "Analyze educational data" steps of teacher inquiry. Table 2.5 and Fig. 2.5 present

Table 2.5 Data analysis methods for learner educational data

#	Data analysis method	Occurrence frequency (N = 52)	Percentage (%)
1	Statistics	43	82.7
2	Clustering	17	32.7
3	Classification	11	21.2
4	Regression	8	15.4
5	Social network analysis (SNA)	8	15.4
6	Association rule mining	8	15.4
7	Text mining	4	7.7

The data analysis methods were not mutually exclusive in each research work

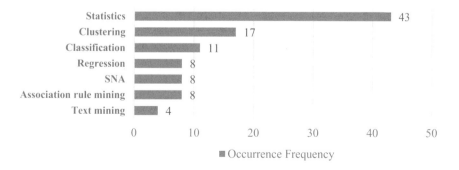

Fig. 2.5 Overview of data analysis methods for learner educational data

Table 2.6 Data analysis methods for teacher educational data

#	Data analysis method	Occurrence frequency (N = 15)	Percentage (%)
1	Statistics	11	73.3
2	SNA	5	33.3
3	Clustering	3	20.0
4	Classification	2	13.3
5	Regression	1	6.7

The data analysis methods were not mutually exclusive in each research work

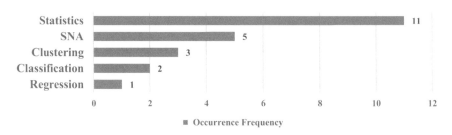

Fig. 2.6 Overview of data analysis methods for teacher educational data

the results of RQ4 regarding the data analysis methods used to process the collected (from Learning Analytics tools) learner educational data, whereas Table 2.6 and Fig. 2.6 present the results of RQ4 regarding the data analysis methods used to process teacher educational data.

As Table 2.5 depicts, a set of seven overarching data analysis methods were employed by the existing TLA research works. Furthermore, it is mentioned that all 52 research works that exploited learner educational data utilized at least one analysis method to process them.

Table 2.6 and Fig. 2.6 present the results of RQ4 regarding the data analysis methods used to process teachers' educational data, which comprised a set of five such methods. As the Table 2.6 depicts, all research works that utilized teachers' educational data (N = 15) also adopted a method (or more) to analyze them.

As both Tables 2.5 and 2.6 depict, the identified data analysis methods used in the existing TLA research works are consistent with the relevant framework proposed by Papamitsiou and Economides (2014). Therefore, these data indicate that TLA approaches have directly built on the existing Learning Analytics methods and tools, simply re-aligning the purpose for which they are exploited (namely, to support teacher inquiry).

2.4.5   Results Related to the Focus of Reflection (RQ5)

The RQ5 was aimed to elicit the aspects of teachers' practice that TLA focused on, namely which was the *TLA task* that aimed to support the final step of the inquiry cycle ("Reflect on data"). Table 2.7 and Fig. 2.7 present the identified set of four overarching TLA tasks. It is mentioned that all 54 research works focused on achieving at least one TLA task.

As Table 2.7 depicts, the elicited TLA tasks are as follows:

- *Evaluation of educational design elements based on educational data* (N = 41, x = 75.9%). This TLA task refers to eliciting evidence from learners' and teachers' educational data in order to evaluate *specific elements* of their educational design. The *main aim* of this TLA task is to evaluate how the learners engaged with each element of the educational design (e.g., Ali et al. 2012) and use these analyses to support teachers answer relevant inquiry questions.
- *Evaluation of overall educational design* (N = 18, x = 33.3%). This TLA task refers to supporting teachers to evaluate the impact of their *overall educational design* to learners. The *main aim* of this TLA task is to allow teachers to reflect on whether the intended educational objectives (e.g., knowledge, skills, attitudes) were successfully met by the learners (e.g., Smolin and Butakov 2012; Jaggars et al. 2016).

Table 2.7 Teaching and learning analytics tasks

#	Focus of reflection (TLA task)	Occurrence frequency (N = 54)	Percentage (%)
1	Evaluation of educational design elements based on educational data	41	75.9
2	Evaluation of overall educational design	18	33.3
3	Reflection on delivery of educational design	15	27.8

The teaching and learning analytics tasks were not mutually exclusive in each research work

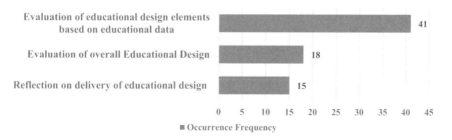

Fig. 2.7 Overview of teaching and learning analytics tasks

- *Reflection on delivery of educational design* (N = 15, x = 27.8%). This TLA task refers to supporting teachers to evaluate how they deliver the educational design. The *main aim* of this TLA task is to process teachers' educational data from the delivery and provide insights for reflection and improvement of the way the teachers deliver their educational designs (e.g., van Leeuwen et al. 2014; Prieto et al. 2016).

2.4.6 Results Related to Whether Teachers Were Provided with Recommendations for Supporting Reflection (RQ6)

The RQ6 was aimed to elicit whether the existing TLA works deployed recommendations to facilitate teachers' reflection and sense-making during the final step of the inquiry cycle ("Reflect on data"). In case that the TLA work did not offer such recommendations, the teacher had to engage in 'ad-hoc' reflective actions on how to utilize the analyses of the educational data. Table 2.8 presents the results of this analysis.

As Table 2.8 depicts, the majority of existing TLA works (N = 50, x = 92.6%) do not support teachers' reflection and sense-making through recommendations for improvement. These works focus on either:

- Providing teachers with the analyses of the collected educational data without further decision support (e.g., Bos and Brand-Gruwel 2016). An example of such analyses can include a Social Network Analysis graph depicting the interactions of learners and teacher in the discussion activities (Dawson et al. 2008).
- Providing teachers with the analyses of the collected educational data and, further allowing the comparison between these analyses (e.g., Kladich et al. 2013; Bakharia et al. 2016). For example, Pardo et al. (2015) used regression analysis to identify the impact of different learner performance indicators

Table 2.8 Analysis of TLA works in terms of whether they provided recommendations for reflection

#	Variables	Occurrence frequency (N = 54)	Percentage (%)
1	Provided recommendations	50	92.6
2	Did not provide recommendations (Ad hoc reflection)	4	7.4

(captured in educational data) on the level of their engagement for better understanding how to improve the educational design.

However, in both cases, the task of translating the results of the analyses or comparisons to actionable insights for improvement is performed by the teacher in an ad hoc manner. On the contrary, very few existing TLA works ($N = 4$, $x = 7.4\%$) support teachers in this final sense-making inquiry step. More specifically, these works mainly focused on either:

- Allowing the teacher to initially define questions on their educational design or delivery, which were answered by the TLA approach based on educational data from the delivery of the educational design. These insights were fed back to the teacher for informing their reflection and improvement actions (Martinez-Maldonado et al. 2016; Rodríguez-Triana et al. 2015).
- Generating textual feedback to the teacher using rule-based, pre-defined feedback templates that were populated based on the analyses of educational data (Kosba et al. 2005; Yen et al. 2015).

In both these cases, teachers received actionable insights, which described specific ways to improve their educational design and delivery. Considering the low number of the TLA works that offer recommendations to teachers, however, it is evident that the TLA state-of-the-art still rely on the teachers' capacity to translate data analyses to actionable insights during the "Reflect on data" inquiry step.

2.5 Discussion and Conclusions

Teaching and Learning Analytics is an emerging research field that aims to combine Teaching Analytics and Learning Analytics in order to support teachers during the process of inquiry. Considering the potential placed on this research field (Mor et al. 2015; Wasson et al. 2016), this book chapter performed the first systematic literature review in order to provide insights on how the state-of-the-art in TLA has realized this potential.

More specifically, using the concept of teacher inquiry and the model of the inquiry cycle as a backbone framework, a set of Research Questions were defined to capture and analyze the TLA research, identify trends (discussed in the previous section) as well as elicit overarching insights and/or shortcomings. The main insights and/or shortcomings from the SLR are as follows:

- The existing TLA works have mainly adopted *basic Teaching Analytics tasks (RQ1)*, which are related to depicting the elements of the educational design in a transparent, but isolated, manner. Furthermore, the interconnections between these elements (e.g., the flow of learning and assessment activities) as well as the actual analysis and classification of these elements (e.g., classify learning activities to specific types) were accommodated in few research works. This insight suggests that existing TLA works provide limited support to teachers in terms of the range of inquiry questions they can potentially investigate, since they afford fragmented analysis of the educational design.

- The existing TLA works have accommodated the collection of a wide range of *learners' educational data (RQ2)*. This diversity suggests that TLA research has largely exploited the potential of Learning Analytics to collect and process diverse types of learner educational data. This is also evident in the *data processing methods (RQ4)* for learner (and teacher) educational data, which are fully aligned with the approaches adopted in the Learning Analytics literature (e.g., Papamitsiou and Economides 2014).

- The aspect of collecting and processing *teachers' educational data (RQ3)* during the educational design and delivery process is addressed by few works. This is consistent with findings from the Learning Analytics field (Dyckhoff et al. 2013). The limited existing work is mainly focused to monitor teachers' contribution in learning and discussion activities. However, this is a significant shortcoming that can hinder teachers' capacity to reflect on their practice in a holistic manner, since it neglects capturing and evaluating their own actions.

- Regarding the *focus of reflection (RQ5)*, the existing TLA works mainly aim to support teachers to target their inquiry in investigating the impact of their educational design to learner, both as a complete product as well as in specific elements of it. This is consistent with the concept of teacher inquiry, which engages teachers to investigate elements of their practice that they consider inefficient. However, few TLA works have explicitly addressed the aspect of supporting *teachers' reflection on the delivery of the educational design*. Following the previously mentioned shortcoming, this can be a hindering factor for holistic inquiry, since it neglects the significant factor of how the teacher actions during the delivery of the educational design can impact its effectiveness to learners.

- Finally, the SLR highlighted that little research attention has been placed on *providing recommendations (RQ6)* to teachers for translating the analyzed data to actionable reflecting actions on their educational design and delivery. This is an important challenge to tackle because the process of eliciting actionable insights for improvement is commonly considered a cumbersome task for teachers (Marsh and Farrell 2014; Mor et al. 2015). Therefore, providing teachers with evidence-based recommendations to translate data analyses to specific reflective insights, can be considered an important need for the TLA research field.

Overall, the contribution of the book chapter was to collect and analyze the existing research woks in the emerging research field of TLA in order to understand and elicit the main trends and limitations. As the above discussion of the results indicated, the field of TLA is still in its infancy, with a heavy reliance on exploiting the existing Teaching Analytics and Learning Analytics methods and tools. However, new methods and tools to explicitly address the scope of TLA are yet scarce. Therefore, future research in the TLA field should build on the aforementioned insights and focus on proposing methods and tools that will address the shortcomings to extend the current state-of-the-art. Additionally, further analyses of the identified pool of research works can also be performed in order to elicit more sophisticated correlations and interconnections between the research focal points, methodologies and outcomes. As a result of the above, new TLA approaches for holistically supporting the full cycle of teacher inquiry can be introduced, aiming to support teachers engage in this important process and improve the teaching and learning conditions for themselves, as well as their learners.

Acknowledgements The work presented in this paper has been partially funded by National Matching Funds 2014–2016 of the Greek Government, and more specifically by the General Secretariat for Research and Technology (GSRT), related to EU project "Inspiring Science: Large Scale Experimentation Scenarios to Mainstream eLearning in Science, Mathematics and Technology in Primary and Secondary Schools" (GA No. 325123).

Appendix

Table 2.9 depicts the full analysis of the 54 identified TLA research works, in terms of the Research Questions of the systematic literature review.

Table 2.9 Analysis of research works included in the SLR

#	Reference	Learner educational data collected (RQ2)	Teacher educational data collected (RQ3)	Data analysis method for learner (RQ4)	Data analysis method for teacher (RQ4)	Teaching analytics method employed (RQ1)	Reflection focus (RQ5)	Recommendations/ad hoc (RQ6)
1	Agudo-Peregrina et al. (2014)	Customizable list of educational data	Engagement in discussion activities	Association rule mining—regression—statistics	Statistics	Analysis of ED in terms of elements	Evaluation of educational design elements based on educational data + reflection on delivery of educational design	Ad hoc
2	El-Bishouty et al. (2015)	Engagement in learning activities—assessment score—engagement in discussion activities	–	Clustering—classification—statistics	–	Analysis of ED in terms of elements	Evaluation of educational design elements based on educational data	Ad hoc
3	Camilleri et al. (2013)	Engagement in discussion activities—assessment score—engagement in learning activities	–	Statistics	–	Analysis of ED in terms of elements + capturing the flow of learning and assessment activities	Evaluation of educational design elements based on educational data	Ad hoc
4	Chounta and Avouris (2016)	Engagement in learning activities—assessment score	–	Classification	–	Analysis of learning and assessment activity types	Evaluation of educational design elements based on educational data	Ad hoc
5	Dawson (2010)	Engagement in discussion activities	Engagement in discussion activities	SNA—statistics	SNA	Analysis of ED in terms of elements	Reflection on delivery of educational design	Ad hoc

(continued)

Table 2.9 (continued)

#	Reference	Learner educational data collected (RQ2)	Teacher educational data collected (RQ3)	Data analysis method for learner (RQ4)	Data analysis method for teacher (RQ4)	Teaching analytics method employed (RQ1)	Reflection focus (RQ5)	Recommendations/ad hoc (RQ6)
6	Fernández-Delgado et al. (2014)	Assessment Score	–	Classification —statistics	–	Analysis of ED in terms of elements	Evaluation of educational design elements based on educational data	Ad hoc
7	Fernández-Gallego et al. (2013)	Engagement in learning activities	–	Classification	–	Analysis of ED in terms of elements	Evaluation of educational design elements based on educational data	Ad hoc
8	Gómez-Aguilar et al. (2015)	Engagement with educational resources/tools— engagement in discussion activities	–	Classification	–	Analysis of learning and assessment activity types	Evaluation of educational design elements based on educational data	Ad hoc
9	Govaerts et al. (2011)	Engagement with educational resources/tools— (time spent) engagement in learning activities —engagement in discussion activities	–	Statistics	–	Analysis of ED in terms of elements	Evaluation of educational design elements based on educational data	Ad hoc

(continued)

Table 2.9 (continued)

#	Reference	Learner educational data collected (RQ2)	Teacher educational data collected (RQ3)	Data analysis method for learner (RQ4)	Data analysis method for teacher (RQ4)	Teaching analytics method employed (RQ1)	Reflection focus (RQ5)	Recommendations/ad hoc (RQ6)
10	Haya et al. (2015)	Engagement with educational resources/tools—engagement in discussion activities—assessment score	–	SNA—text mining—statistics	–	Analysis of ED in terms of elements	Evaluation of educational design elements based on educational data + reflection on delivery of educational design + evaluation of overall educational design	Ad hoc
11	Hernández-García et al. (2015)	Demographics—assessment score—engagement in discussion activities	Engagement in discussion activities	SNA—statistics	SNA	Analysis of ED in terms of elements	Evaluation of educational design elements based on educational data	Ad hoc
12	Howlin and Lynch (2014)	Engagement with educational resources/tools—engagement in learning activities—assessment score—engagement in discussion activities	Engagement in discussion activities	Statistics	Statistics	Analysis of ED in terms of elements + capturing the flow of learning and assessment activities	Evaluation of educational design elements based on educational data + reflection on delivery of educational design	Ad hoc

(continued)

Table 2.9 (continued)

#	Reference	Learner educational data collected (RQ2)	Teacher educational data collected (RQ3)	Data analysis method for learner (RQ4)	Data analysis method for teacher (RQ4)	Teaching analytics method employed (RQ1)	Reflection focus (RQ5)	Recommendations/ad hoc (RQ6)
13	Kladich et al. (2013)	Customizable list of educational data	–	Statistics—association rule mining	–	Analysis of ED in terms of elements + capturing the flow of learning and assessment activities	Evaluation of educational design elements based on educational data + evaluation of overall educational design	Ad hoc
14	Kosba et al. (2005)	Customizable list of educational data	–	Clustering—statistics	–	Analysis of ED in terms of elements	Evaluation of educational design elements based on educational data	Recommendations for feedback
15	Marcos-Garcia et al. (2015)	Engagement with educational resources/tools—engagement in learning activities—engagement in discussion activities	Engagement in learning activities—engagement in discussion activities	SNA	SNA	Analysis of ED in terms of elements + capturing the flow of learning and assessment activities	Evaluation of educational design elements based on educational data + reflection on delivery of educational design	Ad hoc
16	Mazza et al. (2012)	Engagement in discussion activities—assessment scores—engagement with educational resources/tools—engagement in learning activities	Engagement in discussion activities—engagement in learning activities	Statistics	Statistics	Analysis of ED in terms of elements	Evaluation of educational design elements based on educational data + reflection on delivery of educational design	Ad hoc

(continued)

Table 2.9 (continued)

#	Reference	Learner educational data collected (RQ2)	Teacher educational data collected (RQ3)	Data analysis method for learner (RQ4)	Data analysis method for teacher (RQ4)	Teaching analytics method employed (RQ1)	Reflection focus (RQ5)	Recommendations/ad hoc (RQ6)
17	Mazza and Dimitrova (2007)	Engagement in discussion activities—assessment scores—engagement in learning activities	–	Statistics	–	Analysis of ED in terms of elements	Evaluation of educational design elements based on educational data	Ad hoc
18	Mazza and Milani (2005)	Customizable list of educational data	–	Classification	–	Analysis of ED in terms of elements	Evaluation of educational design elements based on educational data	Ad hoc
19	Mendez et al. (2014)	Assessment score	–	Clustering—statistics	–	Analysis of ED in terms of elements	Evaluation of overall educational design	Ad hoc
20	Minović et al. (2015)	Engagement with educational resources/tools—engagement in learning activities—assessment score	–	Statistics	–	Analysis of ED in terms of elements + capturing the flow of learning and assessment activities	Evaluation of educational design elements based on educational data + evaluation of overall educational design	Ad hoc
21	Mirriahi and Dawson (2013)	Engagement with educational resources/tools—engagement in learning activities—assessment score	–	Statistics	–	Analysis of ED in terms of elements	Evaluation of educational design elements based on educational data	Ad hoc

(continued)

Table 2.9 (continued)

#	Reference	Learner educational data collected (RQ2)	Teacher educational data collected (RQ3)	Data analysis method for learner (RQ4)	Data analysis method for teacher (RQ4)	Teaching analytics method employed (RQ1)	Reflection focus (RQ5)	Recommendations/ad hoc (RQ6)
22	Monroy et al. (2015)	Engagement in learning activities	Engagement in learning activities	Statistics—regression	Statistics—regression	Analysis of ED in terms of elements	Reflection on delivery of educational design	Ad hoc
23	Romero et al. (2008)	Customizable list of educational data	–	Statistics, classification, clustering, association rule mining	–	Analysis of ED in terms of elements	Evaluation of educational design elements based on educational data + evaluation of overall educational design	Ad hoc
24	Slotta et al. (2013)	Engagement in learning activities—physical setting—assessment score—engagement with educational resources/tools	–	Classification—clustering	–	Analysis of ED in terms of elements + capturing the flow of learning and assessment activities	Evaluation of educational design elements based on educational data	Ad hoc
25	Smolin and Butakov (2012)	Assessment score	–	Association rule mining—statistics	–	Analysis of ED in terms of elements	Evaluation of overall educational design	Ad hoc
26	Schwarz and Asterhan (2011)	Engagement in discussion activities	Engagement in discussion activities	Classification—SNA	Classification—SNA	Analysis of ED in terms of elements	Reflection on delivery of educational design	Ad hoc

(continued)

Table 2.9 (continued)

#	Reference	Learner educational data collected (RQ2)	Teacher educational data collected (RQ3)	Data analysis method for learner (RQ4)	Data analysis method for teacher (RQ4)	Teaching analytics method employed (RQ1)	Reflection focus (RQ5)	Recommendations/ad hoc (RQ6)
27	van Leeuwen et al. (2014)	Engagement in discussion activities—engagement in learning activities	Engagement in discussion activities	Text mining—statistics	Statistics	Analysis of ED in terms of elements + capturing the flow of learning and assessment activities	Reflection on delivery of educational design	Ad hoc
28	Yen et al. (2015)	Engagement in discussion activities—engagement in learning activities	–	Statistics—text mining	–	Analysis of ED in terms of elements + capturing the flow of learning and assessment activities	Evaluation of educational design elements based on educational data	Recommendations for feedback
29	Zhang et al. (2007)	Engagement in discussion activities—engagement in learning activities—engagement with educational resources/tools	–	Statistics	–	Analysis of ED in terms of elements	Evaluation of educational design elements based on educational data	Ad hoc
30	Bakharia et al. (2016)	Engagement in discussion activities—engagement in learning activities—engagement with educational resources/tools—assessment scores	–	Clustering—association rule mining—SNA	–	Analysis of ED in terms of elements + capturing the flow of learning and assessment activities + analysis of learning and assessment activity types	Evaluation of educational design elements based on educational data + evaluation of overall educational design	Ad hoc

(continued)

Table 2.9 (continued)

#	Reference	Learner educational data collected (RQ2)	Teacher educational data collected (RQ3)	Data analysis method for learner (RQ4)	Data analysis method for teacher (RQ4)	Teaching analytics method employed (RQ1)	Reflection focus (RQ5)	Recommendations/ad hoc (RQ6)
31	Bos and Brand-Gruwei (2016)	Engagement with educational resources/tools—assessment scores—student behavior	–	Statistics—regression	–	Analysis of ED in terms of elements	Evaluation of educational design elements based on educational data	Ad hoc
32	Dawson et al. (2008)	Engagement with educational resources/tools—engagement in discussion activities	Engagement in discussion activities	SNA—Statistics—Clustering	SNA—statistics—clustering	Analysis of ED in terms of elements	Evaluation of educational design elements based on educational data + reflection on delivery of educational design	Ad hoc
33	Dunbar et al. (2014)	Demographics—assessment score—student behavior—engagement in learning activities—engagement with educational resources/tools	–	Regression—clustering—statistics	–	Analysis of ED in terms of elements + capturing the flow of learning and assessment activities + analysis of learning and assessment activity types	Evaluation of educational design elements based on educational data + evaluation of overall educational design	Ad hoc
34	Duque et al. (2015)	Customizable list of educational data	–	Statistics	–	Analysis of ED in terms of elements	Evaluation of educational design elements based on educational data	Ad hoc

(continued)

Table 2.9 (continued)

#	Reference	Learner educational data collected (RQ2)	Teacher educational data collected (RQ3)	Data analysis method for learner (RQ4)	Data analysis method for teacher (RQ4)	Teaching analytics method employed (RQ1)	Reflection focus (RQ5)	Recommendations/ad hoc (RQ6)
35	Elbadrawy et al. (2014)	Assessment scores—engagement in discussion activities—engagement with educational resources/tools	–	Regression—classification—clustering—statistics	–	Analysis of ED in terms of elements + analysis of learning and assessment activity types	Evaluation of educational design elements based on educational data + evaluation of overall educational design	Ad hoc
36	Fritz (2016)	Assessment scores—engagement in discussion activities—engagement in learning activities—engagement with educational resources/tools	–	Statistics	–	Analysis of ED in terms of elements + analysis of learning and assessment activity types	Evaluation of educational design elements based on educational data + evaluation of overall educational design	Ad hoc
37	Hung et al. (2012)	Customizable list of educational data	–	Clustering—association rule mining—statistics	–	Analysis of ED in terms of elements	Evaluation of educational design elements based on educational data + evaluation of overall educational design	Ad hoc

(continued)

Table 2.9 (continued)

#	Reference	Learner educational data collected (RQ2)	Teacher educational data collected (RQ3)	Data analysis method for learner (RQ4)	Data analysis method for teacher (RQ4)	Teaching analytics method employed (RQ1)	Reflection focus (RQ5)	Recommendations/ad hoc (RQ6)
38	Jaggars et al. (2016)	Assessment scores —engagement in learning activities —engagement with educational resources/tools	Engagement in discussion activities	Statistics	Statistics	Analysis of ED in terms of elements + capturing the flow of learning and assessment activities + analysis of learning and assessment activity types	Evaluation of overall educational design	Ad hoc
39	Karkalas et al. (2016)	Engagement in learning activities —engagement with educational resources/tools	–	Clustering—statistics	–	Analysis of ED in terms of elements	Evaluation of educational design elements based on educational data	Ad hoc
40	Martinez-Maldonado et al. (2016)	Assessment scores —engagement in learning activities —engagement with educational resources/tools	Engagement in learning activities	Clustering - Statistics	Statistics	Analysis of ED in terms of elements + capturing the flow of learning and assessment activities	Evaluation of educational design elements based on educational data + reflection on delivery of educational design	Mapping of educational data analysis to initial design questions
41	Merceron (2012)	Customizable list of educational data	–	Clustering—association rule mining—statistics	–	Analysis of ED in terms of elements	Evaluation of educational design elements based on educational data	Ad hoc

(continued)

Table 2.9 (continued)

#	Reference	Learner educational data collected (RQ2)	Teacher educational data collected (RQ3)	Data analysis method for learner (RQ4)	Data analysis method for teacher (RQ4)	Teaching analytics method employed (RQ1)	Reflection focus (RQ5)	Recommendations/ad hoc (RQ6)
42	Ochoa (2016)	Assessment scores	–	Clustering—statistics	–	Analysis of ED in terms of elements	Evaluation of overall educational design	Ad hoc
43	Prieto et al. (2011)	–	Engagement in learning activities	–	Clustering—statistics	Analysis of ED in terms of elements + capturing the flow of learning and assessment activities + analysis of learning and assessment activity types	Reflection on delivery of educational design	Ad hoc
44	Rienties et al. (2015)	Assessment scores —engagement in learning activities —engagement with educational resources/tools	–	Clustering—statistics	–	Analysis of ED in terms of elements + analysis of learning and assessment activity types	Evaluation of educational design elements based on educational data + evaluation of overall educational design	Ad hoc
45	Pardo et al. (2015)	Assessment scores —engagement in learning activities —engagement with educational resources/tools	–	Regression—clustering—statistics	–	Analysis of ED in terms of elements + capturing the flow of learning and assessment activities	Evaluation of educational design elements based on educational data + evaluation of overall educational design	Ad hoc

(continued)

Table 2.9 (continued)

#	Reference	Learner educational data collected (RQ2)	Teacher educational data collected (RQ3)	Data analysis method for learner (RQ4)	Data analysis method for teacher (RQ4)	Teaching analytics method employed (RQ1)	Reflection focus (RQ5)	Recommendations/ad hoc (RQ6)
46	Toetenel and Rienties (2016)	Assessment scores —engagement in learning activities —engagement with educational resources/tools	–	Clustering—statistics	–	Analysis of ED in terms of elements + analysis of learning and assessment activity types	Evaluation of educational design elements based on educational data + evaluation of overall educational design	Ad hoc
47	van Leeuwen et al. (2015)	Engagement in learning activities —engagement with educational resources/tools—engagement in discussion activities	Engagement in discussion activities	Visualization—text mining—regression	Statistics	Analysis of ED in terms of elements + capturing the flow of learning and assessment activities	Reflection on delivery of educational design + evaluation of educational design elements based on educational data	Ad hoc
48	Rodriguez-Triana et al. (2015)	Customizable List of educational data	–	Statistics	–	Analysis of ED in terms of elements + capturing the flow of learning and assessment activities + analysis of learning and assessment activity types	Reflection on delivery of educational design	Mapping of educational data analysis to initial design questions
49	Wong and Lavrencic (2016)	Assessment scores	–	Statistics	–	Analysis of ED in terms of elements	Evaluation of overall educational design	Ad hoc

(continued)

Table 2.9 (continued)

#	Reference	Learner educational data collected (RQ2)	Teacher educational data collected (RQ3)	Data analysis method for learner (RQ4)	Data analysis method for teacher (RQ4)	Teaching analytics method employed (RQ1)	Reflection focus (RQ5)	Recommendations/ad hoc (RQ6)
50	Prieto et al. (2016)	–	Engagement in learning activities —location/physical data	–	Clustering—classification—statistics	Analysis of ED in terms of elements + capturing the flow of learning and assessment activities + analysis of learning and assessment activity types	Reflection on delivery of educational design	Ad hoc
51	Ali et al. (2012)	Assessment scores —engagement in learning activities —engagement with educational resources/tools—engagement in discussion activities	–	SNA—classification —association rule mining—statistics	–	Analysis of ED in terms of elements + Capturing the flow of learning and assessment activities	Evaluation of educational design elements based on educational data	Ad hoc
52	Dyckhoff et al. (2012)	Customizable list of educational data	–	Statistics—regression	–	Analysis of ED in terms of elements + capturing the flow of learning and assessment activities + analysis of learning and assessment activity types	Evaluation of educational design elements based on educational data	Ad hoc

(continued)

Table 2.9 (continued)

#	Reference	Learner educational data collected (RQ2)	Teacher educational data collected (RQ3)	Data analysis method for learner (RQ4)	Data analysis method for teacher (RQ4)	Teaching analytics method employed (RQ1)	Reflection focus (RQ5)	Recommendations/ad hoc (RQ6)
53	Fulantelli et al. (2015)	Customizable list of educational data	–	Statistics	–	Analysis of ED in terms of elements	Evaluation of educational design elements based on educational data	Ad hoc
54	Melero et al. (2015)	Assessment scores—physical setting	–	Statistics	–	Analysis of ED in terms of elements + capturing the flow of learning and assessment activities	Evaluation of overall educational design	Ad hoc

References

Agudo-Peregrina ÁF, Iglesias-Pradas S, Conde-González MÁ et al (2014) Can we predict success from log data in VLEs? Classification of interactions for learning analytics and their relation with performance in VLE-supported F2F and online learning. Comput Hum Behav 31:542–550

Ali L, Hatala M, Gašević D et al (2012) A qualitative evaluation of evolution of a learning analytics tool. Comput Educ 58(1):470–489

Altrichter H, Feldman A, Posch P et al (2008) Teachers investigate their work: an introduction to action research across the professions. Routledge, London

Avramides K, Hunter J, Oliver M et al (2015) A method for teacher inquiry in cross-curricular projects: lessons from a case study. Br J Educ Technol 46(2):249–264

Bakharia A, Corrin L, de Barba P et al (2016) A conceptual framework linking learning design with learning analytics. Paper presented at the 6th international conference on learning analytics & knowledge, Edinburgh, 25–29 April 2016

Bos N, Brand-Gruwel S (2016) Student differences in regulation strategies and their use of learning resources: implications for educational design. Paper presented at the 6th international conference on learning analytics & knowledge, Edinburgh, 25–29 April 2016

Brereton P, Kitchenham BA, Budgen D et al (2007) Lessons from applying the systematic literature review process within the software engineering domain. J Syst Softw 80(4):571–583

Camilleri V, de Freitas S, Montebello M et al (2013) A case study inside virtual worlds: use of analytics for immersive spaces. Paper presented at the 3rd international conference on learning analytics and knowledge, Leuven, 8–12 April 2013

Check J, Schutt R (2012) Research methods in education. Sage Publications, London

Chounta IA, Avouris N (2016). Towards the real-time evaluation of collaborative activities: integration of an automatic rater of collaboration quality in the classroom from the teacher's perspective. Education and Information Technologies, 1–21

Dana NF, Yendol-Hoppey D (2014) The reflective educator's guide to classroom research: learning to teach and teaching to learn through practitioner inquiry. Corwin Press, London

Dawson S (2010) 'Seeing'the learning community: an exploration of the development of a resource for monitoring online student networking. Br J Educ Technol 41(5):736–752

Dawson SP, McWilliam E, Tan JPL (2008) Teaching smarter: how mining ICT data can inform and improve learning and teaching practice. Paper presented at the 2008 ASCILITE conference, Melbourne, 20 Nov–3 Dec 2008

Dunbar RL, Dingel MJ, Prat-Resina X (2014) Connecting analytics and curriculum design: process and outcomes of building a tool to browse data relevant to course designers. J Learn Analytics 1(3):220–240

Duque R, Gómez-Pérez D, Nieto-Reyes A et al (2015) Analyzing collaboration and interaction in learning environments to form learner groups. Comput Hum Behav 47:42–49

Dyckhoff AL (2011) Implications for learning analytics tools: a meta-analysis of applied research questions. Int J Comput Inf Syst Ind Manage Appl 3:594–601

Dyckhoff AL, Zielke D, Bültmann M et al (2012) Design and implementation of a learning analytics toolkit for teachers. Educ Technol Soc 15(3):58–76

Dyckhoff AL, Lukarov V, Muslim A et al (2013) Supporting action research with learning analytics. Paper presented at the 3rd international conference on learning analytics and knowledge, Leuven, 8–12 April 2013

Elbadrawy A, Studham S, Karypis G (2014) Personalized multi-regression models for predicting students performance in course activities. Paper presented at the 5th international conference on learning analytics and knowledge, NY, 16–20 March 2015

El-Bishouty MM, Saito K, Chang T et al (2015) Teaching improvement technologies for adaptive and personalized learning environments. In: Kinshuk, Huang R (eds) Ubiquitous learning environments and technologies, Springer, Berlin, pp 225–242

Emin-Martinez V, Hansen C, Rodrıguez Triana MJ et al (2014) Towards teacher-led design inquiry of learning. eLearning Papers (36)

Fernández-Delgado M, Mucientes M, Vazquez-Barreiros B et al (2014) Learning analytics for the prediction of the educational objectives achievement. Paper presented at the Frontiers in Education Conference, Madrid, 22–25 Oct 2014

Fernández-Gallego B, Lama M, Vidal JC et al (2013) Learning analytics framework for educational virtual worlds. Procedia Comput Sci 25:443–447

Fritz J (2016) LMS course design as learning analytics variable. Paper presented at the 6th international conference on learning analytics & knowledge, 1st learning analytics for curriculum and program quality improvement workshop, Edinburgh, 25–29 April 2016

Fulantelli G, Taibi D, Arrigo M (2015) A framework to support educational decision making in mobile learning. Comput Hum Behav 47:50–59

Gauthier G (2013) Using teaching analytics to inform assessment practices in technology mediated problem solving tasks. In IWTA@ LAK

Gómez-Aguilar DA, Hernández-García Á, García-Peñalvo FJ et al (2015) Tap into visual analysis of customization of grouping of activities in eLearning. Comput Hum Behav 47:60–67

Govaerts S, Verbert K, Duval E (2011) Evaluating the student activity meter: two case studies. In: Leung H, Popescu E, Cao Y et al (eds) Advances in web-based learning-ICWL 2011. Springer, Berlin, pp 188–197

Greller W, Drachsler H (2012) Translating learning into numbers: a generic framework for learning analytics. Educ Technol Soc 15(3):42–57

Greller W, Ebner M, Schön M (2014) Learning analytics: from theory to practice—data support for learning and teaching. In: Kalz M, Marco R (eds) Computer assisted assessment. Research into e-assessment, Springer International Publishing, pp 79–87

Hansen CJ, Wasson B (2016) Teacher inquiry into student learning: the TISL heart model and method for use in teachers' professional development. Nordic J Digital Literacy 10(1):24–49

Haya PA, Daems O, Malzahn N et al (2015) Analysing content and patterns of interaction for improving the learning design of networked learning environments. Br J Educ Technol 46 (2):300–316

Hernández-García Á, González-González I, Jiménez-Zarco AI et al (2015) Applying social learning analytics to message boards in online distance learning: a case study. Comput Hum Behav 47:68–80

Howlin C, Lynch D (2014) Learning and academic analytics in the Realizeit System. Paper presented at the world conference on e-learning in corporate, government, healthcare, and higher education, Louisiana, 2014

Hung JL, Hsu YC, Rice K (2012) Integrating data mining in program evaluation of K-12 online education. Educ Technol Soc 15(3):27–41

Jaggars SS, Xu D (2016) How do online course design features influence student performance? Comput Educ 95:270–284

Karkalas S, Mavrikis M (2016) Towards analytics for educational interactive e-books: the case of the reflective designer analytics platform (RDAP). Paper presented at the 6th international conference on learning analytics & knowledge, Edinburgh, 25–29 April 2016

Kaufman TE, Graham CR, Picciano AG et al (2014) Data-driven decision making in the K-12 classroom. In: Spector JM, Merrill MD, Elen J, Bishop MJ (eds) Handbook of research on educational communications and technology. Springer, New York, pp 337–346

Kitchenham B, Charters S (2007) Guidelines for performing systematic literature reviews in software engineering. Keele University and Durham University

Kladich S, Ives C, Parker N et al (2013) Extending the AAT tool with a user-friendly and powerful mechanism to retrieve complex information from educational log data. In: Human-computer interaction and knowledge discovery in complex, unstructured, big data. Springer, Berlin, pp 334–341

Kosba E, Dimitrova V, Boyle R (2005) Using student and group models to support teachers in web-based distance education. User modeling 2005. Springer, Berlin, pp 124–133

Lockyer L, Heathcote E, Dawson S (2013) Informing pedagogical action: aligning learning analytics with learning design. Am Behav Sci 57(10):1439–1459

Mandinach E (2012) A perfect time for data use: using data driven decision making to inform practice. Educ Psychol 47(2):71–85

Marcos-García JA, Martínez-Monés A, Dimitriadis Y (2015) DESPRO: a method based on roles to provide collaboration analysis support adapted to the participants in CSCL situations. Comput Educ 82:335–353

Marsh JA, Farrell CC (2014) How leaders can support teachers with data-driven decision making a framework for understanding capacity building. Educ Manage Adm Leadersh 1–21

Martinez-Maldonado R, Schneider B, Charleer S et al (2016) Interactive surfaces and learning analytics: data, orchestration aspects, pedagogical uses and challenges. Paper presented at the 6th international conference on learning analytics & knowledge, Edinburgh, 25–29 April 2016

Mazza R, Dimitrova V (2007) CourseVis: a graphical student monitoring tool for supporting instructors in web-based distance courses. Int J Hum Comput Stud 65(2):125–139

Mazza R, Bettoni M, Faré M et al (2012) Moclog–monitoring online courses with log data. Paper presented at the 1st moodle research conference, Heraklion, Crete, 14–15 Sept 2012

Mazza R, Milani C (2005) Exploring usage analysis in learning systems: gaining insights from visualisations. Paper presented at the communication dans: the workshop on usage analysis in learning systems, the twelfth international conference on artificial intelligence in education, pp 65–72

McKenney S, Mor Y (2015) Supporting teachers in data-informed educational design. Br J Educ Technol 46(2):265–279

Melero J, Hernandez-Leo D, Sun J et al (2015) How was the activity? A visualization support for a case of location-based learning design. Br J Educ Technol 46(2):317–329

Mendez G, Ochoa X, Chiluiza K et al (2014) Curricular design analysis: a data-driven perspective. J Learn Analytics 1(3):84–119

Merceron A (2012) Investigating the core group effect in usage of resources with analytics. Paper presented at the 2nd international conference on learning analytics and knowledge, Vancouver, 29 April–02 May 2012

Minović M, Milovanović M, Šošević U et al (2015) Visualisation of student learning model in serious games. Comput Hum Behav 47:98–107

Mirriahi N, Dawson S (2013) The pairing of lecture recording data with assessment scores: a method of discovering pedagogical impact. Paper presented at the 3rd international conference on learning analytics and knowledge, Leuven, 8–12 April 2013

Monroy C, Rangel VS, Bell ER et al (2015) A learning analytics approach to characterize and analyze inquiry-based pedagogical processes. Paper presented at the 5th international conference on learning analytics and knowledge, NY, 16–20 March 2015

Mor Y, Ferguson R, Wasson B (2015) Editorial: learning design, teacher inquiry into student learning and learning analytics: a call for action. Br J Educ Technol 46(2):221–229

Nunn S, Avella JT, Kanai T et al (2016) Learning analytics methods, benefits, and challenges in higher education: a systematic literature review. Online Learning 20(2)

Ochoa X (2016) Simple metrics for curricular analytics. Paper presented at the 6th international conference on learning analytics & knowledge, 1st learning analytics for curriculum and program quality improvement workshop Edinburgh, 25–29 April 2016

OECD (2009) Teacher evaluation: a conceptual framework and examples of country practices. OECD review on evaluation and assessment frameworks for improving school outcomes. http://www.oecd.org/edu/school/44568106.pdf. Accessed 22 Aug 2016

OECD (2013) Teachers for the 21st century: using evaluation to improve teaching. OECD Publishing. http://www.oecd.org/site/eduistp13/TS2013%20Background%20Report.pdf. Accessed 22 Aug 2016

Papamitsiou Z, Economides A (2014) Learning analytics and educational data mining in practice: a systematic literature review of empirical evidence. Educ Technol Soc 17(4):49–64

Pardo A, Ellis RA, Calvo RA (2015) Combining observational and experiential data to inform the redesign of learning activities. Paper presented at the 5th international conference on learning analytics and knowledge, NY, 16–20 March 2015

Prieto LP, Villagrá-Sobrino S, Jorrín-Abellán IM et al (2011) Recurrent routines: analyzing and supporting orchestration in technology-enhanced primary classrooms. Comput Educ 57 (1):1214–1227

Prieto LP, Sharma K, Dillenbourg P et al (2016) Teaching analytics: towards automatic extraction of orchestration graphs using wearable sensors. Paper presented at the 6th international conference on learning analytics & knowledge, Edinburgh, 25–29 April 2016

Rienties B, Toetenel L, Bryan A (2015) "Scaling up" learning design: impact of learning design activities on LMS behavior and performance. Paper presented at the 5th international conference on learning analytics and knowledge, NY, 16–20 March 2015

Rodríguez-Triana MJ, Martínez-Monés A, Asensio-Pérez JI et al (2015) Scripting and monitoring meet each other: aligning learning analytics and learning design to support teachers in orchestrating CSCL situations. Br J Educ Technol 46(2):330–343

Romero C, Ventura S, García E (2008) Data mining in course management systems: Moodle case study and tutorial. Comput Educ 51(1):368–384

Sin K, Muthu L (2015) Application of big data in education data mining and learning analytics—a literature review. J Soft Comput 5(5):1035–1049

Slotta JD, Tissenbaum M, Lui M (2013) Orchestrating of complex inquiry: three roles for learning analytics in a smart classroom infrastructure. Paper presented at the 3rd international conference on learning analytics and knowledge, Leuven, 8–12 April 2013

Smolin D, Butakov S (2012) Applying artificial intelligence to the educational data: an example of syllabus quality analysis. Paper presented at the 2nd international conference on learning analytics and knowledge, Vancouver, 29 April–02 May 2012

Society for Learning Analytics Research (SOLAR) (2011) Proceedings of the 1st international conference on learning analytics and knowledge

Schwarz BB, Asterhan CS (2011) E-moderation of synchronous discussions in educational settings: a nascent practice. J Learn Sci 20(3):395–442

Timperley H, Wilson A, Barrar H et al (2010) Teacher professional learning and development. Report for the New Zealand Ministry of Education. http://www.oecd.org/edu/school/48727127.pdf Accessed 22 Aug 2016

Toetenel L, Rienties B (2016) Analysing 157 learning designs using learning analytic approaches as a means to evaluate the impact of pedagogical decision making. Br J Educ Technol 47 (5):981–992

Van Leeuwen A, Janssen J, Erkens G et al (2014) Supporting teachers in guiding collaborating students: effects of learning analytics in CSCL. Comput Educ 79:28–39

van Leeuwen A, Janssen J, Erkens G et al (2015) Teacher regulation of cognitive activities during student collaboration: effects of learning analytics. Comput Educ 90:80–94

Wasson B, Hanson C, Mor Y (2016) Grand challenge problem 11: empowering teachers with student data. In: Eberle J, Lund K, Tchounikine P, Fischer F (eds) Grand challenge problems in technology-enhanced learning II: MOOCs and beyond. International Publishing, Springer, New York, pp 55–58

Wong WY, Lavrencic M (2016) Using a risk management approach in analytics for curriculum and program quality improvement. In Paper presented at the 6th international conference on learning analytics & knowledge, 1st learning analytics for curriculum and program quality improvement workshop Edinburgh, 25–29 April 2016

Yen CH, Chen IC, Lai SC et al (2015) An analytics-based approach to managing cognitive load by using log data of learning management systems and footprints of social media. Educ Technol Soc 18(4):141–158

Zhang H, Almeroth K, Knight A et al (2007) Moodog: tracking students' online learning activities. Paper presented at the world conference on educational multimedia, hypermedia and telecommunications, Vancouver, 25–29 June 2007

Chapter 3
A Landscape of Learning Analytics: An Exercise to Highlight the Nature of an Emergent Field

Alejandro Peña-Ayala, Leonor Adriana Cárdenas-Robledo and Humberto Sossa

Abstract Before the increasing efforts for understanding, predicting, and enhancing students' learning in educational settings, *learning analytics* (LA) emerges as a candidate research area to tackle such issues. Thus, several work lines have been conducted, as well as diverse conceptual and theoretical perspectives have been arisen. Moreover, quite interesting and useful outcomes have been produced during the LA short lifetime. However, a clear idea of diverse questions is still pending to be given. (e.g., what does learning analytics mean? what are its backgrounds, related domains, and underlying elements? which are the objects of its applications? and what about the trends and challenges to be considered?) This is the reason why the chapter aims at responding those concerns by a sketch of a conceptual scenery that explains the LA background, its underlying domains and nature, including a survey of recent and relevant approaches, and a relation of risks and opportunities.

Keywords Analytics · Learning analytics · Learning settings · Learner · Prediction · Performance · Behavior · Assessment

---

A. Peña-Ayala (✉)
WOLNM: Artificial Intelligence on Education Lab, 31 Julio 1859, No. 1099-B,
Leyes Reforma, 09310 Iztapalapa, Ciudad de México, Mexico
e-mail: apenaa@ipn.mx

A. Peña-Ayala · L.A. Cárdenas-Robledo
Instituto Politécnico Nacional, Escuela Superior de Ingeniería Mecánica y Eléctrica
Unidad Zacatenco, Av Miguel Othón de Mendizabal, S/N, La Escalera,
07738 Gustavo A. Madero, Ciudad de México, Mexico
e-mail: adriposgrado@gmail.com

H. Sossa
Instituto Politécnico Nacional, Centro de Investigación en Computación,
Av. Juan de Dios Bátiz S/N Casi Esq. Miguel Othón de Mendizábal,
Col. Nueva Industrial Vallejo, 07738 Gustavo A. Madero, Ciudad de México, Mexico
e-mail: hsossa@cic.ipn.mx

© Springer International Publishing AG 2017
A. Peña-Ayala (ed.), *Learning Analytics: Fundaments, Applications, and Trends*, Studies in Systems, Decision and Control 94,
DOI 10.1007/978-3-319-52977-6_3

Abbreviations

AA	Academic analytics
CBE	Computer-based education
CBIS	Computer-based information systems
CMS	Courseware management systems
CSCL	Computer-supported collaborative learning
EDM	Educational data mining
IMS	Instructional management systems
ITS	Intelligent tutoring systems
KBS	Knowledge-based systems
LA	Learning analytics
LAK	Learning analytics and knowledge conference
LMS	Learning management systems
m–Learning	Mobile learning
MOOC	Massive open online courses
SRL	Self-regulated learning
SIS	Student information systems
SNA	Social network analysis
SoLAR	Society for learning analytics research

3.1 Introduction

Since computers were used for educational, training, teaching, and learning purposes, diverse sorts of computer-based education (CBE) paradigms have appeared, as the *computer-aided instruction*. In the early 1960s, this paradigm inspired the development of computer programs statically organized to embody both the domain and the pedagogical knowledge of the expert (Wenger 1987). As result of the CBE evolution, its wide application and increment of users, an explosion of *log data* has demanded interdisciplinary views to know *how learning occurs*. The goal is improving and enhancing learning at all stages, as well the study of diverse phenomena that happen in academic settings such as student retention and leaning achievement.

It is because of that since the second mid-2000s several research domains have born and tried to respond diverse issues, as the aforementioned, according to specific perspectives, targets, and frameworks. A sample of them correspond to educational data mining (EDM) (Peña–Ayala 2014a), educational data science (Piety et al. 2014), and LA (Larusson and White 2014). Since then, such domains have been evolving, defining their identity, affecting its domain, and extending their scope.

Thus, with the goal to explain the LA essence, the following research questions are made: (1) What is the LA background? (2) Which domains are related to LA?

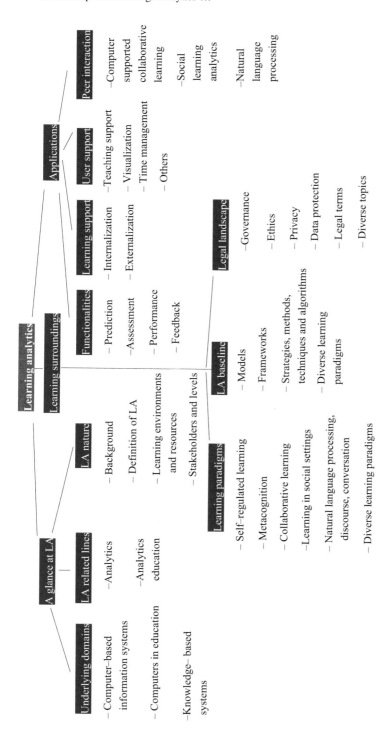

Fig. 3.1 A conceptual landscape of learning analytics

(3) What is the LA sense? (4) What are the learning paradigms related to LA research? (5) How is the LA theoretical baseline composed by? (6) Which are the main application targets for the LA approaches? (7) What to expect from LA?

With the aim at responding those questions, this chapter tailors a conceptual view that privileges essential items. The scene stated since Sect. 3.3 is inspired on a sample of recent works, which has been published in journals from 2014 up to date and pays attention on essential concerns that characterize the LA labor.

Thus, the conceptual LA landscape stated in this chapter embraces three sections, illustrated in Fig. 3.1. In the second section, a profile of LA is stated through its ancestor domains and related fields, as well as the definition of its nature. The third section unveils learning paradigms and theoretical elements, in addition to uncover legal topics. In the fourth section a sample of approaches is organized according to functionalities, peer interaction, learner and user support. Finally, in the fifth section, a vision of LA labor is traced to address future research, as well as a summary of the work and the responses to the research questions are provided.

3.2 A Glance at Learning Analytics

LA is a surfacing field interested in improving learner success, as well as teaching efforts, where research questions, as the ones raised by Ochoa et al. (2014), are made: How do we measure the important characteristics of the learning process, and how do we use those measurements to improve it? Thus, LA aims at developing models, methods, and tools that can be widely used, whose deliverables are reliable and valid at a scale beyond a course or cohort to provide benefits for learners and educators without distracting or misleading them (Ferguson et al. 2014).

With the purpose to orient researchers interested in practicing LA labor, a profile is outlined in this section, where a scenery of both underlying and related lines is stated to provide the background that surrounds LA arena. Moreover, a sketch of the LA nature is drawn, as well as the LA stakeholders and levels are unveiled.

3.2.1 Underlying Domains

The origin, nature, and scenery that surrounds LA can be recognized through three essential domains illustrated in Fig. 3.2. Therefore, this section provides a summary of those domains to frame the LA environment and labor.

3.2.1.1 Computer-Based Information Systems

Since the invention of the computers, *data processing* has been one of the most demanded application targets for computer systems, particularly in business arena

Fig. 3.2 Underlying domains that skirt learning analytics field

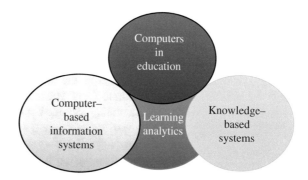

(e.g., government, banks, companies, and schools), where *computer-based information systems* (CBIS) have been built to satisfy such a need (Vlahos 2004).

By means of software engineering methods, techniques, and tools, as well as computer programming languages, CBIS have been built to gather, validate, update, store, and process data with the purpose to generate valuable information that reveals basic knowledge about diverse domains (Leach 2016).

Some of the classic applications of CBIS concern to the administration tasks, including decision-making, that are daily achieved in the organizations. They require suitable management of transactions, accurate data processing to produce useful information, and specialized tasks that discover valuable knowledge to decide action courses oriented to guide the behavior of people and organizations.

Thus, a diversity of CBIS, data models, and computational data processes have emerged to support those duties such as: management information systems (Lucey 2005), decision support systems (Power and Sharda 2009), customer relationship management (Kim 2003), enterprise resource planning (Chand et al. 2005), databases (Harrington 2016), data warehouses (Rahman 2016), online analytical processing (Alkharouf et al. 2005), and business intelligence (Isik et al. 2013).

Inclusive academic CBIS, labeled as *student information systems* (SIS), have been built to facilitate planning, organization, management, control, and monitoring labor in various school areas of concern. SIS provide administrative information and analyzed data (e.g., student profile, enrollment files, courses approved, tuition, scholarships, attendance records, and teaching evaluation …) to the internal and external academic stakeholders (Elouazizi 2014).

3.2.1.2 Knowledge-Based Systems

Another branch of data processing computer systems corresponds to *knowledge-based systems* (KBS) (Flasiński 2016), which are grounded on artificial intelligence and sophisticated computational paradigms that are oriented to acquire, elicit, discover, represent, and exploit knowledge about different domains.

KBS represent a broad diversity of approaches such as: expert systems (Grove 2000), knowledge bases (Laskey 2008), computational intelligence (Azar and

Vaidyanathan 2015), evolutionary computing (Eiben and Smith 2015), neural networks (Schmidhuber 2015), fuzzy logic (Yager and Zadeh 2012), soft computing (Karray and De Silva 2004), and rule-based systems (Ligêza 2006).

In addition, formal disciplines as statistic, probability, and operations research have provided essential basis to underpin the KBS line of *knowledge discovery* that includes: machine learning (Kodratoff and Michalski 2014), knowledge discovery in data bases (Wille 2002), data mining (Shmueli et al. 2016), data sciences (Schutt and O'Neil 2013), as well the assortment of analytics stated in a further section.

The scope and value of knowledge discovery is rising as result of the deployment of CBIS and KBS approaches on the internet. Moreover, its relevance is growing due the tendency to build distributed computing (Powell 2012), parallel processing (Moldovan 2014), and cluster computing (Colmenares et al. 2015) that have produced novel paradigms as big data (Marr 2015) and cloud computing (Chao 2012).

As consequence of the aforementioned trends, astronomical distributed data repositories are object of knowledge discovery applications that unveil valuable descriptive and predictive patterns, as well reveal findings and analysis to support the labor being achieved in diverse sorts of organizations.

3.2.1.3 Computers in Education

Education field has also been an object of research, development, and practice in the computer systems arena. It is because of that CBE has been conceived as *knowledge communication systems* (Wenger 1987), which represent a mixture of CBIS and KBS that also include specific requirements, functionalities, and content. That is why the first CBE applications emerged in the early 60s, as for example the nationwide online system built by Harvey Long to train International Business Machines field engineers (Hunter 2005), and the approach to generate problems on arithmetic and questions about vocabulary recall tailored by Leonard Uhr (1969).

Since then an explosion of teaching and learning systems emerged as the following: *computer-assisted instruction* that identifies and meets the needs of individual learners who interact with the computer to receive content and feedback (Tolman and Allred 1984); *intelligent tutoring systems* (ITS) which behave like problem-solving monitors, coaches, laboratory instructors, and companions (Sleeman and Brown 1982); *computer-assisted learning* to support diverse learning methods, where student is paired with a computer as virtual teacher (Guttormsen and Krueger 2000); and *intelligent learning environments* that include ITS traits and items for student-driven learning and knowledge acquisition (Brusilovsky 1999).

Later with the inclusion of personal computers, local and wide access networks, and the internet, CBE systems extended their scope and included novel functionalities to arise: *computer-supported collaborative learning* (CSCL), which recreates a mediated setting that helps learners to communicate in joint activities and provide assistance in their coordination and application of knowledge (Sinha et al. 2015).

Courseware management systems (CMS) to support the routine of teachers in the classroom (e.g., publishing material, examples, and auto-grade quizzes) (Weber and Brusilovsky 2001). *Learning management systems* (LMS) to allow instructors and students to share materials, submit and return assignments, and communicate online (Lonn and Teasley 2009). *Instructional management systems* (IMS) combine curriculum, assessment, practice tools, and student data for educators and staff into one online system (Ysseldyke et al. 2004). *Web-based educational systems* emerged as an open platform to spread teaching–learning services (Su et al. 2005).

Moreover, a trend to provide learner–centered education inspired the conception of adaptive and intelligent CBE systems such as: *adaptive learning systems* that adapt curricula, content, sequencing, assessment, and assistance according to learners' goals, skills, and progress (Rehak et al. 2000); *intelligent educational systems*, which include curriculum sequencing, solution analysis, and interactive problem solving to provide the student with individually planned series of tasks (Brusilovsky and Peylo 2003); *intelligent and adaptive systems* that combine traits and functionalities of both sorts, adaptive and intelligent systems (Peña–Ayala 2013).

At present, Web 2.0, wireless communication, mobile supplies, smart devices, as well as ubiquitous and pervasive computing inspire CBE approaches, such as: *mobile learning* (m-Learning) that allows access to learning content and information from anywhere and at any time (Ally 2009); *ubiquitous learning* is based on a model of human–computer interaction in which computer processing has been integrated fully into daily activities and objects with which users routinely interact (Peña–Ayala 2015); *pervasive learning* recreates an intelligent environment and context awareness to shape immersive experiences that mediate between the learner's mind, physical objects, and virtual contexts (Laine and Joy 2009).

Moreover, a trend for spreading education has emerged to privilege its *socialization* through two well-known options. The first corresponds to *education based on social networks*, also called *educational networking*, which use social networking technologies for educational purposes (Holcomb et al. 2010). The second delivers education at scale by means of *massive open online courses* (MOOC). Where MOOC teach domain knowledge based on 'connectivism' paradigm for learning (named cMOOC), as well as a hyper-centralized, content-based, and linear sequence of topics, named xMOOC (Margaryan et al. 2015). Inclusive, the opposite option is called *small private online courses* (Hecking et al 2014).

3.2.2   *Learning Analytics Related Lines*

LA is an emerging and expanding field that pertains to a trend labeled by the term *analytics*. This filed pursues to go in depth of huge databases and beyond of the scope reached by classic *knowledge discovery* approaches. Its goal is to draw a wide, profound, and multidimensional viewpoint that under a holistic perspective describes and predicts events and situations of a given domain. Thus, in order to state an overall idea of what analytics is, diverse kinds of analytics are identified, including those oriented to educational settings.

3.2.2.1 Analytics

Analytics is a novel branch that is gaining interest in *knowledge discovery*. Analytics tries to collect valuable data, intelligently examine information, and discover useful knowledge to support decision-making. According to Campbell et al. (2007): "Analytics marries large data sets, statistical techniques, and predictive modeling. It represents the practice of mining institutional data to produce actionable intelligence". Analytics tools provide statistical evaluation of rich data sources to discern patterns that help decision–making in organizations (EDUCASE 2010).

One derivative line corresponds to *web analytics*, which according to McFadden (2005) is: "The collection, analysis and reporting of web site usage by visitors and customers to understand the effectiveness of online …" It involves the compilation of data from many users, where trends are noted, hypotheses are formed, and adaptations to the web site are implemented and tested (Rogers et al. 2010).

Others terms corresponds to: *data analytics*, which is defined as: the application of computer systems to the analysis of large data sets for the support of decision–making; where data are assessed, selected, cleaned, visualized, and analyzed, as well as the outcomes are interpreted (Runkler 2012); *big data analytics* has the goal to analyze massive datasets, which increasingly occur in web-scale business intelligence problems, by distributing the processing utilizing massive parallel analysis systems (Saecker and Markl 2013); *business analytics* is the practice of bringing quantitative data to bear on decision–making by a range of data analysis methods, including data visualization and reporting understanding (Shmueli et al. 2016).

What is more, the variety of terms also includes: ubiquitous analytics (Elmqvist and Irani 2013), social network data analytics (Aggarwal 2011), predictive analytics (Siegel 2013), advanced analytics (Ryza et al. 2015), visual analytics (Ceneda 2016), text analytics (Xiang 2015), and supply chain analytics (Souza 2014).

3.2.2.2 Analytics in Education

Although analytics have arrived later to education than to government (Siemens 2014), analytics has ventured to support academic labor. That is why specialized lines have emerged, since those that are suitable for chancellors of academic institutions, whilst others concentrate on the learning process, as well as the own learner.

One of such lines is *institutional analytics* that focuses on the business side of higher education. Within institutions, units (e.g., finance, advancement, institutional research, and effectiveness) have troves of data related to institutional performance.

The aim is to discover new efficiencies, cost savings, or revenue streams by means of the potential of analytics approaches (Brooks and Thayer 2016).

Another line is *academic analytics* (AA), term coined by WebCT as Katz says in the "preface" section of Goldstein and Katz (2005). According to Goldstein (2005): "AA is the intersection of technology, information, management culture, and the application of information to manage the academic enterprise".

Even though they were interested in the application of information to support decision-making in the academy business functions, they preferred the term AA, instead of business intelligence, because AA initiatives predict which students are in academic difficulty, allowing faculty and advisors to customize learning paths or provide instruction tailored to learning needs (Campbell et al. 2007).

Whereas Norris et al (2008) call to play more action into AA by means of *action analytics*, which define it as: "Learner-centric that focuses on issues related to access, affordability, and success for learners at all stages of their learning lives". Over time, action analytics empowers learners to take greater responsibility for their success, in collaboration with parents, teachers, mentors, and employers.

According to Piety et al (2014), *learner analytics* collects information around differences among learners with regard to cognitive traits like aptitudes, cognitive styles, and the like. It also studies learners' non-cognitive traits, such as: motivation, attitudes toward content, attention and engagement, expectancy and incentive, experiences, extra-curricular interests, socioeconomic status, and family situations.

Social learning analytics refers to the use of social network analysis (SNA) for learning processes, with the goal of understanding, explaining, and improving learning as a means to observe the social learning occurring in classes with a high number of geographically distant and disconnected users (Hernández–García et al. 2015).

In addition, there are some knowledge discovery lines closely related to LA, that although they are not "authentic analytics" (i.e., because hold their own background and purpose), it is convenient to consider such as: EDM (Peña–Ayala 2014b), educational data science (Williamson 2015), science education (Talbot–Smith et al. 2013), learning sciences (O'Neill 2016), and information visualization (Ware 2012).

There are other fields related with LA that are worthy to be mentioned: pedagogy (Knight et al. 2014), psychometrics (Gray et al. 2014), user modeling (Kahraman et al. 2016), human–computer interaction (Falzon 2015), computer-mediated communication (Walther et al. 2015), and recommender systems (Aggarwal 2016).

3.2.3 Learning Analytics Nature

Once diverse domains and lines that ground and shape the LA labor have been stated in the prior subsections, in this new one the nature of LA is outlined. Thus, as first step is pertinent to identify the LA background, which takes into account the foundation of the Society for Learning Analytics Research (SoLAR) and the diffusion of the field through its Learning Analytics & Knowledge Conference

(LAK) and the Journal of Learning Analytics.[1] As a second topic, a sample of relevant definitions for the LA term is given, as well as the academic settings and most common learning environments are recognized.

3.2.3.1 Background

Notwithstanding, the systemic use of analytics for improving teaching and learning is still emerging (Siemens 2014). Thus, it is possible to trace a timeline composed by three eras. Where the first corresponds to the pioneer works published before 2011. Whilst the emergence of SoLAR and the celebration of the first LAK conference, both in 2011, represent the kick off for the second era. In so far as the current one starts in 2014 with the launching of the Journal of Learning Analytics. Thus, a summary of the first two eras is given in this section, and the third is addressed in the rest of the chapter.

In relation to the first era, one of the earliest works that concerns to LA concept was the study of causes and cures of college students' attrition (Tinto 1987), where a theory of student departure from school is given to serve as a barometer of the social and intellectual health of college life as much as of the students' experiences.

Afterwards, the LA term appears in a briefing made by Mitchel and Costello (2000). They mention the report on opportunities in the e-Learning industry stated by SRI Consulting Inc., where the firm asserts: "The e-learning industry, now in its infancy, will grow into adulthood after 2005, driven by advances in e-commerce, according to the latest report, *emerging e-Learning industry*, stated by Language on Demand, Inc. Its adolescence, roughly spanning 2002–05, will see key opportunities in … *Tools for LA* …" Later on, Berk (2004) edits a report about the state of LA, which concerns the effectiveness of corporate training programs.

Other pioneer works correspond to Moore (2005) that explains how LA is able to measure the effectiveness of learning, which is related to the pedagogy. He seeks to understand whether learning effectively meets its original design objectives. While Retalis et al. (2006) work on *networked LA* to discover issues that help students move toward knowledge construction, understanding of topics of the subject domain, and problem-solving skills acquisition, either individually or in groups. In addition, Bach (2010) traces a framework for the development of LA and ethical issues involved in the application of its methods to educational contexts.

In relation to the second era, most of the LA work is published in LAK conferences organized by SoLAR, whose links appear in the first footnote. According to Conol et al. (2011), three indicators were considered to develop LAK conferences: (1) the growth of data surpasses the ability of organizations to make sense of it; (2) learning organizations make little use of the data learners "throw off" in the process of

---

[1]To know more about SoLAR, the Journal of Learning Analytics, and the first conferences, readers should visit: SoLAR: https://solaresearch.org/ Journal of Learning Analytics: http://learning-analytics.info/ LAK'2011: https://tekri.athabascau.ca/analytics/ LAK'2012: http://lak12.sites.olt.ubc.ca/ LAK'2013: https://lakconference2013.wordpress.com/.

accessing learning materials ...; (3) educational institutions are under growing pressure to reduce costs and increase efficiency .... As result, 27 papers make up the LAK'2011 proceedings, where SNA, prediction and tools are the main targets.

The proceedings of the LAK'2012 conference organize 42 papers into the following topics: SNA, adaptive-recommender systems, reflective learning, institutional views, educator interventions, visualization, EDM, and prediction. Whilst, the LAK'2013 conference publishes 40 papers related to next domains: visualization, SNA, collaboration, discourse, affects, prediction, MOOC, assessment, and architectures. The proceedings of these three first LAK conferences, as well as the subsequent ones, LAK'2014–LAK'2016, are worthy to be analyzed in detail by a future work that highlights the relevant findings and uncovers new tendencies.

3.2.3.2 Definitions of Learning Analytics

In order to shape a concept for LA, several definitions given through the time are stated, starting with Moore (2005), who said: "The term LA has been thrown around the training ... industry by everyone from technology heavyweights to analysts ... In short, LA means the study of the impact of learning on its learners". Later, Bach (2010) defines LA as: "The use of predictive modeling and other advanced analytic techniques to help target instructional, curricular, and content resources to support the achievement of learning goals". Whereas, Siemens (2010) asserts: "LA is the use of intelligent data, learner-produced data, and analysis models to discover information and social connections, and to predict and advise on learning".

After, Elis (2011) claims: "LA focuses on building systems able to adjust content, levels of support and other personalized services by capturing, reporting, processing, and acting on data on an ongoing basis in a way that minimizes the time delay between the capture and use of data". Whilst, Siemens et al (2011) say: "LA is the measurement, collection, analysis, and reporting of data about learners and their contexts, for purposes of understanding and optimizing learning and the environments in which it occurs". Whilst, van Barneveld et al. (2012) assert: "LA focuses on the learner, gathering data from CMS and SMS to manage student success, including warning processes where a need for interventions may be warranted".

Recent LA definitions includes the one given by Wise and Shaffer (2015) to state: "LA is the ability to discover patterns and associations across modalities (e.g., coordinating talk), over time (e.g., in the revisiting of studied material), or at a micro-genetic level (e.g., how a teacher uses analytics to monitor student activity)": Another is given by Brooks and Thayer (2016) to mean: "LA is the area of analytics investment and interest that is directly related to the student experience and learning outcomes with the aim at improving students' success and student services".

Before given a new definition for LA, two views of "what learning is" stated by Schmeck (2013) are expressed. One corresponds to the *experiential*, also called *phenomenological*, where learning is defined by individuals engaged in learning (i.e., learners depict their experience of events involved in learning). While, the *behavioral* point means, learning is an observable change in a person's reaction to an equally observable stimulus situation. Thus, in this work LA is defined as: "A research line that pursues to study, understand, describe, explain and predict the learning phenomenon, from both experimental and behavioral views, that happens in CBE settings by the support of CBIS and KBS to enhance the efficacy of teaching–learning experiences and increase the learners' achievements and gratification".

3.2.3.3 Learning Environments and Resources

Even though, the essence of LA concerns to all modalities (e.g., face-to-face, open …), levels (i.e., from pre-primary up to postgraduate), settings (e.g., classroom), and kinds of CBE systems, most of its applications have been oriented to higher distance education based principally on CMS, LMS, IMS, ITS, and MOOC.

Moreover, the main *source of data* represents the information automatically produced by such systems as result of the interaction between users and CBE systems, recording who accessed what, and when; as well as the dialogues expressed by learners in social networks and online forums. Use of this kind of data is *termed usage logs, audit trails, log data, dataset* … (Phillips et al. 2011). The data are used to track, how students use the CBE systems, as well as to reconstruct an individual student's online presence in detail. Therefore, the analysis of the association between variables of learners' behavior extracted from log data attain attention.

A good example of the information stated in log data is given by Joksimovic et al. (2015), who map Moodle logs to the next interaction types: student–student, student–teacher, student–system, and student–content. Whilst, Gašević et al (2015) assert: trace data contain time-stamped events about views of specific resources, attempts and completion of quizzes, or discussion messages viewed or posted. In consequence, they edit the following individual types of trace data recorded by a LMS across diverse courses: assignment, book, chat, course logins, feedback, forum, light box gallery, map, quiz, resource, virtual classroom, and Turnitin (i.e., program that can be integrated with Moodle to detect plagiarism).

In addition to source data, emerges the *synthetic data*, also known as simulated data, with the purpose to avoid accidental disclosure or reconstruction of information. According to Berg et al. (2016), such kind of data can be used as example data to train predictive models. Also, synthetic data are useful for benchmarks.

In another vein, *learning resources* are the main asset of any CBE system because represents the domain knowledge to be taught and learned. Online learning resources are authored through a diversity of multimedia content, such as static and dynamic web pages, quizzes, exercises, wikis, self–tests, as well as tools for manipulation and search to satisfy different learning needs. In this regard, Hecking et al. (2014) apply network analysis methods to investigate the dynamics of relations between students and resources to identify characteristic patterns of the courses.

As supplementary sources for exploitation appear SIS, curriculum, learner models, learning resources metadata, assessments, pedagogical strategies, learning paradigms, and multimodal data (e.g., gesture, eye-tracking, biosensors, object manipulation). This increasing relevance of datasets in LA is: they are information assets that, if used properly, empower many aspects of the academic institution labor.

Nevertheless, diverse issues need to be taken into account such as: the assumption that data stored within institutional SIS can be directly associated with an individual student's identity, something that is not always feasible due to ethical regulations (Ognjanovic et al. 2016). Other issue corresponds to the meaning of the masses of data collected, it is not always clear. One more is pointed out by Ochoa et al. (2014) who assert: A challenge is the collection of datasets that can be shared for research goals in order to collect data captured in real–life settings, from diverse learning settings, and to make such data available for comparison purposes.

In addition, Elouazizi (2014) identifies a couple of issues related to datasets as information assets: (1) they are often the worst governed, least understood, and most poorly utilized key asset; (2) they are dynamic in nature, multifaceted, and increase exposure to security and privacy risks.

3.2.3.4 Stakeholders and Levels

The complexity of LA is not only due its wide and heterogeneous underlying do-mains, it is also because the diversity and dynamic interests of its stakeholders to be met. As this regard, Buerck and Mudigonda (2014) claims: Few studies exist that depict the constraints that preclude a LA initiative from succeeding fully, meeting the criteria of success as defined by involved stakeholders.

The LA labor demands hierarchical, distributed and networked actions, as well as data streams, where stakeholders submit input, provide advice, and demand information, whose outcome represents knowledge, visualization, interpretation and analysis to enrich the learning process and reach the goals of all stakeholders. For instance, Knight and Littleton (2015a) acknowledge a three-tier architecture to collect, analyze, and feedback data to stakeholders at the macro, meso, and micro levels, which respectively correspond to region, state, national and international scope; institution wide, and individual user actions.

A key mapping between stakeholders and data uses is given by Elouazizi (2014), which also includes data sources. Some of its items are used to illustrate a proposed *taxonomy of LA stakeholders*, which also includes other instances provided by Reyes (2015), and Buerck and Mudigonda (2014) as follows:

- *Organizations*: (1) government: educational policies, improving accountability, assess of policies … (2) academic institutions: enrollment, attrition, desertion, graduates impact … (3) training companies: quality, prestige, certifications …
- *Formal areas*: (1) academic boards: instructional practices, assessment criteria … (2) research labs: pedagogy paradigms, learning styles … (3) counselling: students at risk, retention … (4) management: student flow-through and atten-dance …

- *Roles*: (1) students: access to learning resources, self-monitoring … (2) faculty: teaching effectiveness, learning processes, learners collaboration … (3) instructors: monitor student activity, warning signals … (4) researchers: EDM outcomes, visualization hints … (5) community and donors: educational outreach, investment profit, social impact … (6) learning systems, content and support staff: user experience, instructional design, user support services … (7) course coordinator: usefulness of content, problem solving, sequence patterns, learners interaction …
- *Individuals*: (1) department head: teaching effectiveness, program evaluation … (2) dean: enhancing reputation, empowering education; (3) executive officer: process optimization, improving graduation rates … (4) tutor: student behavior and performance … (5) parents: student status, achievements, issues …

3.3   Learning Surroundings

In this section[2] a conceptual, logistic, and juridical surrounding of LA is sketched to illustrate the learning paradigm that grounds LA approaches. Moreover, the theoretical frame that guides the development of LA approaches is outlined. Inclusive, the legal context to be taken into account to regulate LA labor is also exposed.

3.3.1   Learning Paradigms

According to Wise and Shaffer (2015): "There is a danger in thinking that with enough data, the numbers speak for themselves. In fact, with larger amounts of data, learning theory plays an ever more critical role in analysis". Thus, a sample of learning paradigms considered in the surveyed works is stated as follows.

3.3.1.1   Self-regulated Learning

According to Segedy et al. (2015), Self-Regulated Learning (SRL) is an active theory of learning that describes how learners are able to set goals, create plans, monitor their progress, and revise their plans. They develop coherence analysis to analyze students' SRL behaviors, which focus on students' ability to seek for, interpret, and apply information encountered while at problem-solving. Whilst, You (2016) aims at predicting course achievement by using frequencies, duration of

---

[2]Citations stated in Sects. 3.3 and 3.4 pertain to the papers published in journals since 2014, where their statements presented here could correspond to other authors cited in those works. Thus, readers should seek the real author of the exposed definitions in those citations.

study time, and measures that are indicative of SRL practices such as time-management.

In addition, Roll and Winne (2015) claim: Because learners are agentic, LA can inform them about options that may bear on the phases of their SRL. Thus, the goal is to explore how options for SRL are usefully described given the complexity of learning activities patterns. Whilst, Tabuenca et al. (2015) use mobile time–logs to foster SRL in m-Learning, they deliver LA to students via mobile chart-visualizations and notifications to help learners by raising awareness on time management.

Cutumisu et al (2015) deploy a game–based assessment, Posterlet, to measure students' SRL choices, in which learners design posters and apply graphic design principles from feedback. They found evidence that seeking negative feedback and revising are good behaviors for SRL. An additional series of LA and SRL works is: Sonnenberg and Bannert (2015) focus on the effect of metacognitive prompting on the sequential structure of SRL processes using process mining techniques; Timmers et al. (2015) apply SRL to facilitate regulation feedback on problem solving; Nussbaumer et al. (2015) build a competence—based service to support SRL; Colthorpe et al. (2015) evaluate students use of SRL strategies; Siadaty et al. (2015, 2016) measure SRL, as well as scaffold at micro-level processes of SRL.

3.3.1.2 Metacognition

Based on a third-party concept for metacognition (i.e., it is monitoring and regulating one's own knowledge, emotions, and actions), Chiu and Fujita (2014) assert: *Social metacognition* is a group members' monitoring and controlling one another's knowledge, emotions, and actions. In that way, they apply statistical discourse analysis to study how people influence one another through their interactions in online forums. Thus, they test whether three types of cognition and three types of social metacognition (ask for explanation, ask about use, and different opinion) increase the likelihoods of new information or theoretical explanations in next messages.

Furthermore, Segedy et al. (2015) apply coherence analysis to model learning behavior at problem-solving by measuring students' metacognitive behavior (e.g., goal setting, planning, monitoring, and reflection). Sonnenberg and Bannert (2015) provide metacognitive support for learning through inducing regulatory activities by asking students to reflect upon, monitor, and control their own learning process.

In addition, Nussbaumer et al. (2015) aim at fostering SRL and reflection. They also provide support for planning, goal setting, self–monitoring, and self-evaluation by metacognitive strategies (i.e., monitoring, evaluating, and planning the learning). Colthorpe et al. (2015) examine the SRL traits of students, using the evaluation of responses to meta-learning (i.e., awareness of one's own learning that involves various metacognitive aspects of learning) assessment tasks supported by access data from a LMS. As Siadaty et al. (2015, 2016) assert: Traces are the observable indicators about cognition that students create as they engage with a task. They achieve a trace–based micro analytic measurement of learners' metacognitive states.

3.3.1.3 Collaborative Learning

Paraphrasing Kelly et al. (2015), who cite several authors: *Collaborative learning* is ground on the *sociocultural theory* of Vygotsky, communities of practice, conversation analysis, and computers in cognition—including Bandura's social learning theory—where cognition is "distributed across people and tools, situated in contexts, within small groups, involved in activities and across communities of practice". From that view, they develop a tool for orchestration of online collaborative learning. In addition, Schneider and Pea (2015) study the effect of mutual visual gaze perception on student discourse in CSCL. They analyze students' linguistic coherence and coordination, and assess their value for estimating learning gain.

Moreover, van Leeuwen et al. (2015) find that groups of students may face problems concerning cognitive activities in CSCL. Thus, they provide LA adaptive support to teachers for monitoring and solving the occurrence of problems. Van Leeuwen (2015) reports on teacher use of diverse LA tools to diagnose student progress and intervene during student learning activities in CSCL. Whilst, Fidalgo–Blanco et al. (2015) build a LA system as support teachers to evaluate and monitor individual progress within teamwork. The goal is preventing problems, performing corrective measures, and making decisions to improve the learning process of teamwork.

3.3.1.4 Learning in Social Settings

Chung and Paredes (2015) review learning and social network theories to culminate at a social networks model for understanding learning and performance. One of their cited theories is *situated learning theory* that claims: Learning takes place in social situations where individuals develop skills by interacting with others who can provide them with insights about existing knowledge and previous experiences within a community of practice. Another mentioned theory is *social learning*, which assumes that modeling processes (e.g., attention, retention, motoric reproduction, reinforcement, and motivation) generate learning due their informative functions and that observers acquire symbolic representations of modeled associations.

Hernández–García et al. (2015), cite a work where two lines of social analytics are stated, one is labeled as *social learning analytics* (i.e., derived from the analysis of interpersonal relationships), another is called *discourse analytics* (i.e., focused on language–based constructed knowledge). They focus on the first one to study the emerging social network structures from student and teacher interactions in online classrooms and their relation to student performance. Moreover, Hecking et al.

(2014) analyze resource access patterns in a *blended learning* (i.e., provide hybrid approaches that combine the benefits of online learning and face-to-face sessions) course and a MOOC from collaborative wikis, self-tests, and thematic videos. They adapt SNA methods and apply them to dynamic bipartite student–resource networks built from event logs of the students' resource access.

3.3.1.5   Natural Language Processing, Discourse, and Conversation

Knight and Littleton (2015b) explore productive educational dialogue in classroom, or free-chat based environments. The approach lays on the *sociocultural theory*, which explains how individual mental functioning is related to cultural, institutional, and historical context. Their object of enquiry is the dialogue itself (as a representation of, and a tool for learning). They study the ways that dialogue is used to create common knowledge (i.e., a shared understanding built during the time by people involved in a dialogue, and which is an essential part of learning).

Also, Knight and Littleton (2015a) claim: Discourse-centric LA facilitates exploring the ways in which discourse about learning resources and evidences occurs. Their approach supports the study of ways in which small and large groups, as well as individuals make and share meaning together through their language. Moreover, they analyze diverse kinds of language from discipline specific, to argumentative and socio emotional associated with positive learning outcomes. What it is more, Ferguson and Clow (2015) investigate whether the same patterns of active engagement with course (e.g., content, assessment, and discussion) are found in MOOC that employ *social constructivist*. A pedagogy based on the conversational framework, where knowledge is jointly constructed through conversation, and contributing to or reading discussion comments are relevant parts of the learning process.

3.3.1.6   Diverse Learning Paradigms

Gray et al. (2014) show a depth review of psychometric variables to evaluate learners' academic performance, such as: (1) SRL; (2) *expectancy theory of motivation* explores how an outcome is a consequence of person behavior; (3) *goal theory of motivation* associates a student setting with challenging goals about academic achievement (e.g., performance goals—where a person is looking for favorable feedback—and learning goals—where people wish enhancing competency); (4) *self-determination theory* focuses on human innate psychological need for competency and analyzes the kinds of goals learners adopted and the reasons; where intrinsic motivation arises from enjoyment of activity, and extrinsic motivation reveals that the outcome is attractive; (5) *learning style* holds classic constructs known as instructional preference, information processing style, and cognitive personality style, whilst learners are classified as shallow, deep, and strategic.

Tabuenca et al. (2015) aim at facilitating a mobile tool that can be smoothly integrated by any student in his daily learning routine as a way of *seamless learning*, paradigm that makes the transitions between diverse learning situations and context as easy as possible. Whereas Andergassen et al. (2014) investigate potential correlations between learning results and LMS usage during exam preparation in *blended learning* courses, focusing on practice and repetition.

Wanli et al. (2015) based on the *theory of online learning as online participation* operationalize *activity theory* to holistically quantify students' activity in the CSCL. Where the former asserts: online learner participation, which is supported by physical and psychological tools, is a complex process of taking part and maintaining relations with others as well as all kinds of engaging activities. While the later offers a holistic framework that depicts activities in practice while join both individual and social behavior. The aim is to explore the development and representation of prediction models using data from a collaborative problem-solving environment.

In relation to Fulantelli et al. (2015), they shape a framework to facilitate educational decision-making in m-Learning that takes into account the relationships between the different types of interactions occurring in an activity and the tasks which are pedagogically important for learning activity. The framework's baseline includes the task model for mobile learners, which considers the activity theory.

In another vein, Serrano–Laguna et al. (2014) apply LA and game analytics in *game-based learning* scenarios (i.e., paradigm based on the *theory of narrative-centered learning* that holds two principles of cognitive processes, one where learners are transported, the other where learners perform the narrative). They aim at creating reliable assessment systems for videogames that facilitate the diverse forms of assessment. In addition, Gibson and de Freitas (2015) trace an exploratory analysis that summarizes methods, observations, challenges, and implications for games-based virtual performance assessment.

3.3.2 Learning Analytics Baseline

Even though LA is an incipient research line, an increasing and valuable work has been made to outline a conceptual baseline that supports research, development, deployment, and application of diverse kinds of approaches. Therefore, in this section an inventory of several works oriented to introduce some theoretical contribution (e.g., models, frameworks, methods, techniques, and tools) is given.

3.3.2.1 Models

The essential and most common underlying LA element are the models! Where diverse models trace how the LA process should be carry out, whilst others focus on particular targets of description. A sample of them is summarized as follows:

- *Three big data analytical models*: Conceives the intersection of descriptive, predictive, and prescriptive analytics models to support association, triggering, and assessment, and analytical representations; model proposed by Daniel (2015).
- *Learning analytics model*: Depicts the dynamic interaction of stakeholders with their data supported by visual analytics, for instance: self–organizing maps, to generate conversations, and solution-seeking (de Freitas et al. 2015).
- *Model for instructional effect of feedback*: Embraces five stages: learners initial state, search and retrieval strategies, response, evaluation, and adjust initial state, model used in computer-based formative assessment (Timmers et al. 2015).
- *Process of information problem solving*: Models five skills: defining information problem, searching information, scanning information, processing information, and organizing and presenting information, applied in (Timmers et al. 2015).
- *Social networks-based model for understanding learning and performance*: Studies the association between social networks, content richness in academic learning discourse, and performance, stated by Chung and Paredes (2015).
- *Community of inquiry*: Assists the design of educational experiences in communities and networks of learners, by the description of inquiry through the cognitive, social, and teaching presences, model used by Kovanović et al. (2015a).
- *5E+I/A*: Adds the intervention and acceleration steps to the five ones (e.g., engage, explore, explain, elaborate, evaluate, intervention, and acceleration) of the *science inquiry model* to comprise a lesson cycle, stated by Monroy et al. (2014).
- *Data and learning outcomes curriculum development model*: Depicts a learning-outcome-centered curriculum cycle that includes: specification of curricular design, identification of a conceptual model, curricular development, testing and evaluation, and curriculum refinement, described in (Méndez et al. (2014).
- *Technology-enhanced learning technology complex*: Identifies the components of the technology complex: pedagogy, stakeholders, communities, current practices, context, technical, and business model, cited in (Ferguson et al. 2014).
- *Technology-enhanced learning innovation process*: Highlights components to address educational projects, mentioned by Ferguson et al. (2014).
- *Analyze, design, develop, implement, and evaluate model*: Guides curriculum and course design through a cycle make up by those five actions. This former model is cited and used by Dunbar et al. (2014).
- *Structural topic model*: Finds syntactic patterns with semantic meaning in un-structured text, as well as identifies variation in those patterns, and unveils texts that exemplify documents within a topical pattern. Model stated by Reich et al. (2015).
- *LA frameworks and models*: Summarizes five models: (1) knowledge continuum: data, information, knowledge, wisdom; (2) five steps of analytics: capture, report, predict, act, refine; (3) web analytics objectives: define goals, measure, use, share; (4) collective applications model: select, capture, aggregate, process, display; (5) process of LA: select, capture, aggregate and report, predict, use,

refine, share. Knight and Littleton (2015a) cite a third-party work that presents such models, and compare them.

- *Model for retention of coherent understanding of complex phenomena*: Joins the knowledge integration framework and the research on distributed practice sequence in order to guide learning design. Model adopted by Svihla et al. (2015).
- *Course engagement*: Considers five observed behaviors (e.g., counts of logins, days, spent hours, posts viewed, posts authored) that reflect course engagement. Conceptual model sketched by Lowes et al. (2015).
- *RAPID Outcome Mapping Approach*: This model is based on the technology complex model with the goal to deploy LA for supporting technology—enhanced learning. This model has been introduced in the prior section (Ferguson et al. 2014).

3.3.2.2 Frameworks

Other relevant and popular LA ground corresponds to the frameworks! They provide key insights to be considered at designing, or represent a scaffold to orient the development of a LA approach. Diverse frameworks are identified as follows:

- *Framework to describe big data in higher education:* Acknowledges four items: institutional analytics, information technology, AA, and LA (Daniel 2015).
- *Ideal data:* Logically relates six data systems: administrative, research, student, teaching and learning, curriculum, and department (Daniel 2015).
- *Theoretical framework for student performance prediction model*: Joins theory, EDM, application, and LA through computation, optimization, interpretation, and contextualization tasks to predict performance. This framework is cited in (Wanli et al. 2015).
- *Quality indicators*: Standardizes the evaluation of LA tools and advices how to capture evidence for the impact of LA on education (Scheffel et al. 2014).
- *Construction of proxy variables*: Transforms an unstructured data set into proxy variables to define time management strategies, outlined by Jo et al. (2015).
- *Development of predictive models based on student data*: Studies the scaling issues concerned to portability and intervention effectiveness. Framework developed by the Open AA Initiative (Jayaprakash et al. 2014).
- *Online academic support environment design framework*: Considers learner interactions with the aim at creating a compelling online environment in which learners feel part of an academic support community (Jayaprakash et al. 2014).
- *Psycho-pedagogical*: Infers psychological assumptions on underlying skills and competences needed for problem solving, which includes both domain and user models to plan and set goals, use learning resources, assess knowledge and competence, and reflect on learning process. Framework outlined by Nussbaumer et al. (2015).
- *General framework for design of automated discourse analysis instruments and representations from first principles*: Guides the design of tools for assisting

instructors with the orchestration of online groups. Framework stated by Kelly et al. (2015).

- *Knowledge integration*: Advices learners to integrate their diverse and conflicting ideas about core topics by building coherent understanding through elicit, add, distinguish, and link ideas. Framework described and cited by Svihla et al. (2015).
- *Learning awareness tools-user eXperience*: It is an iterative five stages (e.g., problem identification, low and higher fidelity prototyping, pilot studies, and classroom use) workflow to guide the development of awareness tools for technology-enabled learning settings (Martínez–Maldonado et al. 2015).
- *Trace-based microanalytic measurement protocol*: Estimates the effects of scaffolding interventions on SRL processes by the achievement of prerequisites (e.g., formulation of the SRL model, defining micro-level SRL processes ...) and the measurement (e.g., identifying SRL interventions events, translating traces to SRL events, editing contingency records). Framework shaped by Siadaty et al. (2015).

3.3.2.3 Strategies, Methods, Techniques, and Algorithms

Once the first-tier underlying elements to conceive and design a LA approach have been identified, now a sample of strategies, methods, and techniques is going to be stated respectively in Tables 3.1, 3.2 and 3.3 in order to sketch an overview of

Table 3.1 Sample of strategies used in LA field

Name/author	Profile
Muñoz-Merino et al. (2015)	*Precise effectiveness*: analyzes learners' effectiveness with educational resources and activities in MOOC
Kim et al. (2016)	*Facilitative*: encourages potential low achiever students to participate during a course *Structured*: guides novice learners at an initial stage of discussion by means of discussion prompts
Chiu and Fujita (2014)	*Statistical discourse analysis strategies to address analytic difficulties*: contains suitable strategies to deal with diverse difficulties for data set, dependent and explanatory variables
Knight and Littleton (2015a)	*Inductive*: it is a data-driven approach to derive data, from which hypotheses are produced and validated *Deductive*: it is the traditional scientific hypothetic-deductive model that constructs hypotheses to test through data collection and analysis
Kovanović et al. (2015b)	*Time-oriented heuristics*: places an upper limit on the total session time or a single Web page time *Navigation-oriented heuristics*: seeks a web page connectivity to identify user sessions
Siadaty et al. (2015)	*Motivational strategies for learning questionnaire*: assess learners' motivational orientation and use of different learning strategies *Learning and study strategy inventory*: assesses the learning strategies that university students report using

Table 3.2 Sample of methods used in LA arena

Name/author	Profile
Fidalgo–Blanco et al. (2015)	*Comprehensive training model of the teamwork competence*: integrates tools from several LMS and facilitates registration of user interactions, as well as the access to teamwork evidence
Kim et al. (2016)	*Extracting proxy variables*: depicts key factors considered in asynchronous online discussion
Chiu and Fujita (2014)	*Statistical discourse analysis*: modeling online discussion processes to face diverse analytic difficulties involving the data, outcome, and explanatory variables
Knight and Littleton (2015a)	*Traditional and data-driven science*: compares the views of both methods, as well as their steps based on inductive and deductive strategies
Kovanović et al. (2015b)	*Time-on-task estimation*: analyzes learners performance based on two strategies, time-oriented heuristics, and navigation-oriented heuristics
Siadaty et al. (2015)	*Trace-based measurement protocol*: measures the effects of scaffolding interventions on SRL processes

Table 3.3 Sample of techniques used in LA domain

Name/author	Profile
Fidalgo–Blanco et al. (2015)	*Virtual teamwork*: includes forums, wikis, what's up … in a blended learning environment, where teams also have meetings in person
Muñoz–Merino et al. (2015)	*Expert validation*: gets diverse perspectives from the results according to the application context
Kim et al. (2016)	*Random forest*: features random sampling and ensemble strategies
Wise and Shaffer (2015)	*Sequence segmentation*: organizes sequences of user actions at each time a user completes an "evaluate" action
Knight and Littleton (2015a)	*Discourse functions, focus, content, and example discourse-centric learning analytics*: contains a set of suitable techniques devoted to four instances of relations between a specific function, focus, and content
Kovanović et al. (2015b)	*Time-on-task estimation*: calculates time spent reading discussions, time on-task from trace-data, and self-reported data on the amount of time students spent using the system

logistic frames that guide the LA labor. Moreover, in the sample of published journal papers in the LA arena, only three works reveal particular interest in the use of a kind of EDM algorithm. Thus, such works are going to be summarized as follows.

The first is the *tree-based* that uses random subsets of the data where each node in the tree is then split using the best split among all variables. It is used by Kim et al. (2016) to measure how well the proxy variables classify the low and high achievers in terms of accuracy, precision, recall, and specificity. The second algorithm is *k-means* that is oriented to clustering a data set according to a distance criterion of some attributes. It is applied by Wise and Shaffer (2015) with the purpose of using all of the students' action segments to develop a set of generalizable action segments. The third algorithm is pattern—matching, which is designed to generate patterns of traits that are common to a set of items that are used to compare against other instances to try to identify some similarities. Such an algorithm is deployed by Siadaty et al. (2015) to accomplish the goal of first searching for occurrences of all available patterns defined in the pattern library in users' log files.

3.3.2.4 Tools

A valuable resource in LA environments are the tools. A software application, a complete system or environment that facilities the development of specialized tasks to be achieved by learners, academic, and research staff. A sample of tools built, used or cited by the collected sample of LA works is introduced in this section.

The first is the "mobile environment for learning with linked open data", *MeLOD*. The tool provides didactical contents directly embedded into the learner's current situation and deploys m–Learning scenarios (Fulantelli et al. 2015). The second tool is *ALAS–KA*, the "add-on of the LA support of the Khan Academy" developed by Ruipérez–Valiente et al. (2015) as a module to extend the MOOC Khan Academy plaform. It provides visualizations and information for the class, as well as for individual students that help teachers and learners to make decisions. Gómez–Aguilar et al. (2015) develop the "semantic spiral timeline", *SPT*, to group and filter demands by providing a landscape of the overall use of CMS.

In another vein, Epp and Bull (2015) analyze uncertainty representation in visualizations of LA by a review of learning dashboards and visual analytics tools, such as: (1) *SQL-Tutor*, open learner model: shows learners what they have understood and misunderstood; (2) *Narcissus*: enables the monitoring and comparison of individual team member's contributions against one another … (3) *My-Pet*: illustrates learner's affect and motivation, as well as reflects the learner's observed interest in a topic; (4) *Pepper*: shows the relationship between discussion forums and learner interest, as well as depicts group level interest in topics; (5) *ProTutor*: tracks the user's ability to pronounce Russian characters; (6) *Next—TELL* open learner model: shows the relationship between competencies; (7) "student activity monitor", *SAM*: visualizes student activity; (8) *Comtella*: encourages learners to notice both their and their classmates' contributions, and highlights the differences in their contributions.

Sedrakyan et al. (2014) analyze modeling behavior using experimental logging functionality of the *JMermaid* modeling tool. Such a tool is a computer aided software engineering program that assists users in creating and validating models in the requirements analysis phase of an enterprise systems development project according to the method for requirements engineering, labeled *MERODE*. Whilst Reyes (2015) identifies the following LA resources: (1) *Blackboard analytics*: offers self-service analytics applications; (2) *GISMO*: displays data by a graphical interface; (3) *SNAPP*: is a SNA tool; (4) *Meerkat–ED*: analyzes users' participations and their interactions in forums; (5) *SunGard* assessment and curriculum management: facilitates assessment analysis and management, as well as curriculum management; (6) *Desire2Learn*: analyses engagement, retention, and learning outcomes.

Wise et al. (2014) aims at helping learners to monitor and regulate how they speak and listen in online discussions. Thus, with the purpose to foster engagement in discussions, they cite the *Starburst* discussion forum. Such a tool exhibits discussion threads as a radial tree, allowing learners to see the structure of the discussion and the location of their comments within it. Dunbar et al. (2014) build the tool "browser of student and course objects", *BoSCO*, to relate the analytics space and the course—curriculum design environment and encourage faculty to use analytics for course and curricular design. As for Nussbaumer et al. (2015), they tailor a SRL process model to depict learning as a cyclic sequence of four main phases: planning and goal setting, using learning resources, knowledge and competence assessment, and reflecting on learning behavior and progress. For each of these phases, a visual tool is built to support the respective cognitive and metacognitive activities.

Finally, van Leeuwen (2015) study the effects of LA tools and how teachers use them. Thus, she develops a case where collaborating groups have to read and analyze historical sources (using the *Sources* tool), discuss the information (by the *Chat* tool), and write a report (using the *Cowriter* tool). For all, the three activities are all automatically logged by the "Virtual collaborative research institute" CSCL system in the form of opening and closing of tools, messages, and written words.

3.3.3   Legal Landscape

With the goal to frame the LA labor, diverse topics should be considered for researchers, developers, and practitioners. Where the first one corresponds to the governance of the applications, data, and outcomes. Ethical and privacy matters constitute the next topics, which are needed to shape a legal framework to be observed by LA stakeholders. The fourth concerns to data protection matters, whilst legal terms are stated in the fifth topic. The last one describes a couple of subjects, one related to the epistemology nature of LA and the other to vulnerability.

3.3.3.1   Governance

In regards to governance, Elouazizi (2014) cites the definition given by Richard Alfred to say: "Governance is the process for distributing authority, power, and influence for academic decisions among campus constituencies". He also identifies some of the challenges of data governance modeling in the context of LA (e.g., the ownership of LA data sets, the interpretation of the data, and the decision-making based on data), as well as discusses the critical factors for designing data governance models (e.g., unicameral, bicameral, tri-cameral, and hybrid).

In order to inspire a LA data governance model that clearly defines who owns the physical LA data, who holds the interpretation of the LA data, and who is able to make decisions based on LA data, Elouazizi (2014) traces a relationship between LA stakeholders, data uses, and data sources overlaps.

Additionally, Berg et al. (2016) analyze the impact of *synthetic data* on LA infrastructures, with a particular focus on data governance. They consider data governance and ethics issues should be taken into account for any LA framework. Thus, they propose the use of synthetic data to foster the prototyping of services before the real data feed the LA application. Its availability supports proof of concept, security testing, practicing, and training around data governance processes.

3.3.3.2 Ethics

In order to clarify the meaning of ethics, Ferguson et al. (2016) cite the definition given by Drachsler and Greller to mean: "Ethics is the philosophy of morality that involves systematizing, defending, and recommending concepts of right and wrong conduct". In addition, Ferguson et al. (2016) identify 21 LA challenges with ethical dimensions, where six correspond to ethics as a restriction on action. One more is devoted to informed consent, three correspond to the ethical goal that should safeguard. Another concerns to an ethical goal of a society in which everyone has equal access to education, and one more is associated with the purpose of a just society.

Inclusive, Ferguson et al. (2016) suggest nine ethical goals related to: student success, educational institutions, private and group assets, property rights, educators and educational institutions, access to education, laws, freedom from threat, and integrity of self. Additionally, Khalil and Ebner (2016) cite a work that points out diverse categories of ethical issues, such as: transparency of data collection …, anonymization and de-identification of individuals, ownership of data, data accessibility and accuracy of the analyzed results, security of the examined data sets.

Even though, the extraction and analysis of learners and faculty online behavior can uncover useful insights into the learning process, these analyses also raise concerns about the ethics and privacy of these forms of analysis and research (Ognjanovic et al. 2016). Additionally, Elouazizi (2014) claims: Who designs and interprets the ethical guidelines for gathering, using, and purging such data? He also points out that: Some of LA stakeholders may generate or use LA data object under legal and ethical restrictions. Furthermore, LA data governance models should allow for a shared understanding of the ethical and legal aspects of using the data.

Moreover, Andergassen et al. (2014) acknowledge the need of an ethical framework that defines learner rights and data ownership, including opting out of the analytics record and giving informed consent for data usage to researchers. What is more, they cite six principles, given by third-party authors, towards an ethical framework, whose essence is: (1) LA should function as moral practice; (2) students should be seen as agents … (3) data collected about performance should be seen as temporal, dynamic constructs … (4) student success should be seen as a multidimensional phenomenon … (5) transparency about data usage; (6) need to use LA better to understand and develop outcomes for students.

As for the ethics view of Rodríguez–Triana et al. (2016), they depict the ethical issues that emerged in small-scale classroom oriented approaches that aim at supporting teachers in their practice learning scenarios such as: responsibility, transparency, consent, privacy, validity, minimizing impact, and stewardship of data. Also, they propose expanding an ethical framework for LA through a series of recommendations devoted to the following categories: consent, transparency, access, responsibility, stewardship, validity, privacy, and avoiding negative impact.

In regards ethical principles, Steiner et al. (2016) survey the following five series proposed by diverse authors they cited, which are worthy to be analyzed: privacy, consent, location of data, management, ownership, possibility of error, role of knowing, legal requirements, cultural and social norms, risks, safeguards, vulnerable groups, clarity, comfort and care, choice and consent, consequence and complaint, moral practice, students as agents, student identity, students success, use of data, transparency, right to access, student control over data, accountability and assessment, responsibility, transparency, validity, enabling positive interventions, minimizing adverse impacts, and stewardship of data.

What is more, Steiner et al. (2016) define the following eight principles relating to privacy, data protection, and ethics to scaffold their LEA'S Box privacy and data protection framework: data privacy, purpose and data ownership, consent, transparency and trust, access and control, accountability and assessment, data quality, and data management and security.

3.3.3.3 Privacy

Once again, it is convenient to consider another definition given by Drachsler and Greller that is cited in (Ferguson et al. 2016): Privacy is a living concept made out of continuous personal boundary negotiations with the surrounding ethical environment. Moreover, Ferguson et al. (2016) assert: Privacy is understood as a freedom from unauthorized intrusion: the ability of an individual or a group to seclude themselves or the information about them, and thus to express themselves selectively.

Additionally, a study about privacy, as well as a review about theories of privacy are published by Heat (2014), who claims: Privacy is an ill-defined concept and subject to various interpretations and perspectives, including those of philosophers, lawyers, and information systems specialists. Moreover she cites, a person can be said to have privacy if, in a given situation, the individual is offered protection from intrusion, interference, and information access by others.

Also, Heat (2014) traces a view of privacy through four levels: (1) individuals have privacy in a particular situation if they are offered protections from inference, information access, and intrusion; (2) normative and descriptive privacy situations; (3) control and limitation early theories of privacy; (4) recent theories:

ontological theory of information privacy, contextual integrity theory of information privacy, and hybrid Regional Alternative Licensing Centers theory of privacy.

Whereas Buerck and Mudigonda (2014) claim: Eny LA initiative should ensure compliance to various privacy related guidelines and laws, where external constraints concern to conventions, norms, and legal requirements pertaining to data privacy (e.g., the US Family Educational Rights and Privacy Act), research ethics.

In regards the proposal for expanding an ethical framework for LA made by Rodríguez–Triana et al. (2016), they provide a couple of recommendations for privacy: (1) if some data must be anonymous, be sure it cannot be re-identified by contextual information available to the users; (2) if you use data from external sources, be sure you can identify the owner correctly, and that other ethical and privacy issues are not put at risk when using those sources.

As for the data privacy principle that guides the LEA'S Box privacy and data protection framework, Steiner et al. (2016) assert: Collection and use of personal data must be fair and provide appropriate protection of privacy. Thus, they encourage designing and building data-sensitive educational apps equal to the well-established principles of other critical online solutions, including transaction numbers for accessing delicate information.

A key object of privacy corresponds to *personal information*, which gathers the minimal data necessary to identify an individual (e.g., name, address, photograph, Email, phone ...). In the education sector, this kind of information is known as *personal identifiable information*. Particularly, when such an information includes educational details (e.g., courses, qualifications, assessments ...). In contrast, a leak of learners' personal and educational information could induce misuse of data, and unreliable outcomes. Even though academic institutions that are engaged in exploiting LA approaches are demanded to share diverse findings, as well as provide some details about such kind of information (e.g., students' drop out, attrition ...).

Those are some reasons why Khalil and Ebner (2016) proposed a conceptual de-identification LA framework to prevent uncovering individual identity and keeping the personal identifiable information absolutely confidential. The framework begins with students involved in learning environments. In the next step the de-identification process applies techniques (e.g., anonymization, masking, blurring, and perturbation) to convert personal and private information into anonymized data take place. In the final step the de-identified data linked with a unique descriptor that may be examined by LA researchers and benefit stakeholders, but ultimately must be used only to the advantage of students.

On the side of Hoel and Chen (2016), they claim, for privacy-driven design as an essential part of LA applications development. Particularly, they are aware of giving priority to privacy in terms of data exchange and application design. Thus, they propose the "LA design space model", as a conceptual tool to ease the requirement solicitation and design for new LA solutions. Such a tool is organized as a cyclic workflow composed by three construction processes: problem space, solution space, and the design space and selecting a first solution.

According to Prinsloo and Slade (2016), vulnerability in the digital context depends on our understanding of the notion of privacy. Thus, they explore the Solove's taxonomy of privacy to illustrate the potential for increasing individuals' vulnerability, which embraces four categories of privacy problems that contains particular elements as follows: (1) information collection (e.g., surveillance, interrogation); (2) information processing (e.g., aggregation, identification, insecurity, secondary use, exclusion); (3) information dissemination (e.g., disclosure, exposure, accessibility, blackmail, appropriation, distortion); (4) invasion (e.g., intrusion, and interference).

3.3.3.4 Data Protection

A key topic is data protection, which corresponds to the logistic, processes, resources, and security measure to preserve data of authorized access, destruction, and lost. In this regards, Ferguson et al. (2016) define seven LA challenges with ethical dimensions that concern with data protection, and demand a legal response that depends on the area of jurisdiction and regional attitudes towards data security.

However, a "contradictory" statement is cited by Steiner et al. (2016) that says: The right to data protection is not an absolute right; it must be balanced against other rights, it needs to be considered and implemented always in relation to its function in society. However, they aim at defining a *privacy and data protection framework for a LA toolbox*, which considers several ethical and privacy principles for researching and exploring the educational possibilities of benefitting from LA without sacrificing privacy. In consequence, four of the eight principles stated for their framework correspond to data protection, such as: purpose and data ownership, access and control, data quality, and data management and security.

For its part, Cormack (2016) proposes a data protection framework for LA based on an approach used in data protection law, where he advises: An ethical framework should treat LA as two separate stages, using different justifications and their associated ways of protecting individuals. The first is called *analysis*, which is oriented to discover significant patterns treated as a legitimate interest of the organization that must include safeguards for individuals' interests and rights. Whilst the second is labeled *intervention*, which is devoted to the application of those patterns to meet the needs of particular individuals that require their informed consent or, perhaps in future, a contractual agreement.

3.3.3.5 Legal Terms

Ethical and legal objections to LA are barriers against the development of the field. In consequence, Sclater (2016) introduces an initiative to define a code of practice for LA. The code covers the main issues that institutions need to address in order to progress ethically and legally. As result, he tailors a taxonomy of ethical, legal, and logistical issues for LA, which embraces the following groups: ownership and control, consent, transparency, privacy, validity, access, action, adverse impact, and stewardship. In regards the code, it is grouped into the next eight areas: responsibility, transparency and consent, privacy, validity, access, enabling positive interviews, minimizing adverse impacts, and stewardship of data.

In relation to consent, Cormack (2016) reflects: To date LA has largely been a subject for educational research. However, the techniques are increasingly being adopted as part of the routine operation of universities and colleges. Such processes may affect all current and future students and staff, not just those who participate in research studies, through changes to how education is provided in general and through specific individual interventions. With this significantly increased impact, informed consent may no longer provide adequate protection and guidance either for individuals or for organizations.

What is more, Cormack (2016) cites works achieved by diverse authors to state: Law and ethics claim that for *consent* to be valid, it must be both informed and freely provided. Moreover, the law presumes that consent is not freely given in situations where the party requesting consent has significant power over the individual granting it. Inclusive, he states: The use of consent may well bias the outcomes of LA, potentially excluding those who have most to gain from the process. A consent demands individual learners to take responsibility for technologies and business practices that they do not create themselves, but find themselves increasingly dependent upon.

As this regard, Berg et al. (2016) analyze the student consent service that an analytics-based alert and intervention system is able to prompt staff and students in the case of certain specific situations. For instance, such a system offers students some degree of control over what is done with their data by means of a student consent service.

3.3.3.6 Diverse Topics

A couple of supplementary topics are introduced in this section, the first concerns to *epistemology*, known as the nature of knowledge. According to Knight et al. (2014): Epistemology is the philosophical study of what knowledge is, and what it means for someone to "know" something. Central to the field of epistemology, there are questions regarding to the nature of truth, the nature of justification, and types of knowledge (e.g., knowing how—skills—or knowing that—facts). Inclusive, they assert: Epistemology could be seen as driving assessments aimed at uncovering student knowledge, and driving pedagogy to build high quality knowledge to that end. They justify a consideration of epistemology for LA because the assessment ways, the sorts of tasks, and the kinds of learning, and how epistemology relates to assessment regimes. They also introduce epistemic beliefs to relate the intrapersonal and psychological conceptualizations that individuals hold regarding knowledge.

As for the second topic, it corresponds to *vulnerability*, both institutional and individual, which triggers a relevant and useful lens on the collection and use of student data. In this regard, Prinsloo and Slade (2016) adopt the notion of vulnerability as an interpretive lens to consider the control and choices available to users of digital networks. All of this with the aim at engaging with issues surrounding privacy and student privacy, self-management, and agency. They focus particularly on student vulnerability in the nexus between realizing the potential of LA, and the fiduciary duty of academic institutions in the context of their asymmetrical information and power relations with students. Thus, they take into account their former framework for mapping user vulnerability and privacy self-management to design an enhanced version called "framework for learner agency" with the purpose to explore ways to decrease student vulnerability, increase their agency, and empower them as participants in LA.

3.4 Applications

The employment of LA in education is oriented to pursue diverse goals related to the learners, academics, researchers, staff, and their teaching–learning contexts, with the purpose to understand and improve learning. These are the reasons why this section presents a review of relevant functionalities and approaches devoted to apply LA in educational settings. In the first part, the typical LA functionalities oriented to enhance learning processes are described. Whilst in the second subsection, several approaches are briefly described stated.

3.4.1 Functionalities

A key aspect of LA is the ability to provide outcomes and findings to support decision-making for the different entities that intervene in a learning process such as the student and faculty members. Thus, some of the most common functions and features outlined in the works that compose the sample are highlighted as follows.

3.4.1.1 Prediction

LA in education aims at predicting students in academic difficulty in order to provide timely instruction or advise to specific needs supported by actionable intelligence, which is based on the exploration of datasets, statistical techniques, and predictive modeling. In relation to *performance prediction*, its objective is to estimate an unknown value, in this case the final performance of the student (Wanli et al. (2015). It is because of that such authors propose a framework for exploring more understandable prediction based on genetic programming.

Regarding *academic prediction*, Tempelaar et al. (2015) use a range of demographic data, cultural differences, learning styles, learning motivation, engagement, learning emotions, and diverse user behaviour attributes from a LMS. In the same manner, Iglesias–Pradas et al. (2015) predict *teamwork* (i.e., referring to knowledge, skills, and attitudes) and *commitment* (e.g., degree of engagement and students' willingness) based on interactions extracted from a LMS.

In another vein, *temporal prediction* is critical to place at–risk students in a chronological order, so the teachers can provide timely intervention to the students (Xing et al. 2016). They propose a temporal modeling approach for students' dropout behavior based on the principle component analysis and stacking generalization. Other instance is an open source analytical initiative introduced by Jayaprakash et al (2014), where the process and challenges of collecting, organizing, and mining student data to predict academic risk are anticipated. Moreover, the prediction models have been released under the standard *predictive model markup language* to further use and enhancement by the community.

What is more, Ognjanovic et al. (2016) develop an approach for extracting student preferences from institutional data to predict student *course selections* in higher education. They employ a framework based on the analytical hierarchical process to model different preference structures according to their importance.

3.4.1.2 Assessment

Regarding assessment, Timmers et al. (2015) express that the main aim of formative assessment is to support and stimulate student learning. In consequence, they examine the effect of regulation feedback in computer-based *formative assessment* applied on information problem solving.

Related to *teamwork assessment*, Fidalgo–Blanco et al. (2015) propose indicators based on the interaction to assess the individual development in the teamwork context. In the same manner, LA is also applied to assess learners based on their interaction and performance derived from gaming. For instance, Serrano–Laguna et al. (2014) present a two-step approach to define a scalable LA system that supports diverse forms of assessment in *game-based learning* activities. Another instance is presented by Cutumisu et al. (2015), who build an *assessment game*. In this setting, students design posters and learn graphic design principles from feedback. The authors also measure the student behaviors and achieved learning with respect to positive and negative feedback, as well as determine the differences of individuals' SRL skills.

3.4.1.3 Performance

LA has an essential role in predicting student performance by means of extracting key indicators to provide insights to benefit students and institutions. In this context, Aguiar et al. (2014) study the relation *engagement-performance* to predict student attrition. They claim: "… placing disproportional focus on academic performance data can result in warning systems that may fail to identify students losing interest and disengaging from school …" Those authors employ measurements of engagement from students' electronic portfolios and use them to augment the quality of predictions.

In contrast, Lowes et al. (2015) explore the link between *online behaviors* and *course performance* based on LMS data. They find that the higher levels of online behaviors are associated with higher performance, and two types of behavior. Where one is associated with attendance, and the other with interactivity. Nevertheless, both operate separately and differently on gender.

On the other hand, a study that aims to identify significant LMS data indicators, including SRL indicators to predict *course achievement* is conducted by You (2016). The author examines whether the data collected in the middle of the course can successfully predict final course achievement, and studies the relationship between online learning strategies and academic achievement.

Another performance work is developed by (Gray et al. 2014), who assert: Psychometric factors (e.g., ability, personality, motivation, and learning strategies) are taken into account to predict academic performance. Their study emphasizes on factors that can be measured prior to, or during learner enrolment to facilitate and inform early engagement with students potentially at risk of failing.

Furthermore, Chung and Paredes (2015) develop a theoretical model based on social learning and social network theories to understand how knowledge professionals engage in learning and performance, both as individuals and as groups. The model fosters understanding social factors that influence learning and performance in project management.

3.4.1.4 Feedback

In relation to feedback, it provides information to students and teachers about the state and performance of the learning processes. Because timely feedback is essential to regulate learning process, Timmers et al. (2015) focuses on *regulation feedback* in a way that stimulates students to engage in evaluating their performance. The purpose of their work is to examine the effect of regulation feedback on student performance and behavior.

In addition, Tempelaar et al. (2015) explore components from diverse data sources to generate timely feedback and signaling risk of underperformance. They investigate learning dispositions, outcomes of continuous *formative assessments*, and other system generated data in modeling student performance and their potential to generate informative feedback. Other feedback example case is carried out by Cutumisu et al. (2015), where they assess children's choices to seek *informative negative feedback* and to revise their work. The authors express: "… behaviors after feedback, such as revising and help seeking, can be important factors of learning because they enable students to practice the correct skill".

3.4.2 Learner Support

Another component of an effective learning process corresponds to the support that helps learners beyond the primary delivery content. In this section factors related to the inner aspects of the learner are presented, as well as outer features of the learner are point out. A brief sample related with both views is presented as follows.

3.4.2.1 Internalization

Regarding *engagement*, Aguiar et al. (2014) utilize electronic portfolios to measure student engagement by means of capturing and documenting student learning and engagement through their reflection, rationale building, and planning. They suggest that applying EDM techniques to this kind of portfolio could generate a LA approach to improve the understanding of teaching and learning.

In relation to *motivation*, Lonn et al. (2015) focus on students' motivational orientations, as well as how to assess them. Furthermore, the authors believe such orientations can conduct to an intervention based on LA with the aim at supporting data-driven decisions and actions of the academics.

Another example corresponds to the work published by Gray et al. (2014), who review *ability*, *personality*, and *motivation* factors that could be used to predict academic performance. The authors claim that: Models predicting academic performance that include factors of *motivation* with cognitive ability yield a lower error variance than models of cognitive ability alone.

3.4.2.2 Externalization

Online courses generate data from CBE systems, which can be employed to provide insights related to student habits and behavior. In relation to *behavior*, Lowes et al. (2015) express that the literature suggests a link between *online behaviors* and learning outcomes. The authors explore behaviors related to *attendance* such as number of days, number of logins, session duration, as well as *interactivity* (e.g., posts viewed and posts authored). Whilst, Ferguson and Clow (2015) explore patterns of *individual learner* behavior within a MOOC. They express that studying learner behavior might shed light on the associations between MOOC characteristics and learner behaviors.

Concerning *retention*, Jayaprakash et al. (2014) express that: It can be defined as continued enrollment or graduation at a given institution. The authors try to identify students at risk of course failure. They also study the effectiveness of two different interventions, awareness messaging and participating in an online academic support environment, with the aim at improving student outcomes.

For its part, Buerck and Mudigonda (2014) use the *retention* center of Blackboard LMS to implement an early warning system in a selection of courses.

The purpose is to determine which approaches seem to work best and identify students that need intervention and feedback opportunely. Likewise, Svihla et al. (2015) identify analytics that guide the design of learning experiences to support retention. The authors explore *revisiting* previously studied material in different ways, for example: revisiting specific curricular steps, revisiting material on different days, and revisiting specific steps on different days.

In so far as Pardos et al. (2014), they study the correlation between student *affect* and behavioral engagement. The authors focus on detectors that estimate the student state (e.g., boredom, concentration, confusion, frustration) based on data from an ITS. In relation to *advising,* Krumm et al. (2014) depict an early warning system for an advising program. The goal is to identify students who need academic support. Thus, they develop a three-level classification scheme—engage, explore, and encourage—that informs advisors the relationships between academic performance data, longitudinal data, intra-course comparisons, and log-in events. Whilst, Colthorpe et al. (2015) examine the *self-regulatory* traits of health students. They employ the evaluation of responses to meta-learning assessment tasks supported by access data from LMS, and depict student behavior at interacting with course materials.

3.4.3   Peer Interaction

Student success in a course can be facilitated by offering communication channels among peers, instructors, and contents. LA provides tools and facilities to exploit the data to benefit relationships among students during their learning experience. Thus, in this section a paragraph is reserved for describing a CSCL approach, while a sample of works concerned to social LA and natural language processing are summarized, exposed, and cited in the following two subsections.

According to van Leeuwen et al. (2015), CSCL is an instructional strategy supported by technology that fosters collaboration among students, based on the idea that collaboration is beneficial for learning. The authors explore the effect of two learning analytics tools: *concept trail* and *progress statistics* that give information about students' cognitive activities. They examine whether teacher supporting tools could assist teachers by visualizing analyses of students' cognitive activities.

3.4.3.1   Social Learning Analytics

Instead of focusing on learners in an isolated manner, social LA is interested in processes where learners are engaged in a social activity with peers to develop educational networking. A sample of works are briefly described as follows:

- *Message boards in online learning*: Examines the relation between the parameters of both social networks and classroom in regards with student performance (Hernández–García et al. 2015). They also explore the potential of social network visualizations to observe student and teacher behaviors.
- *Analyzing the main paths of knowledge*: Presents a network analysis technique to address learning dynamics in the context of an open learning community. Thus, scientometric methodology is used to analyze biology and electrical engineering domains in wikiversity for collaborative creation of knowledge artifacts (Halatchliyski et al. 2014).
- *Analysis of dynamic resource access patterns*: Investigates characteristic patterns of resource usage of the learners. Hecking et al. (2014) adapt methods from SNA and apply them to dynamic bipartite student-resource networks using event logs. They also outline a method to identify patterns of the cluster evolution over time with the purpose to gain deeper insights into the usage of learning materials.

3.4.3.2 Natural Language Processing

LA supports natural language processing which concerns to the interactions between computers and human natural languages. The aim is to provide evidence to academics about the words and dialogues expressed by students during the development of learning activities. Thus, a set of related works is presented as follows.

- *Student's written language*: Develops an analysis method and tool to track how a student's written language migrates from paraphrase to mastery. The information is provided to instructors, allowing customized visualization and self-monitoring (White and Larusson 2014).
- *Narrative and cohesive linguistic features*: Examines students' essays within an automated writing evaluation system. Natural language processing techniques and entropy analyses are employed to calculate how rigid or flexible students are in their narrative and cohesive linguistic features over time Snow et al. (2015).
- *Additional works*: A series of diverse LA approaches correspond to the following works: Wise et al. (2014) explore asynchronous online discussions applying embedded and extracted analytics; Chiu and Fujita (2014) focus on discourse analysis and propose a method for modelling discussion; Reich et al. (2015) describe an approach to language processing to find syntactic patterns with semantic meaning in unstructured text and identify variations. Whilst supplementary works are accomplished by Kelly et al. (2015), Knight and Littleton (2015a, b), and Schneider and Pea (2015).

3.4.4   User Support

Effective use of LA helps students and academics to enhance learning achievements by providing timely support to increase the efficiency of teaching and learning processes. Thus, a sample of works related to teaching support, visualization, and time management are described and others cited in the next subsections.

3.4.4.1   Teaching Support

LA can provide insights of the students' performance and behavior, where teachers and academics become aware of learners' conduct. In this way, they are able to foster and enhance best practices in teaching and learning resources. For instance, a couple of aspects are described and instantiated by a set of citations.

In relation to *educational content*, Gunnarsson and Alterman (2014) study the practicality of using learner promotions of content, to identify quality content, and implications for students and instructors. The authors identify which students are good and poor predictors of quality content, and what instructors can do with this information in terms of feedback and guidance.

Regarding *curriculum design*, it is essential to plan the sequence of learning experiences taking advantage of LA. In this way, LA empowers the academia to do a better design of the lectures and classes, workgroup, private study, and assessment. A sample of works related to curriculum is published by Monroy et al. (2014), Méndez et al. (2014), Dunbar et al. (2014), and Leeuwen et al. (2015), whose contributions have been introduced in previous sections, where other aspects are exposed.

3.4.4.2   Visualization

In regards visualization, Martinez–Maldonado et al. (2015) express: the achievement of affective visualizations is one of the key aspects to be addressed for advancing the LA field. However, the main issue is to identify the real value of visualizations, when they are used in real learning contexts, and the impact they represent on the learning experience.

The authors offer a set of guiding principles and recommendations derived from a workflow for designing and deploying awareness tools for technology-enabled learning settings. The purpose is to guide the design of LA visualizations that can inform pedagogical decisions or intervention strategies.

In this context, Minovic et al. (2015) develop a tool for visualization of student learning model during a gameplay session. This tool is useful for tracking the game progress by educators and students. Thus, educators receive real-time tracking of students learning. Moreover, the tool enables them to react and influence the overall learning process. Additional visualization related works are reported by Ruipérez–

Valiente et al. (2015) and Epp and Bull (2015), whose characteristics were described in previous sections.

3.4.4.3 Time Management

The application of LA in *time management* helps to deal with analytics related to recording and reporting time. This kind of user support is useful to provide insight into the future and actionable recommendations for actors involved in learning process. One example corresponds to the approach presented by Miyamoto et al. (2015), who examine the relationship between students' allocation of their time in MOOC and their performance. They express that the number of sessions students initiate is correlated with certification rate across students in all courses.

In contrast, *procrastination* refers to put off or defer an action. As this regard, You (2016) examines the effect of academic procrastination on achievement using LMS data in e-Learning. The author focuses on the delays in weekly scheduled learning and late submission of assignments calculated from log data. Additional works concerning time management are found in Ferguson and Clow (2015), Jo et al. (2015), Kovanović et al. (2015b), and Tabuenca et al. (2015), whose description has been outlined in prior sections concerning to supplementary topics.

3.5 Conclusions

Once the essence of the landscape for LA has been unveiled through Sects. 3.2, 3.3, and 3.4, in this last one a couple of topics are stated in order to complement the LA scene. Therefore, the first topic corresponds to a discussion of the current state and future trends for LA arena; while the second is oriented to provide the responses for each research question made in Sect. 3.1.

3.5.1 Discussion

Even though LA is a novel domain of research, a growing and vigorous community of researchers, academics, and practitioners is devoted to deal with demanding challenges, ambitious goals, emergent demands, and diverse issues. For its part, learners, staff, and academic authorities are increasing their demands, requirements, and expectations. Inclusive, the evolution of CBE systems and the spread of their application, as well as the extension of their scope represent additional reasons to consider the need to invest in more and better LA resources. Thus, based on the already traced LA conceptual landscape and the sample of related works already summarized, in this section a sample of strengths, weaknesses, opportunities, and threats is outlined in Tables 3.4, 3.5, 3.6 and 3.7.

Table 3.4 Example of strengths that empowers LA

Id	Strengths
1	A robust background composed by its underlying domains (e.g., CBIS, KBS, and CBE systems) grounds LA as a research and application area
2	Being part of a revolutionary branch, analytics, including the wide sort of diverse analytics, as well as the variety of analytics in education
3	The broad and heterogeneous repertory of learning paradigms that inspire hypothesis-driven research
4	An evolving collection of underlying elements that ground the design, development, and deployment of approaches
5	An incipient legal frame to inspire and regulate the labor and application
6	The improvement and extension of the scope and reliability of the applications that gains diffusion, impact, acceptance, and demand
7	The support of the four classic epistemological items: (1) an incipient, but dynamic growing theory of knowledge about LA; (2) a formal community, SoLAR; (3) a devoted four-monthly Journal of Learning Analytics; (4) a specialized annual LAK conference
8	The alliances and collaborative work for spreading and improving research with other communities specialized in diverse domains (e.g., EDM, User Modeling and Artificial Intelligence in Education societies …)

Table 3.5 Set of weaknesses to deal with in LA

Id	Weaknesses
1	The complexity nature of the LA arena, its approaches, and goals
2	The lack of a well-sounded systematic, cybernetic, and holistic theoretical baseline ground on the theory of systems
3	The plurality of requirements, needs, and interests represented by each kind of stakeholder that demand a holistic solution
4	The low knowledge, consciousness, and acknowledgment of the LA labor between the technical practitioners, as well as user community
5	The need of a new kind of software engineering that merges the mixture of resources, interests, and requirements from the underlying domains and related lines; particularly SIS, knowledge discovery, knowledge communication systems, and the plethora of analytics variations
6	The additional investment that represents for the academic institution the implementation of LA human and computer resources devoted to develop and operate LA approaches to enhance the diversity of CBE systems
7	The viability, convenience, and overhead that represent the application of LA in real-time processing linked or embedding to CBE systems

3.5.2   Final Comments

This work has been conceived as result of becoming aware of the need to know and precise the nature of LA, as well as to spread its research and application. Particularly we found out three prerogatives, where the first corresponds to the need for disseminating the knowledge and awareness of the LA domain through a clear

Table 3.6 Relation of opportunities for LA

Id	Opportunities
1	The maturity in progress of LA opens new targets of study and application such as: active assistance in classrooms, help in informal learning, adaptive functionalities, active support for decision-making …
2	The design and instrumentation of an architecture that organizes the offer that the diverse kinds of educational analytics (e.g., from institutional analytics, AA, LA … up to learner analytics) provide to academic institutions
3	The conception, development, and deployment of a multidimensional layout that meets the requirements of all types of stakeholders with the suitable grey level of services, functionalities, tools, and outcomes
4	To rise LA as a virtual companion of learners, as well as teachers, that facilitates their daily learning and teaching activity respectively
5	To demonstrate the benefits, profits, and improvements produced by the application of LA approaches to enhance the academic institution's life
6	To encourage the LA labor to the best practices, codes, and laws that regulate ethics, privacy, data protection, and user consents
7	The integration of LA functionalities, services, approaches, and complete systems to classic CBE systems

Table 3.7 List of threats to be faced by LA

Id	Threats
1	The misunderstanding of the LA nature (i.e., a classic misconception is to consider LA and EDM are similar fields, inclusive interchangeable terms)
2	The ignorance, unawareness, and doubts of academic authorities about the LA labor, its benefits and return on investment, which avoid or constraint the provision of funding to support research, development, and deployment
3	The revolution in the academic life of the institutions as a result of implementing LA as a regularly service that should be taken into account to supervise, warn, and transform the daily activities
4	The improper use of data, LA approaches, and its outcomes that violate the legal, ethics, privacy, and consent statutes
5	The undue exploitation of LA applications and results to manipulate and distort its pure purposes
6	Potential claims of users that dislike a kind of "big brother" system that coldly monitors and interprets what they do all the time …
7	The technical overhead in the CBE systems that incorporate LA services at real–time, as well as the cognitive load that impose on the stakeholders who are responsible for interpreting the outcomes and making the correspondent decisions

exposition of the LA essence. All of this in order to benefit the quality of education and improve the learning achievements.

As for the second, it represents the growing demands for achieving better results in the use of CBS, as well as innovate diverse learning paradigms, more reliable assessments, improve the predictive power of the approaches, and recreate a proactive attitude of the learning environment that benefits students and academics.

Whilst the third corresponds to the confusion that exists between the nature, purpose, scope, baseline, and differences between LA and other fields, especially EDM, where some practitioners use both terms as if they were synonym.

In consequence, this chapter tries to explain the roots, essence, orientation, and application of the LA field to promote future labor. Therefore as the last task, the responses to the seven research questions that inspired this chapter are unveiled next with the purpose to reinforce the sketched LA landscape that is illustrated in Fig. 3.1.

1. What is the LA background? It is make up by three underlying domains CBIS, KBS, and computers in education (Sect. 3.2.1). These domains represent a plethora of research lines, paradigms, theoretical elements, approaches, and applications that shape a robust, heterogeneous, and rich support for LA.

2. Which domains are related to LA? In addition to the underlying domains, the analytics and analytics in education domains described in Sect. 3.2.2 highlight the diversity of analytics variants that surround LA labor.

3. What is the LA sense? According to the top–down strategy followed in this chapter, Sect. 3.2.1 traces the outer domain for LA as well as its roots, next the related domains stated in Sect. 3.2.2 represent the inner domain that precises the nature of LA, whilst Sect. 3.2.3 constitutes the kernel of LA nature because provides its background, a series of definitions proposed for LA, the classic learning environments and resources, and its stakeholders and levels.

4. What are the learning paradigms related to LA research? Section 3.3.1 identifies diverse learning paradigms and settings that are considered by several LA approaches, where some correspond to soft skills (e.g., SRL and metacognition), others to socialization (e.g., CSCL, social networks, natural language processing), and others are diverse (e.g., m-learning, game-based learning …).

5. How is the LA theoretical baseline composed by? It embraces a variety of models, frameworks, strategies, methods, techniques, algorithms, and tools, whose a sample of them is outlined in Sect. 3.3.2.

6. Which are the main application targets for the LA approaches? They are organized into four categories, such as: functionalities, learner support, peer interaction, and user support, where a review of related works is given in Sect. 3.4.

7. What to expect from LA? The discussion based on the exposition of several strengths, weaknesses, opportunities, and threats exposed in Sect. 8.1 offers a view of diverse objects that demand attention for the LA community.

Acknowledgements The first author gives testimony of the strength given by his Father, Brother Jesus, and Helper, as part of the research projects of World Outreach Light to the Nations Ministries (WOLNM). Moreover, this research holds a partial support from grants given by: CONACYT–SNI-36453, CONACYT 264215, IPN-Sabbatical Leave: DG:2015–118–1–196 y CPE/PIAS/1357–15, IPN–SIP/DI/DOPI/EDI–888/16; IPN-COFAA-SIBE-ID: 9020/2015–2016, IPN–SIP–20160899, IPN-SIP-20171313. The third author received financial partial support from grants given by: CONACYT 289763/CVU360532, CONACYT 264215, IPN–SIP–Thesis-Grants-2016. Inclusive, a special acknowledgment is given to Héctor Gabriel Villegas Berumen for his valuable contribution to the edition of this chapter.

References

Aggarwal CC (ed) (2011) Social network data analytics. Springer Science+Business Media, New York

Aggarwal CC (2016) An introduction to recommender systems. Recommender systems. Springer International Publishing, New York, pp 1–28

Aguiar E, Ambrose GA, Chawla NV et al (2014) Engagement vs performance: using electronic portfolios to predict first semester engineering student persistence. J Learn Anal 1(3):7–33

Alkharouf NW, Jamison DC, Matthews BF (2005) Online analytical processing (OLAP): a fast and effective data mining tool gene expression databases. J Biomed Biotechnol 2:181–188

Ally M (ed) (2009) Mobile learning: transforming the delivery of education and training. AU Press, Athabasca

Andergassen M, Mödritscher F, Neumann G (2014) Practice and repetition during exam preparation in blended learning courses: correlations with learning results. J Learn Anal 1(1):48–74

Azar AT, Vaidyanathan S (2015) Computational intelligence applications in modeling and control. Springer, New York

Bach C (2010) Learning analytics: targeting instruction, curricula and student support. Drexel University. http://www.iiis.org/CDs2010/CD2010SCI/EISTA_2010/PapersPdf/EA655ES.pdf. Accessed 10 Nov 2016

Berg AM, Mol ST, Sclater N (2016) The role of a reference synthetic data generator within the field of learning analytics. J Learn Anal 3(1):107–128

Berk J (2004) The state of learning analytics. Report for American Society for Training & Development

Brooks DC, Thayer TLB (2016) Institutional analytics in higher education. Research report. In: EDUCAUSE_https://library.educause.edu/ ~ /media/files/library/2016/2/ers1504ia.pdf. Accessed 20 Sept 2016

Brusilovsky P (1999) Student model centered architecture for intelligent learning environments. In: Proceedings of the 4th international conference on user modeling, Banff, 20–24 June 1999, p 31

Brusilovsky P, Peylo C (2003) Adaptive and intelligent web-based educational systems. Int J Artif Intell Educ 13:156–169

Buerck JP, Mudigonda SP (2014) A resource-constrained approach to implementing analytics in an institution of higher education: an experience report. J Learn Anal 1(1):129–139

Campbell JP, De Blois P, Oblinger D (2007) Academic analytics: a new tool for a new era. EDUCAUSE Rev 42(4):42–57

Ceneda D (2016) Characterizing guidance in visual analytics. IEEE Trans Vis Comput Graph. doi:10.1109/TVCG.2016.2598468

Chand D, Hachey G, Hunton J et al (2005) A balanced scorecard based framework for assessing the strategic impacts of ERP systems. Comput Ind 56(6):558–572

Chao L (ed) (2012) Cloud computing for teaching and learning: strategies for design and implementation: strategies for design and implementation. IGI Global, Pennsylvania

Chiu MM, Fujita N (2014) Statistical discourse analysis: a method for modelling online discussion processes. J Learn Anal 1(3):61–83

Chung KS, Paredes WC (2015) Towards a social networks model for online learning and performance. Educ Technol Soc 18(3):240–253

Colmenares E, Andersen P, Wei B (2015) An overlap study for cluster computing. In: 2015 International conference on computational science and computational intelligence, IEEE, Las Vegas, 7–9 Dec 2015, p 626

Colthorpe K, Zimbardi K, Ainscough L et al (2015) Know thy student! Combining learning analytics and critical reflections to increase understanding of students' self-regulated learning in an authentic setting. J Learn Anal 2(1):134–155

Conol G, Gašević D, Long P et al (2011) Message from the LAK 2011 general & program chairs. In: Proceedings of the 1st international conference on learning analytics and knowledge, Banff, February 27–March 1 2011, p 3

Cormack A (2016) A data protection framework for learning analytics. J Learn Anal 3(1):91–106

Cutumisu M, Blair KP, Chin DB et al (2015) Posterlet: a game-based assessment of children's choices to seek feedback and to revise. J Learn Anal 2(1):49–71

Daniel B (2015) Big data and analytics in higher education: opportunities and challenges. Br J Educ Technol 46(5):904–920

De Freitas D, Gibson D, Du Plessis C et al (2015) Foundations of dynamic learning analytics: using university student data to increase retention. Br J Educ Technol 46(6):1175–1188

Dunbar RL, Dingel MJ, Prat-Resina X (2014) Connecting analytics and curriculum design: process and outcomes of building a tool to browse data relevant to course designers. J Learn Anal 1(3):223–243

EDUCASE (2010) 7 Things you should know about analytics. In: EDUCASE. http://www.educause.edu/ir/library/pdf/ELI7059.pdf. Accessed 9 Sept 2016

Eiben AE, Smith JE (2015) Introduction to evolutionary computing. Springer, Berlin

Elmqvist N, Irani P (2013) Ubiquitous analytics: interacting with big data anywhere, anytime. Computer 46(4):86–89

Elouazizi N (2014) Critical factors in data governance for learning analytics. J Learn Anal 1(3):211–222

Epp CD, Bull S (2015) Uncertainty representation in visualizations of learning analytics for learners: current approaches and opportunities. IEEE Trans Learn Technol 8(3):242–258

Falzon P (ed) (2015) Cognitive ergonomics: understanding, learning, and designing human–computer interaction. Academic Press, Le Chesnay

Ferguson R, Clow D (2015) Consistent commitment: patterns of engagement across time in massive open online courses (MOOCs). J Learn Anal 2(3):55–80

Ferguson R, Macfadyen LP, Clow D et al (2014) Setting learning analytics in context: overcoming the barriers to large-scale adoption. J Learn Anal 1(3):120–144

Ferguson R, Hoel T, Scheffel M et al (2016) Guest editorial: ethics and privacy as enablers of learning analytics. J Learn Anal 3(1):5–15

Fidalgo-Blanco A, Sein-Echaluce ML, García-Peñalvo FJ et al (2015) Using learning analytics to improve teamwork assessment. Comput Hum Behav 47:149–156

Flasiński M (2016) Defining vague notions in knowledge-based systems. Introduction to artificial intelligence. Springer International Publishing, Switzerland, pp 189–201

Fulantelli G, Taibi D, Arrigo M (2015) A framework to support educational decision making in mobile learning. Comput Hum Behav 47:50–59

Gašević D, Dawson S, Rogers T et al (2015) Learning analytics should not promote one size fits all: the effects on instructional conditions in predicting academic success. Internet Higher Educ 28:68–84

Gibson D, de Freitas S (2015) Exploratory analysis in learning analytics. Technol Knowl Learn 21(1):5–19

Goldstein PJ (2005) Academic analytics: the uses of management information and technology in higher education. ECAR key findings December 2005. In: EDUCASE Center for Applied Research. https://net.educause.edu/ir/library/pdf/ecar_so/ers/ers0508/EKF0508.pdf. Accessed 11 Sep 2016

Goldstein PJ, Katz R (2005) Academic analytics: the uses of management information and technology in higher education. ECAR research study 8, 2005. In: EDUCASE Center for Applied Research. https://net.educause.edu/ir/library/pdf/ERS0508/ekf0508.pdf. Accessed 11 Sep 2016

Gómez-Aguilar DA, Hernández-García A, García-Peñalvo FJ et al (2015) Tap into visual analysis of customization of grouping of activities in eLearning. Comput Hum Behav 47:60–67

Gray G, McGuinness C, Owende P et al (2014) A review of psychometric data analysis and applications in modelling of academic achievement in tertiary education. J Learn Anal 1(1):75–106

Grove R (2000) Internet-based expert systems. J Expert Syst 17(3):129–135

Gunnarsson BL, Alterman R (2014) Peer promotions as a method to identify quality content. J Learn Anal 1(2):126–150

Guttormsen-Schär S, Krueger H (2000) Using learning technologies with multimedia. J IEEE Multimedia 7(3):40–51

Halatchliyski I, Hecking T, Göhnert T et al (2014) Analyzing the main paths of knowledge evolution and contributor roles in an open learning community. J Learn Anal 1(2):73–93

Harrington JL (2016) Relational database design and implementation. Morgan Kaufmann, Maryland

Heath J (2014) Contemporary privacy theory contributions to learning analytics. J Learn Anal 1(1):140–149

Hecking T, Ziebarth S, Hoppe HU (2014) Analysis of dynamic resource access patterns in online courses. J Learn Anal 1(3):34–60

Hernández-García A, González-González I, Jiménez-Zarco AI et al (2015) Applying social learning analytics to message boards in online distance learning: a case study. Comput Hum Behav 47:68–80

Hoel T, Chen W (2016) Privacy-driven design of learning analytics applications: exploring the design space of solutions for data sharing and interoperability. J Learn Anal 3(1):139–158

Holcomb LB, Brady KP, Smith BV (2010) The emergence of educational networking: can non-commercial, education-based social networking sites really address the privacy and safety concerns of educators? MERLOT J Online Learn Teach 6(2):475–481

Hunter B (2005) Learning, teaching, and building knowledge: a forty-year quest for online learning communities. In: Kearsley G (ed) Online learning: personal reflections on the transformation of education. Educational Technology Publications, New Jersey, pp 163–193

Iglesias-Pradas S, Ruiz-de-Azcárate C, Agudo-Peregrina AF (2015) Assessing the suitability of student interactions from moodle data logs as predictors of cross-curricular competencies. Comput Hum Behav 47:81–89

Isik Ö, Jones MC, Sidorova A (2013) Business intelligence success: the roles of BI capabilities and decision environments. Inf Manag 50(1):13–23

Jayaprakash SM, Moody EW, Eitel JML et al (2014) Early alert of academically at-risk students: an open source analytics anitiative. J Learn Anal 1(1):6–47

Jo H-H, Kim D, Yoon M (2015) Constructing proxy variables to measure adult learners' time management strategies in LMS. Educ Technol Soc 18(3):214–225

Joksimovic S, Gasevic D, Loughinet TM et al (2015) Learning at distance: effects of interaction traces on academic achievement. Comput Educ 87:204–217

Karray FO, De Silva, CW (2004) Soft computing and intelligent systems design: theory, tools, and applications. Pearson Education

Kahraman HT, Sagiroglu S, Colak I (2016) Novel user modeling approaches for personalized learning environments. Int J Inf Technol Decis Making 15(03):575–602

Kelly N, Thompson K, Yeoman P (2015) Theory-led design of instruments and representations in learning analytics: developing a novel tool for orchestration of online collaborative learning. J Learn Anal 2(2):14–43

Khalil M, Ebner M (2016) De-identification in learning analytics. J Learn Anal 3(1):129–138

Kim J (2003) A model for evaluating the effectiveness of CRM using the balanced scorecard. J Interact Market 17(2):5–19

Kim D, Park Y, Yoon M et al (2016) Toward evidence-based learning analytics: using proxy variables to improve asynchronous online discussion environments. Internet Higher Educ 30:30–43

Knight S, Littleton K (2015a) Discourse–centric learning analytics: mapping the terrain. J Learn Anal 2(1):185–209

Knight S, Littleton K (2015b) Dialogue as data in learning analytics for productive educational dialogue. J Learn Anal 2(3):111–143

Knight S, Buckingham SS, Littleton K (2014) Epistemology, assessment, pedagogy: where learning meets analytics in the middle space. J Learn Anal 1(2):23–47

Kodratoff Y, Michalski RS (2014) Machine learning: an artificial intelligence approach. Morgan Kaufmann, Massachusetts

Kovanović V, Gašević D, Joksimović S et al (2015a) Analytics of communities of inquiry: effects of learning technology use on cognitive presence in asynchronous online discussions. Internet Higher Educ 27:74–89

Kovanović V, Gašević D, Dawson S et al (2015b) Does time-on-task matter? Implications for the validity of learning analytics findings. J Learn Anal 2(3):81–110

Krumm AE, Waddington RJ, Teasley SD et al (2014) A learning management system-based early warning system for academic advising in undergraduate engineering. In: Larusson JA, White B (eds) Learning analytics: from research to practice, Springer Science + Business Media New York, pp 103–119

Laine TH, Joy MS (2009) Survey on context-aware pervasive learning environments. Int J Interact Mobile Technol 3(1):70–76

Larusson JR, White B (2014) Learning analytics: from research to practice. Springer Science + Business Media, New York

Laskey KB (2008) MEBN: A language for first-order Bayesian knowledge bases. J Artif Intell 172(2–3):140–178

Leach RJ (2016) Introduction to software engineering. CRC Press, Florida

Ligêza A (2006) Logical foundations for rule-based systems. Springer, Berlin Heidelberg

Lonn S, Teasley SD (2009) Saving time or innovating practice: investigating perceptions and uses of learning management systems. Comput Educ 50(3):686–694

Lonn S, Aguilar SJ, Teasley SD et al (2015) Investigating student motivation in the context of a learning analytics intervention during a summer bridge program. Comput Hum Behav 47:90–97

Lowes S, Lin P, Kinghorn B (2015) Exploring the link between online behaviours and course performance in asynchronous online high school courses. Journal of Learning Analytics 2(2):169–194

Lucey T (2005) Management information, 9th edn. Thomson, London

Margaryan A, Bianco M, Littlejohn A (2015) Instructional quality of massive open online courses (MOOCs). Comput Educ 80:77–83

Marr B (2015) Big data: using SMART big data, analytics and metrics to make better decisions and improve performance. John Wiley & Sons, New Jersey

Martínez-Maldonado R, Pardo A, Mirriahi N et al (2015) LATUX: an iterative workflow for designing, validating, and deploying learning analytics visualizations. J Learn Anal 2(3):9–39

McFadden C (2005) Optimizing the online business channel with web analytics http://es. slideshare.net/cmcfadden/optimizing-the-online-business-channel-with-web-analytics. Accessed 9 Sept 2016

Méndez G, Ochoa X, Chiluiza K et al (2014) Curricular design analysis: a data-driven perspective. J Learn Anal 1(3):84–119

Minovic M, Milovanovic´ M, Šoševic´ U et al (2015) Visualisation of student learning model in serious games. Comput Hum Behav 47:98–107

Mitchel J, Costello S (2000) Internationale-VET market research report: a report on international market research for Australian VET online products and services. John Mitchell & Associates, Sydney, New South Wales

Miyamoto RY, Coleman CA, Williams JJ et al (2015) Beyond time-on-task: the relationship between spaced study and certification in MOOCs. J Learn Anal 2(2):47–69

Moldovan DI (2014) Parallel processing from applications to systems. Morgan Kaufmann Publishers, California

Monroy C, Snodgrass RV, Whitaker R (2014) A strategy for incorporating learning analytics into the design and evaluation of a K–12 science curriculum. J Learn Anal 1(2):94–123

Moore C (2005) Measuring effectiveness with learning analytics. Chief Learning Officer. http://clomedia.com/articles/view/measuring-effectiveness-with-learning-analytics. Accessed 10 Nov 2016

Muñoz-Merino PJ, Ruipérez-Valiente JA, Alario-Hoyos C et al (2015) Precise effectiveness strategy for analyzing the effectiveness of students with educational resources and activities in MOOCs. Comput Hum Behav 47:108–118

Norris D et al (2008) Action analytics: measuring and improving performance that matters in higher education. EDUCAUSE Rev 43(1):1–13

Nussbaumer A, Hillemann E, Gütl C et al (2015) A competence-based service for supporting self-regulated learning in virtual environments. J Learn Anal 2(1):101–133

O'Neill DK (2016) Understanding design research-practice partnerships in context and time: Why learning sciences scholars should learn from cultural-historical activity theory approaches to DBR. J Learn Sci. doi:10.1080/10508406.2016.1226835

Ochoa X, Suthers D, Verbert K et al (2014) Epistemology, assessment, pedagogy: where learning meets analytics in the middle space. J Learn Anal 1(2):5–22

Ognjanovic I, Gasevic D, Dawson S (2016) Using institutional data to predict student course selections in higher education. Internet Higher Educ 29:49–62

Pardos ZA, Baker R, San-Pedro M et al (2014) Affective states and state tests: investigating how affect and engagement during the school year predict end-of-year yearning outcomes. J Learn Anal 1(1):107–128

Peña-Ayala A (ed) (2013) Intelligent and adaptive educational-learning systems: achievements and trends. KES-Springer, Berlin

Peña-Ayala A (2014a) Educational data mining: a survey and a data mining-based analysis of recent works. Expert Syst Appl 41(4):1432–1462

Peña-Ayala A (ed) (2014b) Educational data mining: applications and trends. Springer International Publishing, Heidelberg

Peña–Ayala A (ed) (2015) Mobile, ubiquitous, and pervasive learning: fundaments, applications, and trends. Springer, Berling

Phillips et al (2011) Learning analytics and study behavior: a pilot study. In: Proceedings ASCILITE Tasmania, 4–7 Dec 2011

Piety PJ, Hickey DT, Bishop MJ (2014) Educational data sciences—framing emergent practices for analytics of learning, organizations, and systems. In: Pistilli M et al (eds) Proceedings of the 4th international conference on learning analytics and knowledge, Indianapolis, 24–28 Mar 2014, p 193

Powell D (ed) (2012) Delta-4: a generic architecture for dependable distributed computing (Vol 1). Springer Science & Business Media, New York

Power DJ, Sharda R (2009) Decision support systems. In: Shimon NY (ed) Springer handbook of automation. Springer, Heidelberg, pp 1539–1546

Prinsloo P, Slade S (2016) Student vulnerability, agency, and learning analytics: an exploration. J Learn Anal 3(1):159–182

Rahman N (2016) Enterprise data warehouse governance best practices. Int J Knowl-Based Organ 6(2):21–37

Rehak DR, Blackmon B, Humphreys AR (2000) The virtual university: developing a dynamic learning management portal. J IEEE Concurrency 8(3):3–5

Reich J, Tingley D, Leder-Luis J et al (2015) Computer-assisted reading and discovery for student-generated text in massive open online courses. J Learn Anal 2(1):156–184

Retalis S, Papasalouros A, Psaromiligkos Y (2006) Towards networked learning analytics: a concept and a tool. In Proceedings of the 5th international conference on networked learning, Lancaster 10–12 Apr 2006

Reyes AJ (2015) The skinny on big data in education: learning analytics simplified. TechTrends 59 (2):75–80

Rodríguez-Triana MJ, Martínez-Monés A, Villagrá-Sobrino S (2016) Learning analytics in small-scale teacher-led innovations: ethical and data privacy issues. J Learn Anal 3(1):43–65

Rogers PC, McEwen M, Pond S (2010) The use of web analytics in the design and evaluation of distance education. In; Veletsianos G (ed) Emerging technologies in distance education, AU Press, Edmonton, pp 231–248

Roll I, Winne HP (2015) Understanding, evaluating, and supporting self-regulated learning using learning analytics. J Learn Anal 2(1):7–12

Ruipérez-Valiente JA, Muñoz-Merino PJ, Leony D et al (2015) ALAS-KA: a learning analytics extension for better understanding the learning process in the Khan Academy platform. Comput Hum Behav 47:139–148

Runkler TA (2012) Data analytics: models and algorithms for intelligent data analysis. Springer Science & Business Media, Berlin

Ryza S, Laserson U, Owen S et al (2015) Advanced analytics with spark: patterns for learning from data at scale. O'Reilly Media, Inc., Boston

Saecker M, Markl V (2013) Big data analytics on modern hardware architectures: a technology survey. In Business Intelligence. Springer, Berlin Heidelberg, pp 125–149

Scheffel M, Drachsler H, Stoyanov S et al (2014) Quality indicators for learning analytics. Educ Technol Soc 17(4):117–132

Schmeck RR (2013) An introduction to strategies and styles of learning. Learning strategies and learning styles. Springer International Publishing, New York, pp 3–19

Schmidhuber J (2015) Deep learning in neural networks: an overview. Neural Netw 61:85–117

Schneider B, Pea R (2015) Does seeing one another's gaze affect group dialogue? A computational approach. J Learn Anal 2(2):107–133

Schutt R, O'Neil C (2013) Doing data science: straight talk from the frontline. O'Reilly Media Inc, Sebastopol

Sclater N (2016) Developing a code of practice for learning analytics. J Learn Anal 3(1):16–42

Sedrakyan G, Snoeck M, de Weerdt J (2014) Process mining analysis of conceptual modeling behavior of novices: empirical study using JMermaid modeling and experimental logging environment. Comput Hum Behav 41:486–503

Segedy RJ, Kinnebrew JS, Biswas G (2015) Using coherence analysis to characterize self-regulated learning behaviours in open-ended learning environments. J Learn Anal 2(1):13–48

Serrano-Laguna A, Torrente J, Moreno-Ger P et al (2014) Application of learning analytics in educational videogames. Comput Hum Behav 5:313–322

Shmueli G, Patel NR, Bruce PC (2016) Data mining for business analytics: concepts, techniques, and applications in XLMiner. Wiley, New Jersey

Siadaty M, Gašević D, Hatala M (2015) Trace-based microanalytic measurement of self-regulated learning processes. J Learn Anal 3(1):183–214

Siadaty M, Gašević D, Hatala M (2016) Associations between technological scaffolding and micro-level processes of self-regulated learning: a work place study. Comput Hum Behav 55:1007–1019

Siegel E (2013) Predictive analytics: the power to predict who will click, buy, lie, or die. Wiley, New Jersey

Siemens G (2010) What are learning analytics? In: E learnspace. http://www.elearnspace.org/blog/2010/08/25/what-are-learning-analytics/. Accessed 9 Sept 2016

Siemens G (2014) Supporting and promoting learning analytics research. J Learn Anal 1(1):3–4

Siemens G, Gasevic D, Haythornthwaite C et al (2011) Open learning analytics: an integrated and modularized platform: proposal to design, implement and evaluate an open platform to integrate heterogeneous learning analytics techniques. In: Society for learning analytics research. Available via http://www.elearnspace.org/blog/wp-content/uploads/2016/02/ProposalLearningAnalyticsModel_SoLAR.pdf. Accessed 7 Sep 2016

Sinha S, Rogat TK, Adams-Wiggins KR et al (2015) Collaborative group engagement in a computer-supported inquiry learning environment. Int J Comput-Support Collaborative Learn 10(3):273–307

Sleeman D, Brown JS (eds) (1982) Intelligent tutoring system. Academic Press, London

Snow EL, Allen LK, Jacovina ME et al (2015) Keys to detecting writing flexibility over time: entropy and natural language processing. J Learn Anal 2(3):40–54

Sonnenberg C, Bannert M (2015) Discovering the effects of metacognitive prompts on the sequential structure of SRL-Processes using process mining techniques. J Learn Anal 2(1): 72–100

Souza GC (2014) Supply chain analytics. Bus Horiz 57(5):595–605

Steiner CM, Kickmeier-Rust M, Albert D (2016) LEA in private: a privacy and data protection framework for a learning analytics toolbox. J Learn Anal 3(1):66–90

Su B, Bonk CJ, Magjuka RJ et al (2005) The importance of interaction in web-based education: a program-level case study of online MBA courses. J Interact Online Learn 4(1):1–19

Svihla V, Wester MJ, Linn MC (2015) Distributed revisiting: an analytic for retention of coherent science learning. J Learn Anal 2(2):75–101

Tabuenca B, Kalz M, Drachsler H et al (2015) Time will tell: the role of mobile learning analytics in self-regulated learning. Comput Educ 89:53–74

Talbot-Smith M et al (eds) (2013) Handbook of research on science education. Routledge, London

Tempelaar DT, Rienties B, Giesbers B et al (2015) In search for the most informative data for feedback generation: learning analytics in a data-rich context. Comput Hum Behav 47:157–167

Timmers CF, Walraven A, Veldkamp BP (2015) The effect of regulation feedback in a computer-based formative assessment on information problem solving. Comput Educ 87:1–9

Tinto V (1987) Leaving college: rethinking the causes and cures of student attrition, 2nd edn. University of Chicago Press, Chicago

Tolman MN, Allred RA (1984) The computer and education. NEA Professional Library, National Education Association, Washington

Uhr L (1969) Teaching machine programs that generate problems as a function of interaction with students. In: ACM'69 Proceedings of the 24th national conference, p 125

van Barneveld A, Arnold KE, Campbell JP (2012) Analytics in higher education: establishing a common language. EDUCAUSE Learn Initiative 1:1–11

van Leeuwen A (2015) Learning analytics to support teachers during synchronous CSCL: balancing between overview and overload. J Learn Anal 2(2):138–162

van Leeuwen A, Janssen J, Erkens G et al (2015) Teacher regulation of cognitive activities during student collaboration: effects of learning analytics. Comput Educ 90:80–94

Vlahos GE (2004) The use of computer-based information systems by German managers to support decision making. J Inf Manag 41(6):763–779

Walther JB, Van Der Heide B, Ramirez et al (2015) Interpersonal and hyperpersonal dimensions of computer-mediated communication. The handbook of the psychology of communication technology. Wiley Blackwell, Oxford, pp 3–22

Wanli W, Rui G, Eva P, Sean G (2015) Participation–based student final performance prediction model through interpretable genetic programming: Integrating learning analytics, educational data mining and theory. Comput Hum Behav 47:168–181

Ware C (2012) Information visualization: perception for design. Elsevier, London

Weber G, Brusilovsky P (2001) ELM-ART: an adaptive versatile system for web-based instruction. Int J Artif Intell Educ 12:351–384

Wenger E (1987) Artificial intelligence and tutoring systems: computational and cognitive approaches to the communication of knowledge. Morgan Kaufmann Publishers Inc., San Francisco

White B, Larusson JA (2014) Identifying points for pedagogical intervention based on student writing: two case studies for the "Point of Originality". In: Larusson JA, White B (eds) Learning analytics: from research to practice, Springer Science + Business Media New York, pp 157–190

Wille R (2002) Why can concept lattices support knowledge discovery in databases? J Exp Theor Artif Intell 14(2–3):81–92

Williamson B (2015) Learning in the digital microlaboratory of educational data science. Dmlcentral. http://dmlcentral.net/learning-in-the-digital-microlaboratory-of-educational-data-science/. Accessed 25 Oct 2016

Wise AF, Shaffer DW (2015) Why theory matters more than ever in the age of big data. J Learn Anal 2(2):5–13

Wise AF, Zhao Y, Hausknecht SN (2014) Learning analytics for online discussions: embedded and extracted approaches. J Learn Anal 1(2):48–71

Xiang Z (2015) What can big data and text analytics tell us about hotel guest experience and satisfaction? Int J Hospitality Manage 44:120–130

Xing W, Chen X, Stein J et al (2016) Temporal predication of dropouts in MOOCs: reaching the low hanging fruit through stacking generalization. Comput Hum Behav 58:119–129

Yager RR, Zadeh LA (eds) (2012) An introduction to fuzzy logic applications in intelligent systems (Vol 165). Springer Science & Business Media, New York

You JW (2016) Identifying significant indicators using LMS data to predict course achievement in online learning. Internet Higher Educ 29:23–30

Ysseldyke J et al (2004) Use of an instructional management system to enhance math instruction of gifted and talented students. J Educ Gifted 27(4):293–310

Chapter 4
A Review of Recent Advances in Adaptive Assessment

Jill-Jênn Vie, Fabrice Popineau, Éric Bruillard and Yolaine Bourda

Abstract Computerized assessments are an increasingly popular way to evaluate students. They need to be optimized so that students can receive an accurate evaluation in as little time as possible. Such optimization is possible through learning analytics and computerized adaptive tests (CATs): the next question is then chosen according to the previous responses of the student, thereby making assessment more efficient. Using the data collected from previous students in non-adaptive tests, it is thus possible to provide formative adaptive tests to new students by telling them what to do next. This chapter reviews several models of CATs found in various fields, together with their main characteristics. We then compare these models empirically on real data. We conclude with a discussion of future research directions for computerized assessments.

Keywords Latent knowledge extraction · Item response theory · Q-matrix · Cognitive diagnosis models · Adaptive testing · Knowledge space theory

List of Abbreviations

CAT   Computerized adaptive testing
CD   Cognitive diagnosis
DINA   Deterministic input, noisy and
ECPE   Examination for the Certificate of Proficiency in English
GenMA   General Multidimensional Adaptive

J.-J. Vie · F. Popineau (✉) · Y. Bourda
Université Paris-Sud, LRI—Bât. 650 Ada Lovelace, 91405 Orsay, France
e-mail: fabrice.popineau@lri.fr

J.-J. Vie
e-mail: jjv@lri.fr

Y. Bourda
e-mail: yolaine.bourda@lri.fr

É. Bruillard
ENS Cachan—Bât. Cournot, 61 avenue du Président Wilson, 94235 Cachan, France
e-mail: eric.bruillard@ens-cachan.fr

© Springer International Publishing AG 2017
A. Peña-Ayala (ed.), *Learning Analytics: Fundaments, Applications,
and Trends*, Studies in Systems, Decision and Control 94,
DOI 10.1007/978-3-319-52977-6_4

GMAT   Graduate Management Admission Test
GRE   Graduate Record Examination
KC   Knowledge components
LA   Learning analytics
MIRT   Multidimensional item response theory
MOOC   Massive online open course
MST   Multistage testing
PCA   Principal components analysis
TIMSS   Trends in International Mathematics and Science Study

4.1   Introduction

Today, educational assessments are often automatized, so we can store and analyze student data in order to provide more accurate and shorter tests for future learners. The *learning analytics* process consists in collecting data about learners, discovering hidden patterns that can lead to a more effective learning experience, and constantly refining models using new learner data (Chatti et al. 2012). Learning analytics for adaptive assessment have specific and well-defined objectives: they must improve the efficiency and effectiveness of the learning process, and tell learners what to do next by adaptively organizing instructional activities (Chatti et al. 2012). Reducing the test length in needed even more as students today are over-tested (Zernike 2015), leaving less time for instruction.

Traditionally, models used for adaptive assessment have been mostly *summative*: they measure or rank effectively examinees, but do not provide any other feedback. This is in particular the case of models encountered in item response theory (Hambleton and Swaminathan 1985). Recent advances have focused on *formative assessments* (Ferguson 2012; Huebner 2010), providing more useful feedback for both the learner and the teacher; hence, they are more useful to the learning analytics community. Indeed, Tempelaar et al. (2015) have shown that computer-assisted formative assessments have high predictive power for detecting underperforming students and estimating academic performance.

In this chapter, we prove that such adaptive strategies can be applied to formative assessments, in order to make tests shorter and more useful. Our primary focus is the assessment of knowledge and we do not consider dimensions of conscientiousness, i.e., perseverance, organization, carefulness, responsibility. Our second focus is to provide useful feedback at the end of the test. Such feedback can be aggregated at various levels (e.g., at the level of an individual student, of a class, or of a school, district, state, or country) for decision-making purposes (Shute et al. 2016).

We assume that data is provided as *dichotomous response patterns*, i.e., learners answer each question either correctly or incorrectly. A general method is to train user models so they can help uncover the latent knowledge of new examinees using fewer, carefully chosen questions. We here develop a framework that relies solely on dichotomous data in order to compare different adaptive models on the same data. Our approach is thus generic and can be specialized for different environments, e.g., serious games. Based on our analysis, one can choose the best model suitable to their individual needs.

This chapter is organized as follows. First in Sect. 4.2, we present the learning analytics methods that will be used in the chapter. Then in Sect. 4.3, we describe the adaptive assessment models used in diverse fields, ranging from psychometrics to machine learning. Later in Sect. 4.4, we present a protocol to compare adaptive assessment strategies for predicting student performance, and expose our experimental results on real data in Sect. 4.5. In Sect. 4.6, we highlight which models suit which use cases, specify the limitations of our approach in Sect. 4.7, discuss possible directions for the future of assessment in Sect. 4.8 and finally draw our conclusions in Sect. 4.9.

4.2 Learning Analytics

Educational data mining and learning analytics are two research communities that analyze educational data, typically collected in online environments and platforms. The former focuses on automated adaptation while the latter provides tools for human intervention. Indeed, various dashboards, visualizations and analytics packages can help inform pedagogical decision-making. Faculty, instructional designers, and student support services often use data to improve teaching, learning, and course design.

Among the objectives of learning analytics (LA), Chatti et al. (2012) describe the need for intelligent feedback in assessment, and the problem of choosing the next activity to present to the learner. To address these needs, they highlighted the following classes of methods: statistics, information visualization, data mining and social network analysis. In this chapter, we describe the methods used to provide adaptive assessments.

Adaptive assessments can lead to improved personalization, by organizing learning resources. For example, the problem of *curriculum sequencing* studies how we can choose learning paths in a space of learning objectives (Desmarais and Baker 2012). It aims to use skills assessment to tailor the learning content, based on as little evidence as possible. As stated by Desmarais and Baker (2012), "The ratio of the amount of the evidence to the breadth of the assessment is particularly critical for systems that cover a large array of skills, as it would be unacceptable to ask hours of questions before making a usable assessment."

In educational systems, there is an important difference between *adaptivity*, the ability to modify course materials using different parameters and a set of pre-defined rules, and *adaptability*, the possibility for learners to personalize the course materials by themselves. As Chatti et al. (2012) indicate, "more recent literature in personalized adaptive learning have criticized that traditional approaches are very much top-down and ignore the crucial role of the learners in the learning process." There should be a better balance between giving learners what they need to learn (i.e. adaptivity) and giving them what they want to learn (i.e. adaptability), the way they want to learn it (e.g., giving them more examples, or more exercises, depending on what they prefer). In either case, learner profiling is a crucial task.

As a use case scenario, let us consider users who register on a massive online open course (MOOC). As these users may have acquired knowledge from diverse backgrounds, some may be missing some prerequisites of the course, whereas other could afford to skip some chapters of the course. Therefore, it would be useful to adaptively assess user needs and preferences, to filter the content of the course accordingly and minimize information overload. Lynch and Howlin (2014) describe such an algorithm to uncover the latent knowledge state of a learner, by asking a few questions at the beginning of the course. Another lesser-known use case is the automated generation of testlets of exercises on demand, that reduce the costs of practice testing.

In learning analytics, methods in data mining include machine learning techniques such as regression trees for prediction. For instance, gradient boosting trees can be used to highlight which variables are the most informative to explain why a MOOC user obtained a certificate (or failed to obtain it). Gradient boosting trees have also been successful to tackle prediction problems, notably in data science challenges, because they can integrate heterogeneous values (categorical variables and numerical variables) and they are robust to outliers. It is surprising to see that learning analytics methods produced so many models to predict some objective from a fixed set of variables, and so few models to assess the learner about their needs and preferences. We believe that a lot of research can still be done towards more interactive models in learning analytics.

Recommender systems are another tool to aggregate data about users in order to recommend relevant resources (such as movies, products). They are increasingly used in technology-enhanced learning research as a core objective of learning analytics (Chatti et al. 2012; Manouselis et al. 2011; Verbert et al. 2011). Most recommender systems rely on *collaborative filtering*, a method that makes automated predictions about the interests of a user, based on information collected from many users. The intuition is that a user may like items that similar users have liked in the past. In our case, a learner may face difficulties similar to the ones faced by learners with similar response patterns. There are open research questions on how algorithms and methods have to be adapted from the field of commercial recommendations. Still, we believe that existing techniques can be applied to adaptive assessment.

Another approach, studied in cognitive psychology, is to measure the *response time* during an assessment. Indeed, the amount of time needed by a person needs to answer a question can give some clues about the cognitive process. To do so, sophisticated statistical models are needed (Chang 2015); we do not consider them in this chapter.

4.3   Adaptive Assessments

Our goal is to filter the questions to ask to a learner. Instead of asking the same questions to everyone, the so-called computer adaptive tests (CATs) (van der Linden and Glas 2010) select the next question to ask based on the previous answers, thus allowing adaptivity at each step. The design of CATs relies on two criteria: a *termination criterion* (e.g., a number of questions to ask), and a *next item criterion*. While the termination criterion is not satisfied, questions are asked according to the next item criterion, which picks questions, e.g., that are the most informative about the learner's ability or knowledge. Lan et al. (2014) have proven that such adaptive tests needed fewer questions than non-adaptive tests to reach the same prediction accuracy.

This gain in performance is important: shorter tests are better for the system, because they reduce load, and they are better for the learner, who may be frustrated or bored if they need to give too many answers (Lynch and Howlin 2014; Chen et al. 2015). Thus, adaptive assessment is more and more useful in the current age of MOOCs, where motivation plays an important role (Lynch and Howlin 2014). In real-life scenarios, however, more constraints need to be taken into account. First, the computation of criteria should be done in a reasonable time; hence the time complexity of the approaches is important. Second, assessing skills must be performed under uncertainty: a learner may *slip*, i.e., accidentally or carelessly fail an item that they could have solved, or they may *guess*, i.e., correctly answer an item by chance. This is why adaptive assessment cannot simply perform a binary search over the ability of the learner, i.e., asking a more difficult question if they succeed and an easier question if they fail. Thus, we need to use more robust methods, such as probabilistic models for skill assessment.

CATs have been extensively studied over the past years, and they have been put into practice. For instance, the Graduate Management Admission Council has administered 238,536 adaptive tests of this kind in 2012–2013 through the Graduate Management Admission Test (GMAT) (Graduate Management Admission Council 2013). Given a student model (Peña-Ayala 2014), the objective is to provide an accurate measurement of the parameters of an upcoming student while minimizing the number of questions asked. This problem has been referred to as *test-size reduction* (Lan et al. 2014), and it is also related to predicting student performance (Bergner et al. 2012; Thai-Nghe et al. 2011). In machine learning, this approach is known as *active learning*: adaptively query the informative labels of a training set in order to optimize learning.

Several models can be used, depending on the purpose of the assessment, e.g., estimating a general level of proficiency, providing diagnostic information, or characterizing knowledge (Mislevy et al. 2012). At the end of the test, rich feedback can help teachers identify at-risk students. It also protects against perseveration errors when students respond incorrectly on a practice test (Dunlosky et al. 2013). In what follows, we describe those models under the following categories: item response theory for summative assessment (Sect. 4.3.1), cognitive models for formative assessment (Sect. 4.3.2), more complex knowledge structures (Sect. 4.3.3), adaptive assessment and recommender systems (Sect. 4.3.4), exploration and exploitation trade-off (Sect. 4.3.5), and multistage testing (Sect. 4.3.6).

4.3.1 Psychometrics: Measuring Proficiency Using Item Response Theory

The simplest model for adaptive testing is the *Rasch model*, also known as the 1-parameter logistic model: it falls into the data mining category of LA. This model represents the behavior of a learner with a single latent trait, called *ability*, and the items or tasks with a single parameter, called *difficulty*. The tendency for a learner to solve a task only depends on the difference between the difficulty of the task and the ability of the learner. Thus, if a learner i has ability θ_i and wants to solve an item j of difficulty d_j, the probability that the learner i answers the item j correctly is given by Eq. (4.1), where $\Phi : x \mapsto 1/(1 + e^{-x})$ is the *logistic function*:

$$Pr(\text{``learner } i \text{ answers item } j\text{''}) = \Phi(\theta_i - d_j). \tag{4.1}$$

Of course, we cannot specify all difficulty values by hand, as it would be time-consuming and probably inaccurate (i.e., be too subjective, and poorly fit student data). Fortunately, the Rasch model makes it possible to estimate parameters efficiently: using former student data, we can calibrate item difficulties and learner abilities automatically, computing the maximum likelihood estimates. In particular, this estimation process does not depend on any domain knowledge.

When a new user takes a test, the observed variables are its outcomes over the questions that are asked to the user, and the hidden variable we want to estimate is the ability of the user, given the known difficulty parameters. This is usually performed using maximum likelihood estimation: we can easily do this computationally, using Newton's method to find the zeroes of the derivative of the likelihood function. Therefore, the adaptive process can be phrased as follows: given an estimate of the learner's ability, which question outcome will be the most useful to refine this estimate? Indeed, we can quantify the information that each item j provides over the ability parameter: this can be done using *Fisher information*, defined as the variance of the gradient of the log-likelihood with respect to the ability parameter, given by Eq. (4.2):

$$I_j(\theta_i) = E\left(\left(\frac{\partial}{\partial \theta} \log f(X_j, \theta_i)\right)^2 \bigg| \theta_i\right). \tag{4.2}$$

where X_j is the binary outcome of the learner i over the item j and $f(X_j, \theta_i)$ is the probability function for X_j depending on θ_i: $f(X_j, \theta_i) = \Phi(\theta_i - d_j)$.

Therefore, an adaptive assessment can be designed as follows: given the learner's current ability estimate, pick the question which yields the most information about the ability, update the estimate according to the outcome (i.e., whether the user answered correctly or incorrectly), and so on. At the end of the test, one can visualize the whole process like in Fig. 4.1. As we can see, the confidence interval for the ability estimate is refined after each outcome.

As the Rasch model is a unidimensional model, it is not suitable for cognitive diagnosis. Still, it is really popular because of its simplicity, its stability, and its sound mathematical framework (Desmarais and Baker 2012; Bergner et al. 2012). Also, Verhelst (2012) has showed that, if the items are split into categories, we can provide to the examinee a useful deviation profile, specifying the categories where the subscores were higher or lower than expected. Specifically, let us consider that, in each category, an answer gives one point if correct, and no point otherwise. The *subscores* are then the number of points obtained by the learner in each category, which sum up to the total score. Given the total score, we can then compute the expected subscore of each category by simply using the Rasch model. Finally, the *deviation profile*, namely, the difference between the observed and expected subscores, provides a nice visualization of the categories that need further work: see Fig. 4.2 for an example. Such deviation profiles can be aggregated across a country to highlight the strong and weak points of its students, which can help identify deficiencies in the national curriculum. These profiled can then be compared worldwide in studies of international assessments, such as the Trends in International Mathematics and Science Study (TIMSS). For instance, Fig. 4.2 presents the TIMSS 2011 dataset of proficiency in mathematics, highlighting the fact that Romania is stronger in Algebra than expected, while Norway is weaker in

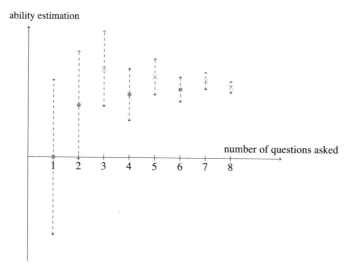

Fig. 4.1 Evolution of the ability estimate throughout an adaptive test based on the Rasch model. *Filled circles* denote correct answers while *crosses* denote incorrect answers

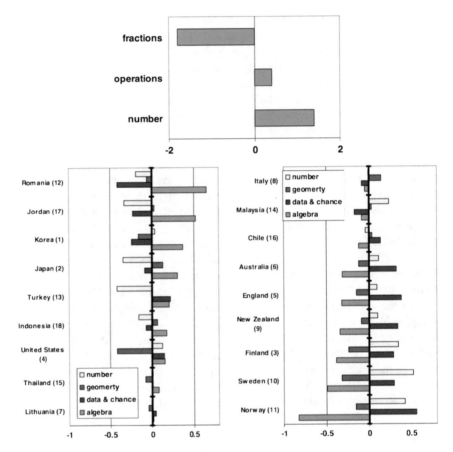

Fig. 4.2 *Above* the deviation profile of a single learner. *Below* the deviation profile of different countries on the TIMSS 2011 math dataset, from the presentation of N.D. Verhelst at the workshop Psychoco 2016

Algebra than expected. This belongs to the information visualization class of learning analytics methods, and shows what can be done using the simplest psychometric model and the student data only.

In adaptive testing, however, we do not observe all student responses, but only the answers to the subset of questions that we asked, and these may differ from a student to another. It is still possible to compute the deviation profile within this subset, but it cannot be aggregated to a higher level in this fashion, because of the bias induced by the adaptive process.

A natural direction to extend the Rasch model is to study multidimensional abilities. In Multidimensional Item Response Theory (MIRT) (Reckase 2009), both learners and items are modeled by vectors of a certain dimension d, and the tendency for a learner to solve an item depends only on the dot product of those vectors. Thus, a learner has a greater chance to solve items that are correlated with their ability vector, and asking a question brings information in the direction of its item vector.

Thus, if learner $i \in \{1, \ldots, n\}$ is modelled by vector $\theta_i \in \mathbf{R}^d$ and item $j \in \{1, \ldots, m\}$ is modelled by vector $d_j \in \mathbf{R}^d$, the probability that the learner i answers the item j correctly is given by Eq. (4.3):

$$Pr(\text{``learner } i \text{ answers item } j\text{''}) = \Phi(\vartheta_i \cdot d_j). \tag{4.3}$$

Using this model, the Fisher information becomes a matrix. When trying to ask the most informative questions, we may either choose to maximize the determinant of this matrix ("D-rule"), or choose to maximize the trace ("T-rule"). The D-rule chooses the item that provides the maximum volume of information, and hence the largest reduction of volume in the variance of the ability estimate. By contrast, the T-rule chooses an item that attempts to increase the average information about each component of the ability, ignoring the covariance between components.

MIRT can be restated as a matrix factorization problem, given by Eq. (4.4):

$$M \simeq \Phi(\Theta D^T) \tag{4.4}$$

where M is the $n \times m$ student data, Θ is the $n \times r$ *learner matrix* composed of the vectors of all learners, and D is the $m \times r$ *item matrix* which contains of all item vectors.

Nevertheless, those richer models involve many more parameters: d parameters are estimated for each of the n learners, and d parameters are estimated for each of the m items. Thus, this model is usually much harder to calibrate (Desmarais and Baker 2012; Lan et al. 2014).

4.3.2   Cognitive Diagnosis: Adaptive Assessment with Feedback

In cognitive diagnosis models, we assume that we can explain a student's success or failure on a learning task, based on whether they master (or fail to master) some *knowledge components* (KC). The point of these knowledge components is that they allow a transfer of evidence from one item to another. For instance, to evaluate correctly the sum $1/7 + 8/9$, a learner needs to know how to add numbers, and how to convert two fractions to the same denominator. By contrast, a learner that solves $1/7 + 8/7$ only needs to know how to add. To use these cognitive models, we need to specify, for each item proposed in the test, which KCs are required to solve it: this information is represented as a binary matrix, called the *q-matrix*. The q-matrix simply maps items to KCs: it is a transfer model. See Fig. 4.3 for a real-world example of a q-matrix.

	Knowledge components							
	1	2	3	4	5	6	7	8
Item 1	0	0	0	1	0	1	1	0
Item 2	0	0	0	1	0	0	1	0
Item 3	0	0	0	1	0	0	1	0
Item 4	0	1	1	0	1	0	1	0
Item 5	0	1	0	1	0	0	1	1
Item 6	0	0	0	0	0	0	1	0
Item 7	1	1	0	0	0	0	1	0
Item 8	0	0	0	0	0	0	1	0
Item 9	0	1	0	0	0	0	0	0
Item 10	0	1	0	0	1	0	1	1
Item 11	0	1	0	0	1	0	1	0
Item 12	0	0	0	0	0	0	1	1
Item 13	0	1	0	1	1	0	1	0
Item 14	0	1	0	0	0	0	1	0
Item 15	1	0	0	0	0	0	1	0
Item 16	0	1	0	0	0	0	1	0
Item 17	0	1	0	0	1	0	1	0
Item 18	0	1	0	0	1	1	1	0
Item 19	1	1	1	0	1	0	1	0
Item 20	0	1	1	0	1	0	1	0

Description of knowledge components:
1. convert a whole number to a fraction
2. separate a whole number from a fraction
3. simplify before subtracting
4. find a common denominator
5. borrow from whole number part
6. column borrow to subtract the second numerator from the first
7. subtract numerators
8. reduce answers to simplest form

Fig. 4.3 The q-matrix corresponding to Tatsuoka's (1984) fraction subtraction data set of 536 middle school students over 20 fraction subtraction test items. The matrix has 8 knowledge components, which are described on the *right*

The DINA model ("Deterministic Input, Noisy And") assumes that the learner will solve a certain item i with probability $1 - s_i$ if they master every required KC, and will solve it with probability g_i otherwise. The parameter g_i is called the *guess parameter* of item i, and it represents the probability of guessing the right answer to item i without being able to solve it. The parameter s_i is called the *slip parameter* of item i: it represents the probability of slipping on item i, i.e., failing to answer it even when the correct KCs are mastered. By contrast, in the DINO model ("Deterministic Input, Noisy Or"), the learner solves an item with probability $1 - s_i$ whenever it masters one of the KCs for this item; if the learner masters none of them, the probability of solving the item is g_i.

The latent state of a learner is represented by a vector of K bits $(c_1, \ldots, c_K)$ where K is the total number of KCs. The vector indicates which KCs are mastered: for each KC k, the bit c_k is 1 if the learner masters the k-th KC, and 0 otherwise. Each time the learner answers an item, we obtain more information about their probable latent state. Xu et al. (2003) have used adaptive testing strategies in order to infer the latent state of the learner using few questions: this is called *cognitive diagnosis computerized adaptive testing* (CD-CAT). Knowing the mental state of a learner, we can infer their behavior over the remaining questions in the test; we can then use this information to choose which questions to ask, as we will now describe. At each point in time, the system keeps a probability distribution over the 2^K

possible latent states: this distribution is refined after each question, using Bayes'
rule. A usual measure of uncertainty on the distribution is *entropy*, defined by
Eq. (4.5):

$$H(\mu) = - \sum_{c \in \{0,1\}^k} \mu(c) \log \mu(c). \tag{4.5}$$

Hence, to converge quickly into the true latent state, the best item to ask is the
one that reduces average entropy the most (Doignon et Falmagne 2012; Huebner
2010). Other criteria have been proposed: for instance, we can ask the question that
maximizes the *Kullback-Leibler divergence*, which measures the difference
between two probability distributions (Cheng 2009). It is given by Eq. (4.6):

$$D_{KL}(P \| Q) = \sum_i P(i) \log \frac{P(i)}{Q(i)}. \tag{4.6}$$

As Chang (2015) states, "A survey conducted in Zhengzhou found that CD-CAT
encourages critical thinking, making students more independent in problem solving,
and offers easy to follow individualized remedy, making learning more interesting."

For large values of K, it may be intractable to maintain a probability distribution
over the 2^K states. Hence, in practice, we often take $K \leq 10$ (Su et al. 2013). We can
also reduce the complexity by assuming prerequisites between KCs: if mastering a
KC implies that the student must master another KC, the number of possible states
decreases, and so does the complexity. This approach is called the *Attribute
Hierarchy Model* (Leighton et al. 2004): it can be used to represent knowledge
more accurately and fit the data better (Rupp et al. 2012).

The q-matrix may be costly to build. Thus, devising a q-matrix automatically has
been an open field of research. Barnes (2005) used a hill-climbing technique while
Winters et al. (2005) and Desmarais et al. (2011) tried non-negative matrix factorization techniques to recover q-matrices from real and simulated multidisciplinary assessment data. Experimentally, these approaches can efficiently separate
items in categories when the topics are clearly separated, e.g., French and
Mathematics. Formally, *non-negative matrix factorization* tries to devise matrices
with non-negative coefficients W and Q such that the original matrix M verifies
$M \simeq WQ^T$. Additional constraints can be made: for instance, *sparse PCA* (Zou et al.
2006) looks for a factorization of the form $M \simeq WQ^T$ where Q is sparse, following
the assumption that only few knowledge components are required for any one task.
On the datasets we described in Sect. 4.5, the expert-specified q-matrix fitted the
data better than a q-matrix devised automatically using sparse PCA. Further, even if
we could fit the data better with an automatically devised q-matrix, this would not
allow us to deduce human-readable names for the knowledge components. Lan
et al. (2014) tried to circumvent this issue, by studying how to interpret a posteriori
the columns of a q-matrix devised by an algorithm, with the help of expert-specified
tags. A more recent work from (Koedinger et al. 2012) managed to combine

q-matrices from several experts using crowdsourcing in order to find better cognitive models that are still understandable for humans.

A natural goal is then to design a model that combines the best of both worlds, and represent which knowledge components are required for tasks, as well as some notion of the difficulty of tasks. Unified models have been designed towards this end, such as the *general diagnostic model for partial credit data* (Davier 2005), which generalizes both MIRT and some other cognitive models. It is given by Eq. (4.7):

$$Pr(``\text{learner } i \text{ answers item } j") = \Phi\left(\beta_i + \sum_{k=1}^{K} \theta_{ik} q_{jk} d_{jk}\right) \tag{4.7}$$

where K is the total number of KCs involved in the test, β_i is the main ability of learner i, θ_{ik} is its ability for KC k, d_{jk} the difficulty of item j over KC k, and q_{jk} is the (j, k) entry of the q-matrix: 1 if KC k is involved in the resolution of item j, and 0 otherwise. Intuitively, this model is similar to the MIRT model presented above, but the dot product is computed only on part of the components. In other words, we consider a MIRT model where the number of dimensions is the number of KCs of the q-matrix: $d = K$. When we calibrate the feature vector of dimension d of an item, only the components that correspond to KCs involved in the resolution of this item are taken into account: see Fig. 4.4. This model has one important advantage: as few KCs are usually required to solve each item, this allows the MIRT parameter estimation to converge faster. Vie et al. (2016) used this model in adaptive assessment under the name GenMA (for General Multidimensional Adaptive). Another advantage of this model is that, at any point in the test, the ability estimate represents degrees of proficiency for each knowledge component. The GenMA model is therefore a hybrid model that combines the Rasch model and a cognitive model.

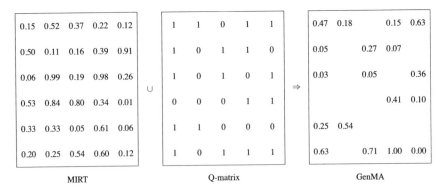

Fig. 4.4 The GenMA hybrid model, combining item response theory and a q-matrix

4.3.3 Competence-Based Knowledge Space Theory and Applications

Doignon and Falmagne (2012) have developed *knowledge space theory*, an abstract theory that relies on a partial order between subsets of a discrete knowledge space. Formally, let us assume that there is a certain number of KCs to learn, following a dependency graph specifying which KCs needs to be mastered before learning a certain KC. We present an example of dependency graph in Fig. 4.5. From this graph, one can compute the feasible knowledge states, i.e., the KCs that are actually mastered by the learner. For example, $\{a, b\}$ is a feasible knowledge state while the singleton $\{b\}$ is not, because a needs to be mastered before b. Thus, for this example there are 10 feasible knowledge states: $\emptyset$, $\{a\}$, $\{b\}$, $\{a, b\}$, $\{a, c\}$, $\{a, b, c\}$, $\{a, b, c, d\}$, $\{a, b, c, e\}$, $\{a, b, c, d, e\}$, $\{a, b, c, d, e, f\}$. An adaptive assessment can then uncover the knowledge state of the examinee, in a similar fashion to the Attribute Hierarchy Model described above at Sect. 4.3.2. Once the knowledge subset of a learner has been identified, this model can suggest to him the next knowledge components to learn in order to help them progress, through a so-called *learning path*. For instance, from the knowledge state $\{a\}$ on Fig. 4.5, the learner can choose whether to learn the KC b or the KC c first.

Falmagne et al. (2006) provide an adaptive test in order to guess effectively the knowledge space using entropy minimization, which is however not robust to careless errors. This model has been implemented in practice in the ALEKS system, which is used by millions of users today (Kickmeier-Rust and Albert 2015; Desmarais and Baker 2012).

Lynch and Howlin (2014) have implemented a similar adaptive pretest at the beginning of a MOOC, in order to guess what the learner already masters, and help them jump directly to useful materials in the course. To address slip and guess parameters, they combine models from knowledge space theory and item response theory.

Another line of work has developed more fine-grained models for adaptive testing, by considering even richer domain representations such as an ontology

Fig. 4.5 *On the left* an example of precedence diagram. *On the right* the corresponding learning paths

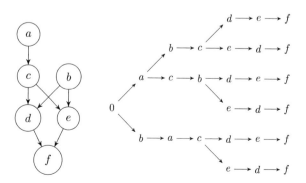

(Mandin and Guin 2014; Kickmeier-Rust and Albert 2015) of the domain covered by the test. However, such knowledge representations are costly to develop.

4.3.4 Adaptive Assessment and Recommender Systems

We now describe how two well-known problems from recommender systems find their counterparts in adaptive assessment. Recommender systems can recommend new items to a user based on their preferences on other items. Two approaches are used:

- *content-based recommendations*, that analyze the content of the items in order to devise a measure of similarity between items;
- *collaborative filtering*, where the similarity between items depends solely on user preferences, i.e., items that are liked by the same people are considered to be similar.

Overall, the aim of these approaches is to predict the preference of a user over an unseen item, based on their preferences over a fraction of the items that they know. In our case, we want to predict the performance of a user over a question that we did not ask yet, based on the previous performance of the user. Collaborative filtering techniques have been applied on student data in an user-to-resource fashion (Manouselis et al. 2011; Verbert et al. 2011) and in an user-to-task fashion (Toscher and Jahrer 2010; Thai-Nghe et al. 2011; Bergner et al. 2012).

All recommender systems face the *user cold-start problem*: given a new user, how to quickly recommend new relevant items to them? In technology-enhanced learning, the problem becomes: given a new learner, how to quickly identify the resources that they will need? To the best of our knowledge, the only work that references the cold-start problem in educational environments is (Thai-Nghe et al. 2011): "In the educational environment, the cold-start problem is not as harmful than in the e-commerce environment where [new] users and items appear every day or even hour, thus, the models need not to be re-trained continuously." However, this article predates the advent of MOOCs, therefore this claim is no longer true.

Among the most famous approaches to tackle the cold-start problem, one method of particular interest is an adaptive interview that presents some items to the learner, and asks the learner to rate them. Golbandi et al. (2011) build a decision tree that starts an interview process with the new user in order to quickly identify users similar to them. The best items are the ones that bisect the population into roughly two halves, and are in a way similar to discriminative items in item response theory. If we transfer this problem to adaptive assessment with test-size reduction, it can be phrased as follows: what questions should we ask to a new learner in order to infer their whole vector of answers? The core difference with an e-commerce environment is that learners might try to game the system more than in a commercial environment, thus their answers might not fit their ability estimate.

Most collaborative filtering techniques assume that the user-to-item matrix M is of low rank r, and look for a low-rank approximation under the matrix factorization $M \simeq UV^{\mathrm{T}}$ where U and V are assumed of width r. Note that, if M is binary and the loss function for the approximation is the logistic loss, we get back to the MIRT model (as a generalized linear model) described in Sect. 4.3.1.

Diversity Recommender systems have been criticized because they "put the user in a filter bubble" and harm serendipity. But since then, there has been more research into diversity (i.e., finding a set of diverse items to recommend), and into explained recommendations. More recently, there has been a need for more inter-active recommender systems, giving more power to users by allowing them to steer the recommendations towards other directions. The application to learner systems is straightforward: this could help the learner navigate the course.

Implicit feedback In e-commerce use cases, recommender systems differentiate explicit feedback given willingly by the user, such as "this user liked this item", from implicit feedback resulting from unintentional behavior, such as "this user spent a lot of time on this page", which may imply that they are interested by the contents of this page. Such implicit feedback data is therefore used by e-commerce websites in order to know their clients better. In technology-enhanced learning use cases, explicit feedback data is often sparse; thus, implicit feedback techniques are attractive candidates to improve recommender performance. For instance, these techniques could use the time spent on a page, the search terms provided by the user, information about downloaded resources, and comments posted by the user (Verbert et al. 2011). Such data may also be useful if they are recorded while the test is administered, e.g., some course content a learner is browsing while attempting a low-stakes adaptive assessment might be useful for other learners.

Adding external information Some recommender systems embed additional information in their learning models: for instance, the description of the item, or even the musical content itself in the scope of music recommendation. In order to improve prediction over the test, one could consider extracting additional features from the problem statements of the items, and incorporate them within the feature vector.

4.3.5 Adaptive Strategies for Exploration–Exploitation Tradeoff

In some applications, one wants to maximize a certain objective function while asking questions. This leads to an exploration–exploitation trade-off: we can increase our knowledge of the user more, by exploring the space of items, or we can exploit what we know in order to maximize a certain reward. Clement et al. (2015) applied these techniques to intelligent tutoring systems: they personalize sequences of learning activities in order to uncover the knowledge components of the learner while maximizing the user's learning progress, as a function of the performance over the latest tasks. They use two models based on multi-armed bandits: the first one relies on Vygotsky's zone of proximal development (Vygotsky 1980) under the form of a dependency graph, the second one uses an expert-specified q-matrix. They tested both approaches on 400 real students between 7 and 8 years old. Quite surprisingly, they discovered that using the dependency graph yielded better performance than using the q-matrix. Their technique helped improve learning for populations of students with larger variety and stronger difficulties.

4.3.6 Multistage Testing

So far, we always assumed that questions were asked one after another. However, the first ability estimate, using only the first answer, has high bias. Thus, ongoing psychometrics research tends to study scenarios where we ask pools of questions at each step, performing adaptation only once sufficient information has been gathered. This approach has been referred to as *multistage testing* (MST) (Yan et al. 2014). After the first stage of k_1 questions, according to their performance, the learner moves to another stage of k_2 questions that depend only on their performance, and so on, see Fig. 4.6. MST presents another advantage: the learner can revise their answers before moving to the next stage, without the need of complicated models for response revision (Han 2013; Wang et al. 2015). In the language of clinical trials, MST design can be viewed as a *group sequential* design, while a CAT can be viewed as a *fully sequential* design. The item selection is performed automatically, but all stages of questions can be reviewed before administration (Chang 2015). Wang et al. (2016) suggest to ask a group of questions at the beginning of the test, when little information about learner ability is available, and progressively reduce the number of questions of each stage in order to increase opportunities to adapt. Also, asking questions in pools means that we can do content balancing at each stage, instead of jumping from one knowledge component to the other after every question.

Fig. 4.6 In multistage testing, questions are asked in a group sequential design

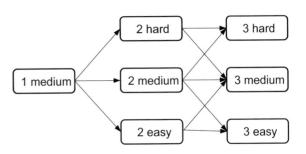

4.4   Comparison of Adaptive Testing Models

Adaptive assessment models need to be validated on real data, in order to guarantee that the model accurately assesses the constructs that it is supposed to assess (Desmarais and Baker 2012). A common way to validate a model is to measure how well the assessment can predict future performance within the learning system.

To evaluate on real data the models that we presented above, we can embed them in a unified framework: all of them can be seen as decision trees (Ueno and Songmuang 2010; Yan et al. 2014), where nodes are possible states of the test, and edges are followed according to the answers provided by the learner, like a flow-chart. Thus, within a node, we have access to an incomplete response pattern, and we want to use our student model and infer the behavior of the learner over the remaining questions. The best model is the one that classifies the remaining outcomes with minimal error.

Formally, let us consider a set I of students who answer questions from a set Q. Our student data is a binary matrix D of size $|I| \times |Q|$, where D_{iq} is 1 if student i answered question q correctly, 0 otherwise. An adaptive test can be formalized as follows.

> TEST(student $i \in I$):
> > **While** some questions still need to be asked
> > > ASK to student i the next question

We want to compare the predictive power of different adaptive testing algorithms that model the probability of student i solving question j. Thus, for our cross-validation, we need to define:

- a student training set $I_{train} \subset I$;
- a student testing set $I_{test} \subset I$;
- a question validation set $Q_{val} \subset Q$.

We use the same sets for all the models that we study. Model evaluation is performed using the EVALUATEMODEL function:

> EVALUATEMODEL(model M, students I_{train}, students I_{test}, questions Q_{val}) :
> > TRAIN model using lines I_{train} of D
> > **For each** student i of I_{test} **do**
> > > **While** not all questions $\in Q \setminus Q_{val}$ have been asked
> > > > CHOOSENEXTITEM and ask it to student i
> > > > Evaluate predictions of model M over questions Q_{val}.

We make a cross-validation of each model over 10 subsamples of students and 4 subsamples of questions (these constant values are parameters that may be changed). Thus, if we number student subsamples I_i for $i = 1, \ldots, 10$ and question subsamples Q_j for $j = 1, \ldots, 4$, experiment (i, j) consists in the following steps:

Fig. 4.7 Cross-validation
over 10 student subsamples
and 4 question subsamples.
Each case (i,j) contains the
results of the experiment (i,j)
for student test set $(I_{test} = I_i)$
and question validation set
$(Q_{val} = Q_j)$

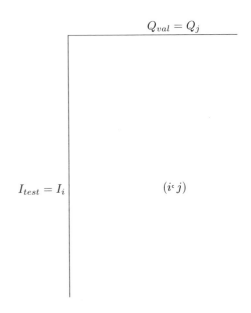

- train the evaluated model over all student subsamples except the i-th, i.e., $I_{train} = I\backslash I_i$;
- simulate adaptive tests on the i-th student subsample (i.e., $I_{test} = I_i$) using all question subsamples except the j-th (namely, Q_j), and evaluate after each question the error of the model over the j-th question subsample (i.e., $Q_{val} = Q_j$).

The error is given by Eq. (4.8), called *score* or *log loss*:

$$e(p,t) = \frac{1}{|Q_{val}|} \sum_{k \in Q_{val}} t_k \log p_k + (1 - t_k)\log(1 - p_k) \qquad (4.8)$$

where p is the predicted outcome over all $|Q|$ questions and t is the true response pattern.

In order to visualize the results, errors computed during experiment (i,j) are stored in a matrix of size 10×4. Thus, computing the mean error for each column, we can see how models performed on a certain subset of questions, see Fig. 4.7.

4.5   Results

For our experiments, we used three real datasets. The models considered were the Rasch model, the DINA model with an expert-specified q-matrix, and the GenMA model with the same q-matrix.

We now describe the results of our cross-validation, for different sizes of training and testing sets. For each dataset, the mean error of each model has been computed over all experiments.

4.5.1 ECPE

This student dataset is a 2922×28 binary matrix representing the results of 2922 learners over 28 English questions from the Examination for the Certificate of Proficiency in English (ECPE). The ECPE purports to measure three attributes, therefore the corresponding q-matrix has only 3 skills: knowledge of morphosyntactic rules, cohesive rules, and lexical rules. This dataset is featured in (Templin and Bradshaw 2014).

For this dataset, there were 5 student subsamples, and 4 question subsamples, i.e., the student training set was composed of 80% of the students, and the validation question sets were composed of 7 questions. The results are given in Table 4.1 and Fig. 4.8.

The GenMA model outperforms the Rasch and DINA models. The estimated slip and guess parameters of the DINA model for this dataset are reported in Table 4.2.

4.5.2 Fraction

This student dataset is a 536×20 binary matrix representing the results of 536 middle school students over 20 fraction subtraction questions. The corresponding q-matrix has 8 skills, described in Fig. 4.3 and can be found in (DeCarlo 2010).

There were 5 student subsamples, and 4 question subsamples, i.e., the student training set was composed of 80% of the students, and the validation question sets were composed of 5 questions. For this dataset only, we compared two occurrences of the GenMA model, one with the original expert q-matrix, the other one with a different q-matrix which was computed automatically, using sparse PCA. The results are given in Table 4.3 and Fig. 4.9.

The best model is GenMA + expert: 4 questions over 15 are enough to provide a feedback that predicts correctly 4 questions over 5 in average in the validation set.

Table 4.1 Mean error of the different models considered for the ECPE dataset

Model	After 5 questions	After 10 questions	After 15 questions
Rasch	0.533 ± 0.010	0.522 ± 0.009	0.514 ± 0.009
DINA	0.535 ± 0.008	0.527 ± 0.008	0.521 ± 0.008
GenMA	**0.515 ± 0.008**	**0.484 ± 0.008**	**0.480 ± 0.008**

The lowest values are denoted in bold

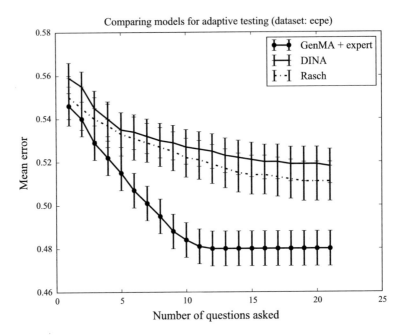

Fig. 4.8 Mean error (negative log-likelihood) over the validation question set as a function of how many questions have been asked, for the ECPE dataset

Table 4.2 The q-matrix used for the ECPE dataset, together with the guess and slip parameters, and the success rate for each question

q-matrix					Success rate (%)
Entries			Guess	Slip	
1	1	0	0.705	0.085	80
0	1	0	0.724	0.101	83
1	0	1	0.438	0.266	57
0	0	1	0.480	0.162	70
0	0	1	0.764	0.040	88
0	0	1	0.717	0.066	85
1	0	1	0.544	0.085	72
0	1	0	0.802	0.040	89
0	0	1	0.534	0.199	70
1	0	0	0.483	0.163	65
1	0	1	0.556	0.099	72
1	0	1	0.195	0.305	43
1	0	0	0.633	0.122	75
1	0	0	0.517	0.212	65
0	0	1	0.749	0.040	88
1	0	1	0.549	0.126	70
0	1	1	0.816	0.058	88
0	**0**	**1**	**0.729**	**0.086**	**84**

(continued)

Table 4.2 (continued)

q-matrix						Success rate (%)
Entries			Guess	Slip		
0	0	1	0.473	0.150		71
1	0	1	0.239	0.295		46
1	0	1	0.621	0.097		75
0	0	1	0.322	0.188		63
0	1	0	0.637	0.075		81
0	1	0	0.313	0.322		53
1	0	0	0.512	0.272		61
0	0	1	0.555	0.211		70
1	0	0	0.265	0.369		44
0	0	1	0.659	0.086		81

The highest guess value is represented in bold

Table 4.3 Mean error of the different models considered for the fraction dataset

Model	After 4 questions	After 10 questions	After 15 questions
Rasch	0.461 ± 0.028	0.420 ± 0.027	0.413 ± 0.027
GenMA + expert	**0.454 ± 0.022**	**0.357 ± 0.018**	**0.322 ± 0.017**
GenMA + auto	0.544 ± 0.013	0.447 ± 0.012	0.393 ± 0.011
DINA	0.578 ± 0.030	0.429 ± 0.027	0.414 ± 0.029

The lowest values are denoted in bold

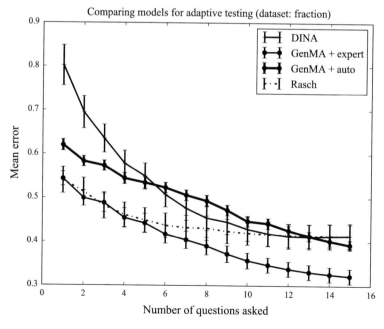

Fig. 4.9 Mean error (negative log-likelihood) over the validation question set as a function of how many questions have been asked, for the fraction dataset

Table 4.4 Mean error of the different models considered for the TIMSS dataset

Model	After 4 questions	After 8 questions	After 11 questions
Rasch	0.576 ± 0.008	0.559 ± 0.008	0.555 ± 0.008
DINA	0.588 ± 0.005	0.570 ± 0.006	0.566 ± 0.006
GenMA	**0.537 ± 0.006**	**0.505 ± 0.006**	**0.487 ± 0.006**

The lowest values are denoted in bold

As an example, for one of the test students, GenMA chooses 4 questions to ask in an adaptive way, then predicts that the student will correctly answer the questions from the validation question set with probabilities [61.7, 12.3, 41.8, 12.7, 12%]. Actually the true performance of the student over the validation question set is [correct, incorrect, correct, incorrect, incorrect], so the mean error is 0.350, according to Eq. (4.8).

4.5.3 TIMSS

This student dataset is a 757×23 binary matrix representing the results of 757 students over 23 questions from the Trends in International Mathematics and Science Study (TIMSS) 2003, an U.S. eighth grade mathematics test. The corresponding q-matrix has 13 skills over the 15 specified in (Su et al. 2013), i.e., all skills except the 10th and the 12th.

There were 4 student subsamples, and 2 question subsamples, i.e., the student training set was composed of 75% of the students, and the two validation question sets were composed of 11 and 12 questions. The results are given in Table 4.4 and Fig. 4.10. The best model is GenMA: after having asked 4 questions, GenMA outperforms the other models.

4.6 Discussion

In all experiments, the hybrid model GenMA with the expert q-matrix performs the best.

In the ECPE dataset, DINA and Rasch have similar predictive power, which is quite surprising given that Rasch does not require any domain knowledge. This may be because, in this dataset, there are only 3 skills: thus, the number of possible states for a learner is $2^3 = 8$, for many possible response patterns (2^{28}). Consequently, the estimated guess and slip parameters are very high (see Table 4.2), which explains why the information gained at each question is low. Indeed, the item which requires KC 2 and 3 is really easy to solve (88% success rate), even easier than items that require only KC 2 or only KC 3. Hence, the only way for the DINA model to express this behavior is to boost the guess parameter.

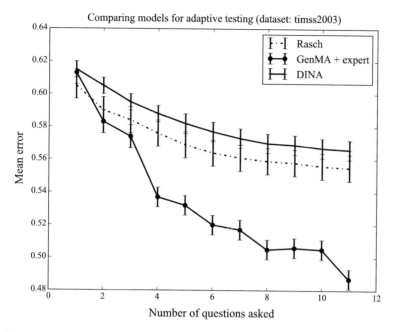

Fig. 4.10 Mean error (negative log-likelihood) over the validation question set as a function of how many questions have been asked, for the TIMSS dataset

On the contrary, GenMA calibrates one difficulty value per knowledge component, so it is a more expressive model. The same reason may explain why the mean error of GenMA converges after 11 questions: this 3-dimensional model may not be rich enough to understand the dataset, while in the Fraction dataset, the 8-dimensional GenMA model can learn after every question.

In the Fraction dataset, the DINA model tries to identify the latent state of the learner over 2^8 possible states, asking questions over few KCs at each step. This may explain why DINA requires many questions in order to converge. Rasch and GenMA-expert have similar predictive power in the early questions, but at least GenMA-expert can provide useful feedback, whereas Rasch cannot. The automatically generated q-matrix used in GenMA-auto has lower predictive power. Hence, for this dataset, Rasch provides a better adaptive assessment model than a q-matrix that is computed automatically.

In the TIMSS dataset, the DINA model tries to identify the latent state of the learner over 2^{13} possible states, which is why it needs many questions in order to reduce the prediction error. Similarly, the unidimensional Rasch model might not be enough to comprehend this multidimensional dataset. The hybrid model GenMA outperforms the other models, and can provide a feedback over 13 dimensions that achieves a mean accuracy of 77% over the validation question set of 12 questions, after 4 questions have been asked in an adaptive way.

4.6.1 Adaptive Pretest at the Beginning of a Course

At the beginning of a course, we have to fully explore the knowledge of the learner, in order to identify their latent knowledge using as few questions as possible. This is a cold-start problem, where we have to identify whether the learner holds the prerequisites of the course, and possibly their weak and strong points. If a dependency graph is available, we suggest to use Doignon and Falmagne's adaptive assessment model (see Sect. 4.3.3). If a q-matrix is available, we suggest to use the GenMA model (see Sect. 4.3.2). Otherwise, the Rasch model at least provides a way to measure the level of the learner in order to detect students that will require more attention.

4.6.2 Adaptive Test at the Middle of a Course

Learners often wish to have a taste of the tasks they will be expected to solve in the final test, in the form of a mock self-assessment that does not count towards their final results. There are several scenarios to consider. If learners have access to the course while taking this low-stakes test, an adaptive assessment should take into account the fact that the level of learners may change while they are taking the test, for example because they are checking the course material during the test. Hence, this is a good use case for models that measure the progress of the learner, such as multi-armed bandits (Clement et al. 2015), mentioned in Sect. 4.3.5. Recall that such models need either a dependency graph or a q-matrix. If learners do not check the course material while taking the test, for example because they have limited time, the GenMA model can ask them a few questions and provide feedback, under the condition that a q-matrix is available.

Depending on the context, students should be tracked from one occurrence of the test to the next one, or not. If the test is fully anonymous, a student might get the same item twice when taking the test twice. Also they will have to record their progress themselves, e.g., by exporting their results. If students are tracked, the teacher can be notified whenever a student struggles at obtaining some KC.

Whenever students want to practice specific KCs, they can filter at the beginning of the test the KCs for which they will be assessed. This is an example of the adaptability of such models instead of pure adaptivity, as stated in Sect. 4.2. Students can therefore learn at their own pace.

4.6.3 Adaptive Test at the End of a Course

A high-stake test at the end of the course might rely on the usual adaptive assessment strategies in item response theory, in order to measure examinees effectively and grade them. On this last examination, we assume that feedback is not so useful, so any model will be suitable. Examples include the GMAT and GRE standardized tests.

4.6.4   Other Applications

Adaptive assessment cognitive models such as DINA or GenMA can provide feedback under the form of degrees of proficiency over several KCs. Whenever a learner wants to sit for an anonymous test, he can understand what he did wrong. Combined with a recommender system, our model could automatically suggest lessons based on the KCs that need further work. A teacher can map student learning outcomes to KCs and KCs to items in order to be notified whenever a student is experiencing difficulty at attaining a concept. All the data collected by tests can be embedded in dashboards for visualization, in order to figure out what KCs are the most difficult to obtain for a population of students, and possibly suggest grouping students with similar difficulties, or at the contrary with disjoint difficulties.

4.7   Limitations

Here we only considered assessment of knowledge and no other dimension such as perseverance, organization, carefulness, responsibility. By reducing items, we reduce the time spent by students being assessed, which prevents boredom and leaves more time for other activities.

In our case, within a test our models never ask the same question twice. In many scenarios though, presenting the same item several times is better, for example in vocabulary learning. Spaced repetition systems based on flashcards such as Anki have been successfully used for vocabulary learning (Altiner 2011). In our case, we prefer to ask different items that need similar KCs (knowledge components), e.g., variants of a same exercise in mathematics. Such an approach has been referred to as interleaved practice (Dunlosky et al. 2013) and reduces the risks of guessing the correct answer.

Our approach is mainly static, which means we assume that the knowledge of the student does not increase within a test, even while he gets several opportunities of being assessed on the same KCs. This assumption can be made because the learner receives feedback only at the end of the test. Thus, our diagnostic test provides a snapshot of the student's knowledge at a certain time. Students can record these snapshots in order to visualize their own progress.

For simplicity, we do not consider learner metadata in our experiments, such as demographic information. This allows us to provide an anonymous test, i.e., the results are stored anonymously. This prevents stress from the examinee and helps them jump more easily into practice testing, which is useful for their learning (Dunlosky et al. 2013).

4.8   The Future of Assessment

We presented several models that could be used for adaptive assessment. A promising application is low-stakes adaptive formative assessments: before high-stakes assessments, learners like to train and to measure what they must know to complete the course. Such adaptive tests would be able to quickly identify the components that need further work and help the learner prepare for the final high-stakes test. It would be interesting to combine this work with automatic item generation. Learners could obtain as many variants of the same problem as they need so as to master the skills involved. The results of these adaptive tests may be recorded anonymously, so that the student can start over "with a clean state", without any tracking. Indeed, no learner would like their mistakes to be recorded for their entire lives (Executive Office of the President 2014).

With the help of learning analytics, *explicit testing* may be progressively replaced with *embedded assessment*, using multiple sources of data to predict student performance and tailor education accordingly (Shute et al. 2016; Redecker and Johanessen 2013). Indeed, if the learner is continuously monitored by the platform and if a digital tutor can answer their questions and recommend activities, they can be full actors of their continually changing progress and there is no need for an explicit test at the end of the course.

Even in such cases, however, we will still need adaptive pretests for specific uses, e.g., for international certifications (GMAT, GRE), or for newcomers at the beginning of a course, in order to identify effectively the latent knowledge they acquired in their past experience (Baker and Inventado 2014; Lynch and Howlin 2014).

Note that the only input to our adaptivity rules are the answers given so far by the learner, not their previous performance: this allows a learner to start from scratch whenever they wish. Using profile information such as the country to select the questions may lead to more accurate performance predictions: for example, from one country to another, the way to compute divisions is not the same. However, if we bias the assessment by sensitive information of this kind, we may inadvertently discriminate against some students.

In the future, an online platform could first ask the learner about their presumed knowledge. The platform could then verify if the self-assessment holds, and, if needed, explain the discrepancy. The learner could then possibly correct this assessment by proving that they actually master the knowledge components required: this could also allow them to learn more material.

4.9 Conclusion

We presented several recent student models that can be used to leverage former assessment data in order to provide shorter, adaptive assessments. As Rupp et al. (2012), rather than attempting to determine the best model for all uses, we have compared them in terms of brevity and predictive power, to see which model is better suited to which use. Note that, throughout this chapter, we have focused on the assessment on a single learner. Readers interested in computer-supported collaborative learning in group assessments may consider reading (Goggins et al. 2015).

Models which use q-matrices are usually validated using simulated data. In this chapter, we compared the strategies on real data. Our experimental protocol could be tried on yet other adaptive assessment models. It could also be generalized to evaluate multistage testing strategies.

According to the purpose of the test (e.g., beginning, middle or end of term), the most suitable model is not the same. In order to choose the best model, one should wonder: What knowledge do we have about the domain (dependency graph, q-matrix)? Is the knowledge of the learner evolving while they are taking the test? Do we want to estimate the knowledge components of the learner or do we want to measure their learning progress while they are taking the test?

The models we described in this chapter have been introduced in several lines of work which are mostly independent. In our opinion, this implies that experts should communicate more across fields, in order to avoid giving different names to the same model. There is a need for more interdisciplinary research, and methods from learning analytics and CAT should be combined in order to get richer and more complex models. Also, crowdsourcing techniques could be applied in order to harvest more data. One might imagine the following application of implicit feedback: "In order to solve this question, you seem to have spent a lot of time over the following lessons: [the corresponding list]. Which ones helped you answer this question?" Such data can help other learners who may experience difficulties over the same questions in the future.

As we stated in Sect. 4.2, we think more research should be done in interactive learning analytics models, giving more control back to the learner. In this chapter, we took a first step in this direction, being inspired by CAT strategies.

The focus on modern learning analytics for personalization does not only lead to automated adaptation: it can also increase the engagement and affect of learners in the system. This raises an open question on whether the platform should let users access everything it knows about them. One advantage would be to leverage trust and engagement, one risk would be that learners may change their behavior accordingly, to try to game the system.

There exist different interfaces for assessment such as serious games or stealth assessment, which lead to more motivation and engagement from the students, e.g., Packet Tracer for learning network routing (Rupp et al. 2012), or Newton's Playground for learning physics (Shute et al. 2013). We believe our approach is more generic: it only needs student data under the form of 1 and 0's and may also be applied to these serious-game scenarios. We leave this for further research.

Acknowledgements We thank Antoine Amarilli and Ryan Lahfa for their most valuable comments. This work is supported by the Paris-Saclay Institut de la Société Numérique funded by the IDEX Paris-Saclay, ANR-11-IDEX-0003-02.

References

Altiner C (2011) Integrating a computer-based flashcard program into academic vocabulary learning. Master's thesis, Iowa State University

Baker RS, Inventado PS (2014) Educational data mining and learning analytics. In: Learning analytics. Springer, New York, pp 61–75

Barnes T (2005) The q-matrix method: mining student response data for knowledge. In: American Association for Artificial Intelligence 2005 educational data mining workshop

Bergner Y, Droschler S, Kortemeyer G, Rayyan S, Seaton D, Pritchard DE (2012) Model-based collaborative filtering analysis of student response data: machinelearning item response theory. International Educational Data Mining Society

Chang HH (2015) Psychometrics behind computerized adaptive testing. Psychometrika 80(1):1–20

Chatti MA, Dyckhoff AL, Schroeder U, Thüs H (2012) A reference model for learning analytics. Int J Technol Enhanced Learning 4(5–6):318–331

Cheng Y (2009) When cognitive diagnosis meets computerized adaptive testing: CD-CAT. Psychometrika 74(4):619–632

Chen S, Choi A, Darwiche A (2015) Computer adaptive testing using the same-decision probability. In: 12th Annual Bayesian modeling applications workshop (BMAW)

Clement B, Roy D, Oudeyer PY, Lopes M (2015) Multi-armed bandits for intelligent tutoring systems. JEDM-J Educ Data Mining 7(2):20–48

Graduate Management Admission Council (2013) Profile of GMAT candidates—executive summary. Technical report, Graduate Management Admission Council. http://www.gmac.com/market-intelligence-and-research/research-library/gmat-test-taker-data/profile-documents/2013-profile-of-gmat-candidates-executive-summary.aspx. Accessed 1 Apr 2016

Davier M (2005) A general diagnostic model applied to language testing data. ETS research report series 2:i–35

DeCarlo LT (2010) On the analysis of fraction subtraction data: the DINA model, classification, latent class sizes, and the q-matrix. Appl Psychol Measure 35(1):8–26

Desmarais MC, Baker RS (2012) A review of recent advances in learner and skill modeling in intelligent learning environments. User Model User-Adap Inter 22(1–2):9–38

Desmarais MC et al (2011) Conditions for effectively deriving a q-matrix from data with non-negative matrix factorization. In: 4th international conference on educational data mining, EDM, pp 41–50

Doignon JP, Falmagne JC (2012) Knowledge spaces. Springer Science & Business Media, Berlin

Dunlosky J, Rawson KA, Marsh EJ, Nathan MJ, Willingham DT (2013) Improving students learning with effective learning techniques promising directions from cognitive and educational psychology. Psychol Sci Public Interest 14(1):4–58

Executive Office of the President, Podesta J (2014) Big data: seizing opportunities, preserving values. Technical report, The White House

Falmagne JC, Cosyn E, Doignon JP, Thiéry N (2006) The assessment of knowledge, in theory and in practice. In: Formal concept analysis. Springer, Heidelberg, pp 61–79

Ferguson R (2012) Learning analytics: drivers, developments and challenges. Int J Technol Enhanced Learning 4(5–6):304–317

Goggins SP, Xing W, Chen X, Chen B, Wadholm B (2015) Learning analytics at "small" scale: exploring a complexity-grounded model for assessment automation. J UCS 21(1):66–92

Golbandi N, Koren Y, Lempel R (2011) Adaptive bootstrapping of recommender systems using decision trees. In: Proceedings of the fourth ACM international conference on Web search and data mining, ACM, pp 595–604

Hambleton RK, Swaminathan H (1985) Item response theory: principles and applications, vol 7. Springer Science & Business Media, New York

Han KT (2013) Item pocket method to allow response review and change in computerized adaptive testing. Appl Psychol Meas 37(4):259–275

Huebner A (2010) An overview of recent developments in cognitive diagnostic computer adaptive assessments. Pract Assess Res Eval 15(3):7

Kickmeier-Rust MD, Albert D (2015) Competence-based knowledge space theory. In: Measuring and visualizing learning in the information-rich classroom. Routledge, New-York and London, p 109

Koedinger KR, McLaughlin EA, Stamper JC (2012) Automated student model improvement. International Educational Data Mining Society

Lan AS, Waters AE, Studer C, Baraniuk RG (2014) Sparse factor analysis for learning and content analytics. J Mach Learn Res 15(1):1959–2008

Leighton JP, Gierl MJ, Hunka SM (2004) The attribute hierarchy method for cognitive assessment: a variation on tatsuoka's rule-space approach. J Educ Meas 41(3):205–237

Lynch D, Howlin CP (2014) Real world usage of an adaptive testing algorithm to uncover latent knowledge. In: 7th international conference of education, research and innovation (ICERI2014 proceedings). IATED, Seville, Spain, Nov 2014, pp 504–511. https://library.iated.org/publications/ICERI2014

Mandin S, Guin N (2014) Basing learner modelling on an ontology of knowledge and skills. In: 2014 IEEE 14th international conference on advanced learning technologies (ICALT). IEEE, pp 321–323

Manouselis N, Drachsler H, Vuorikari R, Hummel H, Koper R (2011) Recommender systems in technology enhanced learning. In: Recommender systems handbook, Springer Science & Business Media, New-York, pp 387–415

Mislevy RJ, Behrens JT, Dicerbo KE, Levy R (2012) Design and discovery in educational assessment: evidence-centered design, psychometrics, and educational data mining. JEDM-J Educ Data Mining 4(1):11–48

Peña-Ayala A (2014) Educational data mining: a survey and a data mining-based analysis of recent works. Expert Syst Appl 41(4):1432–1462

Reckase M (2009) Multidimensional item response theory, vol 150. Springer, New York

Redecker C, Johannessen O (2013) Changing assessment towards a new assessment paradigm using ICT. Eur J Educ 48(1):79–96

Rupp A, Levy R, Dicerbo KE, Sweet SJ, Crawford AV, Calico T, Benson M, Fay D, Kunze KL, Mislevy RJ et al (2012) Putting ecd into practice: the interplay of theory and data in evidence models within a digital learning environment. JEDM—J Educ Data Mining 4(1):49–110

Shute VJ, Ventura M, Kim YJ (2013) Assessment and learning of qualitative physics in Newton's playground. J Educ Res 106(6):423–430

Shute VJ, Leighton JP, Jang EE, Chu MW (2016) Advances in the science of assessment. Educ Assess 21(1):34–59

Su YL, Choi K, Lee W, Choi T, McAninch M (2013) Hierarchical cognitive diagnostic analysis for TIMSS 2003 mathematics. Centre Adv Stud Meas Assess 35:1–71

Tempelaar DT, Rienties B, Giesbers B (2015) In search for the most informative data for feedback generation: learning analytics in a data-rich context. Comput Hum Behav 47:157–167

Templin J, Bradshaw L (2014) Hierarchical diagnostic classification models: a family of models for estimating and testing attribute hierarchies. Psychometrika 79(2):317–339. doi:10.1007/s11336-013-9362-0, URL http://dx.doi.org/10.1007/s11336-013-9362-0

Thai-Nghe N, Drumond L, Horváth T, Schmidt-Thieme L et al (2011) Multi-relational factorization models for predicting student performance. In: Proceedings of the KDD workshop on knowledge discovery in educational data, Citeseer

Toscher A, Jahrer M (2010) Collaborative filtering applied to educational data mining. KDD Cup 2010

Ueno M, Songmuang P (2010) Computerized adaptive testing based on decision tree. In: 2010 10th IEEE international conference on advanced learning technologies. IEEE, pp 191–193

van der Linden WJ, Glas CA (2010) Elements of adaptive testing. Springer Science & Business Media, New-York. http://www.springer.com/la/book/9780387854595

Verbert K, Drachsler H, Manouselis N, Wolpers M, Vuorikari R, Duval E (2011) Dataset-driven research for improving recommender systems for learning. In: Proceedings of the 1st international conference on learning analytics and knowledge. ACM, pp 44–53

Verhelst ND (2012) Profile analysis: a closer look at the PISA 2000 reading data. Scandinavi J Educ Res 56(3):315–332

Vie JJ, Popineau F, Bourda Y, Bruillard (2016) Adaptive testing with a general diagnostic model. In: Design for teaching and learning in a networked world: 11th European conference on technology enhanced learning, EC-TEL 2016, Lyon, France, September 12–16, Proceedings (Springer, to appear)

Vygotsky LS (1980) Mind in society: the development of higher psychological processes. Harvard university press, Cambridge

Wang S, Fellouris G, Chang HH (2015) Sequential design for computerized adaptive testing that allows for response revision. arXiv preprint arXiv:150101366

Wang S, Lin H, Chang HH, Douglas J (2016) Hybrid computerized adaptive testing: from group sequential design to fully sequential design. J Educ Meas 53(1):45–62

Winters T, Shelton C, Payne T, Mei G (2005) Topic extraction from item-level grades. In: American Association for Artificial Intelligence 2005 workshop on educational datamining, Pittsburgh, PA, vol 1, p 3

Xu X, Chang H, Douglas J (2003) A simulation study to compare CAT strategies for cognitive diagnosis. In: Annual meeting of the American Educational Research Association, Chicago

Yan D, von Davier AA, Lewis C (2014) Computerized multistage testing. CRC Press, Boca Raton

Zernike K (2015) Obama administration calls for limits on testing in schools. The New York Times. http://www.nytimes.com/2015/10/25/us/obama-administration-calls-for-limits-on-testing-in-schools.html. Accessed 2 Apr 2016

Zou H, Hastie T, Tibshirani R (2006) Sparse principal component analysis. J Comput Graphical Stat 15(2):265–286

Chapter 5
Data-Driven Personalization of Student Learning Support in Higher Education

Danny Yen-Ting Liu, Kathryn Bartimote-Aufflick, Abelardo Pardo and Adam J. Bridgeman

Abstract Despite the explosion of interest in big data in higher education and the ensuing rush for catch-all predictive algorithms, there has been relatively little focus on the pedagogical and pastoral contexts of learning. The provision of personalized feedback and support to students is often generalized and decontextualized, and examples of systems that enable contextualized support are notably absent from the learning analytics landscape. In this chapter we discuss the design and deployment of the Student Relationship Engagement System (SRES), a learning analytics system that is grounded primarily within the unique contexts of individual courses. The SRES, currently in use by teachers from 19 departments, takes a holistic and more human-centric view of data—one that puts the relationship between teacher and student at the center. Our approach means that teachers' pedagogical expertise in recognizing meaningful data, identifying subgroups of students for a range of support actions, and designing and deploying these actions, is facilitated by a customizable technology platform. We describe a case study of the application of this human-centric approach to learning analytics, including its impacts on improving student engagement and outcomes, and debate the cultural, pedagogical, and technical aspects of learning analytics implementation.

D.Y.-T. Liu · A.J. Bridgeman
Faculty of Science, The University of Sydney, Sydney, NSW 2006, Australia
e-mail: danny.liu@sydney.edu.au

A.J. Bridgeman
e-mail: adam.bridgeman@sydney.edu.au

K. Bartimote-Aufflick
Quality and Analytics Group, The University of Sydney,
Sydney, NSW 2006, Australia
e-mail: kathryn.aufflick@sydney.edu.au

A. Pardo (✉)
Faculty of Engineering and Information Technology,
The University of Sydney, Sydney, NSW 2006, Australia
e-mail: abelardo.pardo@sydney.edu.au

© Springer International Publishing AG 2017
A. Peña-Ayala (ed.), *Learning Analytics: Fundaments, Applications, and Trends*, Studies in Systems, Decision and Control 94,
DOI 10.1007/978-3-319-52977-6_5

143

Keywords Actionable intelligence · Implementation · Intelligence amplification · Learning analytics · Personalization · Student engagement

Abbreviations

EDM   Educational data mining
EWS   Early warning system
LA   Learning analytics
LMS   Learning management system
SRES   Student Relationship Engagement System

5.1 Introduction

5.1.1 The State of Data-Driven Student Support

The rise in use of technology mediation in learning scenarios is providing unprecedented amounts of data about how educational institutions work and how students participate in learning experiences. At the same time, learning scenarios are becoming increasingly diverse and complex. The areas of educational data mining (EDM) and learning analytics (LA) have emerged to address the issue of how to use data to improve our understanding of learning, and enhance the overall quality of the learning experience for the student. Although EDM and LA researchers and practitioners maintain a similar focus (Baker and Siemens 2014), they differ in their approach to data generated in educational settings. Researchers in EDM frequently focus their analyses on the formulation or improvement of data mining algorithms designed to detect and predict important factors in a learning scenario. LA, on the other hand, focuses on how these algorithms can be deployed and integrated in learning designs, used by teachers, and provide tangible improvements for students. However, in their initial stages, both disciplines placed their emphasis mostly on how data can be collected and used by algorithms and not so much on how these data can then lead to actions that have a positive effect on students.

Prior to the availability of massive amounts of data, the areas of intelligent tutoring systems (Corbett et al. 1997), educational hypermedia (Bra 2002), and adaptive hypermedia (Kobsa 2007; Brusilovsky 1996) used technology mediation to increase the support students receive while participating in a learning experience. But this recent increase in the number of data sources about events and information produced while students learn has prompted the use of new types of algorithms and techniques to achieve these improvements through more comprehensive understanding of how students work in these contexts.

The first initiatives in the LA space were conceived by comparing education with other fields such as business intelligence in which massive data sets were processed by algorithms to discover knowledge in a specific context. The term academic analytics was used initially to describe the application of business intelligence techniques to analyze the admission process in higher education institutions (Goldstein and Katz 2005). The objective was to use information about high school transcripts and previous tests to better understand student enrolment and retention during their first year at an institution. Campbell et al. (2007) later defined the steps involved in using student data, the stakeholders in this process, and the ensuing support that could be provided to students. Shortly after these initiatives, numerous decision-making processes in higher education institutions were reconsidered in the presence of data and algorithms. Long and Siemens (2011) further divided this area and provided the name "learning analytics" to those initiatives targeting improvements at the departmental or course level that specifically target learners.

One of the challenges addressed by early initiatives was the detection of so-called students at risk. These students are enrolled in an educational institution but are likely to fail, drop a course, or abandon their studies entirely. Numerous institutions have deployed LA initiatives to detect these students and offer additional support before they decide to abandon their studies (see Norris et al. 2008 for a review), thereby reducing the dropout rate and improving retention.

In more recent years these initial support actions have been extended to address other common difficulties faced by students while participating in a course. These systems are generically known as early warning systems (EWSs) and usually rely on a combination of demographic datasets and data derived from academic environments to identify students who need extra support (Lonn et al. 2012; Jayaprakash et al. 2014). The output from EWSs typically include notifying teachers which students are at risk (and perhaps suggesting a range of ways they could further support these students to stay at university), as well as actions directly proposed to the students (Krumm et al. 2014). Nowadays, this application of LA has grown to encompass a wide variety of sub-areas to provide student support through a variety of methods (Ferguson 2012b). For example, some initiatives provided the information derived from predictive algorithms directly to students to alert them about the possibility of failing a course (Tanes et al. 2011).

Other initiatives consider the social dimension of learning using data retrieved from discussion forums to deduce patterns of interaction among students. These patterns are represented as networks, and social network analysis algorithms used to derive certain features and to visualize their topology (Dawson 2010; Dawson et al. 2010). Students can then be advised to re-assess their participation, or simply to reflect on their position in the network. The text exchanged by students in discussion forums is also a valuable data source for more recent techniques known as discourse-centric analytics that seek to detect evidence of learning, and language usage patterns that are associated with positive academic outcomes (Ferguson and Buckingham Shum 2011; De Liddo et al. 2011; Knight and Littleton 2015). The characterization of these discussions offers the possibility to provide highly detailed and potentially effective feedback for students to increase their performance.

Making data available to teachers can assist them in better understanding and designing learning. For example, data visualizations are often used as artifacts to either help teachers gain insight about how a learning environment unfolds (Verpoorten et al. 2011; Verbert et al. 2014), but can also be offered directly to students to help them reflect on their approach to learning (Kahn and Pardo 2016; Corrin and de Barba 2015). Also, some authors have identified the need to consider LA techniques during the learning design stages and propose how to integrate the data collection, analysis, reporting, and interventions in a unified workflow (Lockyer et al. 2013; Bakharia et al. 2016). In this case, increasing the quality of learning designs indirectly supports students.

Although these initiatives can all be connected to improvements that affect students, their focus is primarily on the steps to collect, analyze, and report data. Wise (2014) identified the need for the LA community to focus more precisely on the actions derived from the use of data. We argue that considering these interventions as personalized learning support actions is a very effective approach that connects the collection of data to tangible and effective changes in learning experiences, which then translate into quantifiable improvements. For example, dashboards that are available to teachers may well provide valuable insight about aspects of a learning experience that were never observed. However, the benefit of the initiative is only realized when teachers deploy actions derived from these observations. Indeed, using technology only for the steps of collection and analysis, and ignoring the actions, may have a serious impact on the overall effectiveness of LA initiatives (Clow 2012).

Many existing approaches to driving actions in response to student data tend to take a one-size-fits-all approach (e.g. Jayaprakash et al. 2014), building models to predict student engagement and success and then applying these models to detect and contact aberrant students. To increase predictive power, these approaches typically seek out large datasets from a range of courses or even across institutions. The innately contextualized nature of different courses means that the variables that are common across courses and institutions (and therefore able to be used in such models) are predominantly based on demographics and educational background. At best, this risks limiting our view of students' ability to their past performance and, at worst, perpetuates stereotypes (Slade and Prinsloo 2013). Further, such analyses ignore the more granular nature of ongoing learning processes. Even when current learning data such as interactions with the learning management system (LMS) are available, the highly contextualized nature of learning environments and instructional designs emphasizes the risks with one-size-fits-all data-driven approaches (Gašević et al. 2016). Therefore, a key argument of this chapter is that the data that drive support actions must be locally contextualized.

5.1.2 Local Contexts Influencing Data-Driven Student Support

With the costs for students of higher education increasing, and participation widening, there has been an increased and understandable focus by institutions, as well as government scrutiny, of dropout and attrition rates. In the Australian context a decrease in government funding to the higher education sector has meant that universities themselves see increasing retention rates as a financial necessity, in addition to the moral imperative most feel to maximize the learning experience and success of all the students they enroll.

Concurrent with these sector-wide structural changes, there has been an increase in the range of available data sources and computational methodologies, which has led many institutions to identify LA as a strategic priority and to invest, sometimes heavily, in software solutions (Colvin et al. 2016). At the time the system reported in this chapter was initially developed, LA was not a priority at our institution but there was already an active network of teaching leaders and central student support staff concerned with the experience of first year students and their transition to university. Their efforts to improve the experience and outcomes of first year students had been informed by the notion of the transition pedagogy.

Building upon extensive research into students' social and learning experiences by researchers including Tinto (2006), Kift (2009), and Nelson and Clarke (2014), the transition pedagogy articulates the importance of a unified design of the undergraduate first year curriculum and co-curriculum and stresses the role of engaging teachers in proactive, just-in-time academic and pastoral support. It thus highlights the need for a whole-of-institution approach where student success and retention are "everybody's business" (Kift 2008) including support staff, teachers, and institutional leaders. Unconnected work from any one single area may be un- or even counter-productive. For example, excellent institution-wide support services may be underused or wasted if students are disengaged by impersonal teaching or swamped by poorly designed or aligned assessment regimes. However, timely and personalized feedback and support, directly connected to each student's own learning data, can positively influence student engagement (Bridgeman and Rutledge 2010).

The transition pedagogy promotes the value of learning communities with active teacher-student interaction. In addition, it highlights the role of formative evaluation, feedback, monitoring, and timely interventions. Given the increasing role of online learning, this requires engagement with data by teachers and course coordinators—those most experienced with the particular stress points in their courses and able to intervene during semester. It also requires this work to be joined up with institutional support and wellbeing frameworks and services. Particularly when enrolments are large and students are taking a wide variety of subjects including electives and service courses out of the enrolling faculty, ready access to relevant engagement and success data enables effective and personalized interventions at the point needed.

Divorcing the teachers from the process through an overly centralized approach has the potential to lead to the usage of easily-obtainable but generic data. As well as excusing or even excluding the teachers from the analysis, such an approach is unlikely to reflect the importance and unevenness of the learning experience. Similarly, without some degree of central coordination, efforts can be duplicated or unaligned with each other and the support systems. For students, this can cause frustrations and disengagement.

5.1.3 Our Approach to Data-Driven Student Support

Here, we present a case study of an LA platform, the Student Relationship Engagement System (SRES), at The University of Sydney that is centered on student-teacher interactions in an attempt to connect teachers with their students through data. We describe the design and development of the SRES, which enables teachers to leverage data that are meaningful to them to provide scalable *and* contextualized personalized learning support with students in large cohorts. These cohorts typically consisted of 600–1800 students in a single course, which contributed to our desire to not only reduce the substantial amounts of money lost to attrition but also improve students' learning experiences in a normally highly depersonalizing environment (Krause 2005).

In the rest of this chapter, we outline the needs, principles, and philosophies that guided its development, and then provide a description of the system itself. We then highlight some real applications of the SRES and the impact it has had on students. Finally, we conclude the chapter with a discussion of potential limitations and affordances of the current system, and avenues for wider institutional impact and development.

5.2 The Student Relationship Engagement System

The SRES started as a small-scale initiative in 2012 that initially sought to improve the efficiency and accuracy of in situ data collection during face-to-face staff-student interactions. At the time, the LA field was in its infancy and was primarily on a different trajectory; that is, finding algorithmic meaning in masses of pre-existing data. Although our approach also involved data, it was starkly contrasted because it presumed that teachers would know the most appropriate data and their meaning, and they needed a platform to collect, analyze, and perform actions on these data at scale. As such, the SRES started with relatively small datasets that were created by teachers, and has gradually expanded to provide for more 'traditional' learning analytics functionality as the data appetites and capabilities of teachers have grown.

5.2.1 Supporting Pressing Needs in Local Contexts

The SRES was initially developed to address a simple need to which most teachers in face-to-face and blended environments are resigned: the perennial scraps of paper or malformed spreadsheets for attendance gathering and grading. These are usually followed by manual transcription and collation into a central spreadsheet, a process that usually ranges from non-existent to error-prone. Even then, teachers could do little with the spreadsheet apart from providing simple numerical grades to students.

An argument could be made that these data are perfunctory as opposed to pedagogically meaningful (and by extension, valuable for LA). Although interim grades and other performance data are often ignored by, or unavailable to, LA systems (Clow 2012), large-scale analyses have shown that they can be one of the most important predictive variables in models of academic risk (Jayaprakash et al. 2014). Similarly, in the context of face-to-face education, class attendance has been positively associated with improved student outcomes (Rodgers 2001; Massingham and Herrington 2006; Superby et al. 2006), and although being a frequently requested data source for teachers, it is notoriously difficult to collect (Shacklock 2016; Dyckhoff et al. 2012). Additionally, a large proportion of meaningful student-teacher interaction and assessment may occur outside of the LMS, which is a blind-spot for typical LA approaches (West et al. 2015).

Beyond data collection, interventions are a key part of LA (Clow 2012), and it is important that affordances for such actions are closely associated (Jones et al. 2013). In this chapter, we adopt a high-level understanding of intervention, involving "any change or personalization introduced in the environment to support student success, and its relevance with respect to the context" (Macfadyen et al. 2014). While direct student contact is certainly not the only intervention that should arise from LA, the affordances of an electronic system to accelerate this process was critical in our context.

5.2.2 Approach and Philosophy for Design and Development

There appears to be a lack of connection between the capabilities of extant LA tools (which, as we have argued, focus on data collection, analysis, and reporting), and the data needs of teachers to act (for example, by connecting with their students at scale). In light of this, a pressing and tangible need for our teachers was therefore a platform capable of allowing efficient and accurate collection of desirable data, and action based on these data.

To address this, we took a participatory design approach similar to that of others working to design and develop LA that would be practically useful and meaningful for teachers and other staff (Lonn et al. 2013; Dyckhoff et al. 2012). From 2012 on, a basic platform that recorded attendance via a web-based, mobile-friendly interface and saved data to a central database was iteratively designed and refined based on user feedback to become the SRES.

Throughout this process, we followed a set of basic design philosophies to guide development. These were fundamentally LA-contextualized reflections of the attributes of diffusible innovations, in particular the notions of relative advantage, compatibility, complexity, trialability, and observability from Rogers (2003), and we ground the following discussion on these attributes.

Teacher-centered. A truism is that "faculty have, for the most part, relied on their intuition and hunches to know when students are struggling, or to know when to suggest relevant learning resources, or to know how to encourage students to reflect on their learning … these hunches are not going to disappear with the advent of learning analytics, nor are the actions derived from them" (Dietz-Uhler and Hurn 2013). Additionally, given that (i) LA is "not an elixir for ineffective teaching, nor does it reveal an ideal pedagogy" (Pistilli et al. 2014), (ii) teachers have pre-conceived notions of meaningful data about their students (as argued above), and (iii) compatibility of innovations with existing ideas and felt needs are positively related to adoption (Rogers 2003), an LA innovation needs to address the contexts that real-world teachers face.

Part of a possible solution lies in a system architecture that corresponds with teachers' conceptualization of data and how to work with it, helping to address issues around compatibility with existing ideas. Additionally, a solution should be cognizant of, and tangibly address, concerns around academic workload (Macfadyen and Dawson 2012), yielding a high level of relative advantage and being compatible with felt needs.

Human-centered. Despite the LA field claiming to differentiate itself from EDM by highlighting the centrality of leveraging human judgment (Siemens and Baker 2012), a large proportion of LA work appears to focus on propelling data at algorithms in order to extract meaning. Ryan Baker's recent propositions of moving these fields towards the amplification of human intelligence are instructive here: "Humans are flexible and intelligent. Humans cannot sift through large amounts of information quickly… But once informed, a human can respond effectively" (Baker 2016).

Lack of human-centeredness in LA also extends beyond approaches to analyses and pervades implementation. A concern that should be raised more frequently is that "the focus of LA appears fixed to an institutional scale rather than a human scale" (Kruse and Pongsajapan 2012). These somewhat condemning perspectives remind us of one of the seminal principles of good practice in higher education, namely encouraging the human relationship between teachers and students (Chickering and Gamson 1987). Solutions addressing this problem must keep humans using the system at the center instead of data and analytics.

Customizable, flexible, and scalable. A substantial amount of learning interactions and data exist outside traditional sources (typically LMS and SIS databases) that LA systems can and do interrogate (West et al. 2015). Beyond the obvious challenges around data warehousing and integration (Bichsel 2012) and despite the best intentions of designers and developers, there may be several pieces of offline or other system data that cannot be automatically integrated. Additionally, teachers often demand the freedom to teach how they wish, which has important implications for the affordances of LA tools (West et al. 2015). Therefore instead of coercing teachers into a system with pre-defined (and possibly limited) data, a different solution lies in building avenues that allow teachers to define and bring in their own local and contextualized data (Graf et al. 2011).

Transparent. In this age where opaque algorithms run so many aspects of our lives, algorithmic accountability has become an important ethical challenge (Diakopoulos 2015). Learning analytics is not immune to this trend. Distrust of data and their analyses can lead to significant barriers for LA adoption (Bichsel 2012). Nevertheless, large-scale deployments of LA systems have typically relied on opaque algorithms to predict student performance (e.g. Arnold 2010; Jayaprakash et al. 2014). A possible solution to avoid such algorithmic black boxes lies in simplifying (perhaps even oversimplifying) the analytics to the extent that it is completely controlled by teachers (Liu et al. 2015). This may help to reduce perceived system complexity, and enhance the ability for teachers to experiment with analytics.

Actionable. In keeping with a human focus, the predominant avenue of intervention arising from LA appears to still be teachers or other staff interacting with students. In an Australian study, personal responses through emails, phone calls, and consultations were the preferred mechanism of data-driven actions (West et al. 2015). In another study, tools that "manage data inputs and generate outputs in the form of actionable feedback" were found to be the most adoptable (Colvin et al. 2016). Even large-scale implementations that involve opaque algorithms eventually involve teachers contacting students based on the outputs of these algorithms (e.g. Arnold 2010; Jayaprakash et al. 2014). In an exemplar of intelligence amplification (Baker 2016), LA provided the means to focus discussions that students had with their academic advisors, and to target help where it was most needed (Lonn et al. 2012). A possible solution includes the provisioning of customizable actions to promote and support teacher-student interactions. These tangible outputs may also help to promote the observability of any LA innovation.

Ethical and secure. An LA system that augments the ability of teachers to provide data-driven student support can help to simultaneously balance ethical and operational issues around irrelevance and intrusiveness. Decontextualization of data and consequent generalizations about students can lead to invalid assumptions and unhelpful data-driven support (Slade and Prinsloo 2014). One possible solution is to leverage the data on students' studies that teachers already have access to and use (perhaps in an inefficient, distributed fashion). If these data were easy to curate and act upon at scale, such an LA solution may not overstep students' existing expectations of privacy. This is in keeping with our design philosophy of augmenting teachers' intelligence and abilities.

Data protection must be a core value in any LA venture and helps to build trust in LA systems (Drachsler and Greller 2016). This may involve ensuring that all student data are encrypted during transit, and stored on secure university-owned and -controlled servers (Slade and Prinsloo 2013). Removing identifiable records after a set timeframe in line with university record retention policies may also help address some concerns over data security. Beyond the critical ethical and legal issues surrounding data security, any negative occurrences could have severe repercussions for the adoption of future innovations (Rogers 2003).

Working with these philosophies, we sought to design and develop an LA system that met real and pressing needs of teachers in our contexts. Our approach was to build a platform that required active input from teachers but provided them the ability to personalize student support at scale and gain insight into their cohorts while saving time in the execution of these processes. We purposely designed the data and system architecture to support these goals.

From a teacher's perspective, an electronic spreadsheet is one of the most common ways to handle student data—it is inherently customizable and extensible, has no hidden algorithms, and typically represents rows of students with corresponding columns of data. The issue with spreadsheets is that they are not immediately actionable, and deriving meaning at scale is difficult. Nevertheless, this matrix structure of student data seems to be eminently accessible and understandable by teachers and other staff. As Rogers (2003) points out, "[o]ld ideas are the main mental tools that individuals utilize to assess new ideas and give them meaning. Individuals cannot deal with an innovation except on the basis of the familiar. Previous practice provides a standard against which an innovation can be interpreted, thus decreasing its uncertainty" (p. 269). Since this matrix structure of data is familiar and flexible, we opted to ground the data architecture on the idea of students (in rows) and data (in columns representing different variables or features) belonging to courses (in tables).

5.2.3   Flexibility in Importing Students and Data

While connection with enterprise student information systems have allowed some LA developers to leverage institutional data warehouses (Lonn et al. 2013), in our context this was not possible, which encouraged us to design an interface that allowed teachers to import student enrolment information semi-automatically. This required them to download an enrolment list from another (enterprise) university system and upload it to the SRES; we made this process as streamlined as possible in the importer interface (Fig. 5.1). The benefits of this included that staff (i) could combine lists from different courses, (ii) could add non-regular students (such as those from outside the university as often exists in bridging courses), (iii) could record other details such as a preferred name which was not possible using enterprise systems, (iv) could have as many lists (tables) with as many students as they liked, and (v) could work in the system safely without affecting data on other enterprise systems. Obvious drawbacks included the need for semi-manual updating of course lists when enrolments changed, and duplication of some data on multiple university systems.

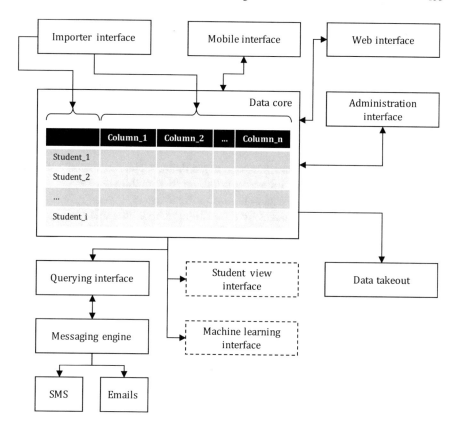

Fig. 5.1 Data and system architecture of the SRES. The SRES core data model is based on familiar tabular student records. Various user interfaces can bring data into the SRES, such as an importer, mobile application, and a web-based interface. Flexible querying and messaging interfaces and engines power the majority of the data-driven student support, allowing teachers to build and deploy highly personalized interventions. Experimental features (shown as *dashed boxes*) take advantage of the data that are already stored in the SRES. Teachers can build customized student views to conditionally show students their own data and other information. A machine learning interface guides teachers through preparing and analyzing data within the SRES using various machine learning algorithms in order to uncover hidden patterns

Once the students (rows) were in place in a list (table), teachers could update these as necessary and also specify an unlimited number of columns. The data in these columns could also be brought into the system through an importer interface (Fig. 5.1), which accepted comma- or tab-delimited plain text files [e.g. comma-separated value (CSV) files] and guided the teacher through mapping a student identifier column and data columns.

5.2.4 Mobile Interface for Staff

One of the primary drivers for developing the SRES was the need for live, in situ, data storage and retrieval in face-to-face learning scenarios. To enable this process, we built a mobile-friendly web-based interface that would allow staff members interacting with students to select columns for which they wanted to input data, specify the data entry pattern, and save data for each row (Fig. 5.2). An example of one of the workflows for this mobile interface is:

1. A teacher authenticates and accesses a pre-defined column in the SRES by scanning a 2D barcode or following a link on their mobile device.
2. The column receiving the data is shown to the teacher.
3. The teacher identifies a student by (i) scanning a code that uniquely identifies the student (e.g. a student card, or a 2D barcode produced by the SRES), or (ii) searching for the student using identification number, email, or name.
4. Once the student is identified, the interface displays a set of values pre-defined by the mentor for that column, and/or allows entry of a custom value.
5. The value selected or entered is saved to the table.

Fig. 5.2 An example of the SRES mobile interface, at step 4 of a timestamp data entry pattern (see text). The *upper section* of the screen is fully customizable and can display data from other columns. The *lower section* provides alternative means to identify students, such as scanning 1D or 2D barcodes, and searching

Aside from the benefit of capturing information in an efficient and secure way in face-to-face scenarios, the mobile interface allows a coordinating teacher to choose the information displayed after a student is identified (Fig. 5.2). This display could include any user-editable hypertext, as well as data drawn from other columns in the table, and identifying information about the student (such as their preferred name). This functionality proved powerful because it allowed teachers to (i) define the important data related to a column, and (ii) have immediate visibility of these data when interacting with students. In a similar way, Lonn et al. (2012) found it powerful to provide mentors with up-to-date data to inform in-person discussions.

By providing data in situ (indeed, as data were being collected), the SRES can support teachers to build better relationships with their students, and engage with them on a deeper, data-driven level. In a case of data systems that augment and leverage the intelligence of humans (Baker 2016), teachers can initiate the necessary conversations and bring in contextual and other factors to which data may be blind. As Rogers (2003) points out, "[w]ords are the thought units that structure perceptions" (p. 276), leading us to name the system the Student *Relationship Engagement* System, in order to emphasize this criticality of engagement and relationships in data-driven student support.

5.2.5   Querying Interface and Messaging Engine

In addition to the face-to-face data-driven support catalyzed through providing pertinent and accurate data, the SRES was also built to be massively scalable and allow teachers of even very large courses to personalize interactions with their students. As we have described, providing a mechanism for efficient data-driven actions addresses a felt need for such teachers. The course size does not necessarily need to number in the hundreds or thousands; it is already a significant workload imposition to personalize regular electronic contact with a cohort of 70. To be effective, the SRES needed to provide a relative advantage for these teachers compared to the alternatives of manual efforts or even not contacting their students. Indeed, relative advantage is one of the strongest positive predictors of whether an innovation will be adopted, and its contributing factors include the saving of time and effort, immediacy of reward, low initial costs, and economic profitability (Rogers 2003).

To provide this, we built a fully customizable querying interface and messaging engine into the SRES (Fig. 5.1). This allowed teachers to use Boolean and other operators to combine condition statements on data stored in the SRES, similar to advanced search engine queries. To increase compatibility with existing ideas, we mimicked the filtering terminology of spreadsheet applications, and built a graphical user interface where teachers could select any column, choose a comparison operator (e.g. less than, contains, not equals to, is empty, etc.), and a comparison value (Fig. 5.3). These conditions could then be combined to form a complex query. For example, a teacher could query the SRES to find students who had a low

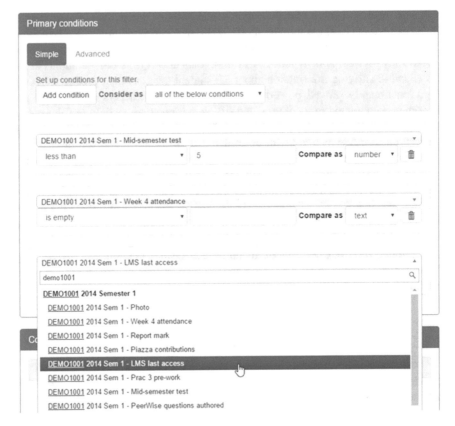

Fig. 5.3 Screenshot of the simple querying interface showing how filter conditions are built by selecting columns and specifying comparisons

performance on a test, and had not attended class, and had not logged into the LMS for a while (Fig. 5.3), while another teacher in a different context could ask different questions.

Another advantage was that a teacher (with the appropriate system permissions) could query across columns from multiple tables; for example, allowing a program coordinator to quickly find high-performing students across a number of courses.

The querying interface was closely linked with a messaging interface and engine, where teachers could compose a personalized message to selected students (Fig. 5.4). This interface allowed the user to bring in any data from the SRES database, including user information (e.g. preferred name) to assist in composing the message to each individual student, drawing on information personally relevant to them. The messaging engine was connected to an email server as well as an SMS service; the former is common practice in LA interventions, while the efficacy of the latter is starting to be explored (Goh et al. 2012).

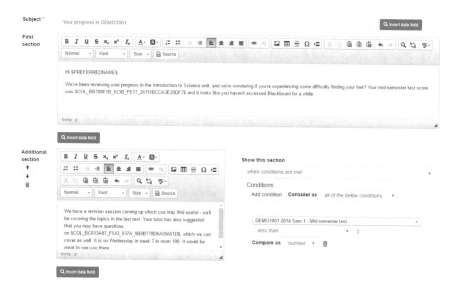

Fig. 5.4 Screenshot of part of the messaging interface and engine. Fully customizable messages could be personalized using students' own data, and sections could be variably included based on conditions in the available data

To help teachers gain confidence in this entire process, we built preview functionality into the SRES so that before anything was committed (e.g. messages sent to students), users could preview and download tabulated results of their query, as well as preview each individual personalized message. This also served to enhance the trialability of the system (Rogers 2003) in that users could safely experiment.

5.2.6 System Adoption, Usage, and Impact

From early on in the LA story, one of three key factors proposed for successful analytics initiatives was a flexible technology platform that allowed users to warehouse data from a variety of sources into a common structure and to perform analyses on these data (Campbell et al. 2007). While the architecture of the SRES is not a data warehouse in the traditional sense and as intended by Campbell et al., our argument here is that the nature of allowing teachers to efficiently select, combine, and apply data of their choice that is relevant to their contexts can be a powerful alternative for LA.

As Colvin et al. (2016) noted, "implementers require an analytic tool or combination of tools that manage data inputs and generate outputs in the form of actionable feedback. The capacity to implement is crucially linked to the quality of these tools and the data they rely on and generate… As these increasingly meet the 'real' needs of learners and educators, organizational uptake is accelerated" (p. 2).

Additionally, West et al. (2015) highlighted an instructive comment that underlines the need to be flexible and context-sensitive: "a lot of the things that you have to do at the moment have to be done manually and they all take time. Anything that can automate the process of that information is beneficial. I suppose there also needs to be some ability to modify it to your own requirements because each course and each cohort of students may differ" (p. 20).

By designing the SRES to encompass the philosophies of flexibility, customizability, and scalability, we have seen considerable uptake in the University of Sydney community, with a variety of applications and impacts. Some representative examples are presented below.

Undergraduate laboratories in the pure sciences. An administrative function of the SRES was to print identity cards, which could be customized to the needs of each course. At the start of each semester, students were given an SRES-generated sticker or card with their unique barcode and other personalized laboratory information (e.g. session, venue, and schedule). This was scanned before or during each laboratory to record attendance and also to initiate conversation between teachers (who perform the scanning) and students (whose relevant data are displayed in the customizable display region of the mobile interface; Fig. 5.2). Marks for laboratory notebooks and reports were also saved directly using the SRES by scanning barcodes on SRES-generated personalized coversheets that students downloaded and printed. Students were typically intrigued by the efficiency and reliability of this approach, which saved hours of staff time in transcribing and correcting records and tracking attendance patterns.

At-scale student support in health and pure sciences. The SRES was used to build and send regular, personalized emails to segments of each cohort. The data that were used to filter and segment the cohorts included attendance recorded through the mobile interface, data imported from the LMS grade book, as well as data imported from third-party adaptive tutorial systems (outside the LMS). One teacher reported that efficiently recording attendance using the SRES was associated with increased attendance at Friday afternoon lectures. Teachers also used the filtering interface to segment cohorts (e.g. into no-, low-, and high-risk categories), and used the messaging engine to send regular personalized emails to all students in each category.

These helped to keep students on track, feel connected to their teacher, and gave students an easy way to contact the teacher by simply replying to the email. One teacher reported that most students identified as high-risk early in the semester ended up passing the course, with a considerable reduction in students who did not complete compulsory work, in comparison to previous cohorts. Other teachers reported reduced attrition rates and improved distributions of students towards higher grades.

Heavy personalization in philosophy of science. To personalize messages with a cohort of students with lower average university entrance scores, the teacher used the SRES to import quiz scores from an LMS-exported CSV file, as well as other custom fields that the teacher had generated in an offline spreadsheet. Multiple complementary conditions were generated for each of a number of filters to differentiate the emails that different segments of the cohort would be sent. Students received specific feedback based on their up-to-date achievement in the course, and suggestions on how best proceed in the course. Using the SRES, the teacher also identified students who he considered were most at risk, who were then followed up with a phone call from central student support services. The teacher reported a substantial reduction in attrition.

Feedback and follow-up in clinical laboratories. A proposed use of the SRES in clinical settings is for a teaching assistant to record feedback for an individual student as a short piece of text into the SRES, which can then be automatically emailed to the student as part of a customizable message triggered upon saving data. This feedback can then be seen the following week by another teaching assistant working with the same student, via the customizable display in the mobile interface. The teacher suggesting this envisages that students will be more likely to act on feedback if there is an expectation of specific follow-up.

Adoption of the SRES, since its initial pilot in one department and four courses in 2012, has grown to 78 units of study over 19 departments (Fig. 5.5). We believe this successful wider adoption, a result of recommendations by colleagues, is a reflection of the observability of the operation and impacts of the SRES.

A number of factors have likely contributed to this: (i) the SRES was designed from the ground up as a teacher-focused platform that addressed a felt need and offered tangible relative advantages compared to existing methods; (ii) its architecture was compatible with how teachers commonly use and manipulate data; (iii) it sought to reduce complexity and enhance trialability; (iv) regular communication between the developers (who are also teachers) with users meant the

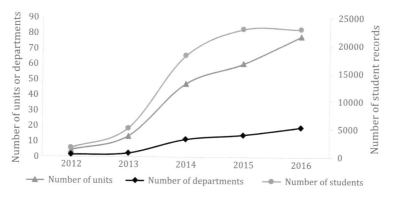

Fig. 5.5 Uptake of the SRES at the University of Sydney. Since an initial pilot in 2012, the SRES has been adopted by more teachers in more units of study (courses) and departments, and is being used to provide data-driven personalized student support data for an increasing number of students

system was able to be updated or extended relatively quickly in response to user feedback.

5.2.7 Experimental Interfaces

As the SRES has expanded in reach, the data appetite of staff using the system has grown. For example, we have seen this in terms of how data may be reported, as well as how data may be analyzed. In keeping with our iterative and teacher-centered design philosophies, we have gradually added new experimental features to the SRES to address emergent needs.

The existing paradigm of delivering data-driven support to students through personalized messages from the SRES characterized a 'push' mechanism; that is, teachers set up and activated a personalized message to students. We have recently been trialing a 'pull' mechanism where teachers set up a customized 'student view' of students' own data. In a similar way to the personalized messages, teachers use a graphical interface to easily write and format information to be shown to students, and use data that exists within the SRES database to either conditionally display relevant information and show the data itself. These 'student view' pull-type interventions can then be embedded into the LMS so that students see pertinent information from their teachers when they log in. We see this approach as a rudimentary but highly customizable reporting engine that could help to put learners back in control of their data (Drachsler and Greller 2016).

As the SRES encourages and makes it more efficient for teachers to curate local data in one place, i.e. within the SRES, the issues with disconnected data silos start to be eroded. As a result, there are more locally meaningful data available on which to perform analyses. Leveraging this situation, we have started exploring various machine learning approaches to help teachers analyze data stored in the SRES and uncover hidden patterns that may influence their curriculum and teaching approaches. Our philosophy is to provide teachers with an easy-to-use interface to perform fully customizable explorations themselves (Liu et al. 2016).

Specifically, we are leveraging web-based machine learning application programming interfaces (initially experimenting with www.bigml.com) to build decision trees, mine for association rules, and cluster students based on data in the SRES. For example, cluster analysis of data may indicate that attendance is not substantially different in clusters with lower-performing students, while concurrently highlighting that early summative quizzes and exams may be important in identifying these cohorts (Fig. 5.6). In a similar example, decision tree analyses may reveal particular characteristics of students with different outcomes; for example, demonstrating that online formative quizzes may differentiate students who fail with those who pass (Fig. 5.7). These analyses open a new dimension to the SRES, as the system in its entirety gives teachers the flexibility to bring in data they want, query and action these data, and now interrogate these data using typical LA and EDM techniques to uncover hidden patterns.

	Cluster 0 (n = 6)	Cluster 1 (n = 239)	Cluster 2 (n = 22)	Cluster 3 (n = 217)	Cluster 4 (n = 248)
Early_prework_avg	4.944	3.732	4.386	4.063	4.569
C_COURSEACTIVITYINHOURS	79.843	26.625	62.914	31.317	48.570
Piazza_answers	34.000	0.303	4.211	0.229	0.585
Formative_quizzes	13.208	2.683	8.382	8.196	12.983
Early_attendance	3.000	2.921	2.947	2.752	2.962
Enzyme_short_communication	2.000	1.809	1.961	1.884	1.926
Test_1	19.500	12.850	17.684	16.131	17.875
Piazza_questions	8.333	0.700	16.421	0.608	1.429
	PA: 1 DI: 3 HD: 2	PA: 120 FA: 98 CR: 17 DI: 3 HD: 1	PA: 6 FA: 2 CR: 8 DI: 4 HD: 2	PA: 114 FA: 23 CR: 52 DI: 27 HD: 1	PA: 70 FA: 4 CR: 76 DI: 77 HD: 21

Fig. 5.6 Example screenshot of the output of a cluster analysis from within the SRES on data that a teacher has brought into the SRES. The numbers represent the cluster centroids. *Test_1* is the first of three mid-semester exams, *formative_quizzes* are non-compulsory online quizzes, *early_attendance* is the count of attendance at the first three practical classes, *early_prework_avg* is the average mark in the first three compulsory online pre-work quizzes, and *Piazza* is the online discussion forum

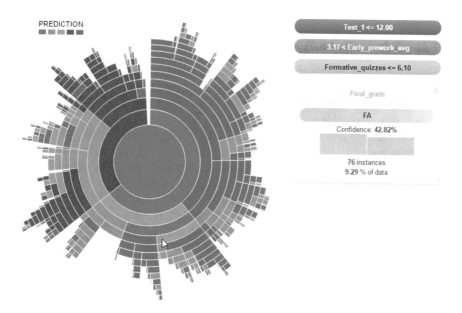

Fig. 5.7 Example screenshot of the output of a decision tree analysis from with the SRES, generated through the BigML application programming interface, on data that a teacher has brought into the SRES. This interactive interface, powered by BigML, allows teachers to hover over subsets of their cohort (*cursor* shown hovering over a group of students who achieved a fail grade) and examine the decision points that the algorithm has identified

5.3 Discussion

The decisions made during the design process of the platform presented in the previous sections have provided deep insights into two very relevant aspects of LA. The first one is the need to explore the space between fully manual and fully automated solutions, addressing the real needs of teachers and focusing on the human elements of learning. The second one concerns the measures that can be adopted at the institutional level to foster the use of these types of platforms and make sure they provide a tangible improvement to all stakeholders. In the remainder of this section we further elaborate on these two areas.

5.3.1 Amplifying Teacher Intelligence

Increasingly, researchers and practitioners in LA are identifying the importance of context, especially when addressing the factors and interventions that impact student success. This context includes aspects such as educational history, instructional conditions, and course revisions, as well as the complex and largely hidden (to machines) realm of students' cognitive, emotional, and social processes (Gašević et al. 2015, 2016). At the same time, thought leaders in these fields are turning to the idea of amplifying and supporting human intelligence, as opposed to blindly following machine outputs (Baker 2016).

Could it be that in the age of big data, we are becoming tantalized by data and potentially neglecting the personal teacher-student interactions that are so crucial to learning and teaching? To purposely contort an idiom, are we missing the trees for the forest? To reconcile these, we have described in this chapter an LA approach to addresses teachers' real needs that aligns with their understanding of their students, courses, data, and student support.

At the same time, there are pressing needs driven from institutional contexts that cannot be ignored. As massification in higher education continues, one-on-one teacher–student interactions have become less common, and personalized student support increasingly challenging. Many LA efforts to date have been focused on trying to algorithmically triage support resources to the most 'at risk' students. This focus on maximizing retention in LA has meant there is a real risk that we lose the human element in higher education and replace it with predictive models based on large datasets but perhaps a limited number of variables with dubious meaningfulness. Further, by retaining a narrow focus on the at-risk portion of the student population, we fail to support and maximize the learning experience and outcomes of all students. In LA, we need to continue to push ourselves to develop and use systems that take research-informed actions to support and challenge *all* students.

We see SRES as a first step towards a possible solution. Data-driven personalized learning support may be positioned between machine models and personal relationships; by leveraging machines and humans, it can capitalize the best of both worlds and be scalable as well as contextualized. To do this well, teachers need a system capable of scaling their natural workflows (which are exceedingly context-dependent) to large student cohorts while providing a level of student personalization. The SRES, which we have presented here, addresses teachers' needs of efficient and centralized data curating, augmenting their abilities to personalize student support using these data. This is afforded through providing highly customizable push and pull intervention mechanisms.

Arguably, the SRES currently deals with small data as opposed to the traditional view of big data (Berman 2013), and the data points that lie within its matrix-like data architecture may represent aggregated data and therefore mask rich complexity. Other associated risks with this design include the oversimplification of metrics and the possibility of missing potentially meaningful data that the teacher did not consider including, both of which can be partly alleviated through the sharing of good practice. As long as some meaningful data are collected, teachers could use the nascent machine learning interfaces to uncover hidden meaning and possibly use this to inform intervention or learning design decisions. However, all of these are teacher-facing in terms of data collection and reporting, and providing affordances for action.

We envisage that future developments would also include student-facing interfaces that could be customized by the teacher so that their students could input data (e.g. 'pulse' data, psychosocial variables, self-assessment of skill attainment, self-reports of perceptions) and visualize data (e.g. performance compared to the cohort, self-progression through tasks) directly to and from the SRES table(s) via a web or mobile interface. This way, the data outputs could be contextualized by the teacher instead of relying on a one-size-fits-all dashboard across all courses. Further, building application programming interfaces into the SRES itself would allow easier data interoperability with other systems, and potentially be able to expose limited datasets for interested students to analyze themselves. Together, these emphasize our focus on practical LA systems that are customizable, flexible, scalable, actionable, and human-centered.

5.3.2 Enabling Scaling-Up of Data-Driven Student Learning Support

In the Australian higher education context, Colvin et al. (2016) noted that sustainable LA adoption relies on (i) organization strategy that lays the groundwork for LA, (ii) implementation capability that integrates teachers' practices with data and tools, (iii) tools that address real needs, and (iv) a capacity to monitor and improve the quality and usability of implementations. Our journey with the SRES has serendipitously approached this from the bottom up and nevertheless has seen increasing adoption at our institution.

Now, the increasing data appetite of our colleagues, catalyzed through the use of LA tools such as the SRES, are fueling a number of top-level challenges including organizational culture and complexity (Macfadyen and Dawson 2012; Macfadyen et al. 2014), automating data workflows and interoperability (Lonn et al. 2013), stakeholder engagement and expectation management (Ferguson et al. 2014), connecting LA with learning design (Lockyer et al. 2013), and developing an ethical framework to benefit learners (Ferguson 2012a). Additionally, as the user base of a technology innovation expands, expectations for enterprise-standard system reliability, user experience design, and user support and training begin to grow. To address these challenges associated with scaling up LA innovations, the framework applied by Arnold et al. (2014) is instructive and corresponds with institution-wide steps that we are taking as LA becomes a strategic priority for the University of Sydney.

Technology infrastructure, analytics tools, and applications. We are brokering connections between data owners and users and central information technology and business intelligence units, looking to characterize and integrate data that are currently collected, and identify gaps in collection of meaningful data. At the same time, we are working towards tools and business processes that allow LA to be embraced by the academic masses, in a range of roles, and making LA not just the domain of data and technology enthusiasts. Part of this involves creating a space for bespoke software development by LA researchers and practitioners.

Policies, processes, practices, and workflows. More widespread use of data to drive timely interventions understandably causes anxieties in staff and students. Alongside issues of invasions of privacy and even surveillance, real-time data are necessarily incomplete and potentially inaccurate. There are therefore legitimate concerns in the ways that data are obtained, held, and used that must be addressed in parallel to the development of software and data collection tools. To address this, we are establishing LA principles that align with legal requirements for student privacy and the values of the institution.

Values and skills. We are working to connect people across the institution with relevant skills—both academic and professional staff, and those in departments and central portfolios. This will aid the evaluation of LA, particularly the technology and methodologies, the allied support services, and learning support actions used in individual courses. Supporting this will be professional learning around using LA systems effectively and clarifying roles for faculty and staff. This will become increasingly important as more agile access to relevant learning data allows teaching staff to personalize and target support *en masse*. Unsupportive teachers can cause large-scale damage and disengagement if personalized student support is delivered in ill thought-out or destructive ways, or if based on an overreliance on data which are messy or not meaningful.

Culture and behavior. Connected with building values, we are working to inspire and support LA research and innovation by funding EDM and LA projects, establishing networks and research groups, brokering research ethics arrangements, and connecting the institution with groups such as international societies (e.g. the Societies for LA Research and EDM) and local interest groups.

Leadership. To provide strategic support for all of this, we are establishing governance and strategy groups, as well as providing coordination, leadership, and advocacy for LA at the whole-of-institution level.

Using the SRES as a case for scaling up a bespoke LA innovation, we have already started to think about how organizational resources could be exploited, such as enhancing the connectivity of the SRES with data warehouses, growing its institutional profile, providing professional learning opportunities for teachers about effective pedagogical strategies and learning support actions, and fostering an SRES community of practice.

As an institution, we are seeking to actively encourage innovation in EDM and LA and let a thousand flowers bloom. To support subsequent scaling, we need to have a process to identify which new innovations hold promise for wider use, and how to further support, develop, and implement these at the enterprise level by engaging with institutional infrastructure, resources, and personnel.

5.4 Conclusion

In this chapter we have suggested that the EDM and LA communities have to reflect on how to better achieve the ultimate goal of improving students' overall learning experiences. We propose increasing the focus on systems that enhance the decision-making skills and support abilities of humans (i.e. teachers, students, and support staff) and truly achieve personalized learning scenarios. The rich set of existing data sources, sophisticated algorithms to discover knowledge, and complex visualizations still need to be considered under a lens that brings the human to the center of the design and formally leverages the effect of technology in their day-to-day operations. Humans are in the best position to establish the required connection between data, algorithms, and educational underpinnings. We propose the notion of personalized learning support actions as the focal point where contributions should aim in order to make a quantifiable difference. Our argument is that this approach allows for greater relational connection between students and teachers.

The SRES has been presented as an example of a tool that seeks to connect the existing expertise of teachers with their students' data-rich learning environments. Our approach centers on the relationship between teachers and their students, both in terms of collecting and curating meaningful local data as well as supporting actions based on these data. This is in stark contrast to prevailing approaches to learning analytics. These predominantly focus on warehousing a plethora of existing data such as from learning management, student information, and library systems, followed by applying statistical and other modeling approaches in order to predict student performance. We posit that these approaches can potentially miss out on the rich pedagogical expertise of teachers, ignore the relationships between teachers and students, and fail to encapsulate local data that teachers may find more meaningful.

Based on these needs, we have discussed the application of a series of human-centered design philosophies, rooted in the notion that teachers need decision-support tools that can accommodate the diversity of context-specific data sources present in learning environments, and the variety of possible vehicles to provide personalized support. The SRES has been deployed at a large higher-education institution in Australia with a significant uptake. Its trajectory has served to highlight the main adoption barriers at both staff and institutional levels, and how these may be addressed.

The future avenues to explore offer a promising landscape in which data, algorithms, staff, and students all interact to effectively combine data richness and algorithmic efficiency with human intelligence to yield tangible improvements in the overall learning experience.

Acknowledgements The authors wish to thank the many teaching and support staff who have patiently implemented the SRES in their units of study, supported its use, and provided valuable feedback.

References

Arnold KE (2010) Signals: applying academic analytics. Educause Q 33(1):n1

Arnold KE, Lynch G, Huston D, Wong L, Jorn L, Olsen CW (2014) Building institutional capacities and competencies for systemic learning analytics initiatives. Paper presented at the international conference on learning analytics and knowledge. Indianapolis, IN, USA

Baker R (2016) Stupid tutoring systems, intelligent humans. Int J Artif Intell Educ 1–15

Baker R, Siemens G (2014) Educational data mining and learning analytics. In: Sawyer RK (ed) The Cambridge handbook of the learning sciences, 2nd edn. Cambridge University Press

Bakharia A, Corrin L, de Barba P, Kennedy G, Gašević D, Mulder R, et al (2016) A conceptual framework linking learning design with learning analytics. Paper presented at the international conference on learning analytics and knowledge. Edinburgh, UK

Berman JJ (2013) Principles of big data: preparing, sharing, and analyzing complex information. Morgan Kaufmann, Waltham, MA, USA

Bichsel J (2012) Analytics in higher education benefits, barriers, progress, and recommendations. EDUCAUSE Center for Applied Research, pp 1–31

Bra PD (2002) Adaptive educational hypermedia on the web. Commun ACM 45(5):60–61

Bridgeman AJ, Rutledge P (2010) Getting personal: feedback for the masses. Synergy 30 (July):61–68

Brusilovsky P (1996) Methods and techniques of adaptive hypermedia. User Model User-Adap Inter 6:87–129

Campbell JP, DeBlois PB, Oblinger DG (2007) Academic analytics: a new tool for a New Era. EDUCAUSE Review, vol 42. EDUCAUSE White Paper, pp 40–57

Chickering AW, Gamson ZF (1987) Seven principles for good practice in undergraduate education. AAHE Bull 39(7):3–7

Clow, D. (2012) The Learning Analytics Cycle: Closing the Loop Effectively. Paper presented at the international conference on learning analytics and knowledge. New York, NY, USA

Colvin C, Rogers T, Wade A, Dawson S, Gašević D, Buckingham Shum S et al (2016) Student retention and learning analytics: a snapshot of Australian practices and a framework for advancement. Aust Gov Off Learn Teach, Canberra, ACT

Corbett AT, Koedinger KR, Anderson JR (1997) Intelligent tutoring systems. In: Heander M, Landauer TK, Prabhu P (eds) Handbook of human-computer interaction, 2nd edn. Elsevier Science B. V, pp 849–870

Corrin L, de Barba P (2015) How do students interpret feedback delivered via dashboards? Paper presented at the international conference on learning analytics and knowledge. Poughkeepsie, NY, USA

Dawson S (2010) 'Seeing' the learning community: an exploration of the development of a resource for monitoring online student networking. Br J Educ Technol 41(5):736–752. doi:10.1111/j.1467-8535.2009.00970.x

Dawson S, Bakharia A, Heathcote E (2010) SNAPP: realising the affordances of real-time SNA within networked learning environments. In: Dirckinck-Holmfeld L, Hodgson V, Jones C, Laat MD, McConnell D, Ryberg T (eds) International conference on networked learning, pp 125–133

De Liddo A, Buckingham Shum S, Quinto I (2011) Discourse-centric learning analytics. Paper presented at the international conference on learning analytics and knowledge. Banff, Canada

Diakopoulos N (2015) Algorithmic accountability. digital. Journalism 3(3):398–415. doi:10.1080/21670811.2014.976411

Dietz-Uhler B, Hurn JE (2013) Using learning analytics to predict (and improve) student success: a faculty perspective. J Interact Online Learn 12(1):17–26

Drachsler H, Greller W (2016) Privacy and analytics: it's a DELICATE issue a checklist for trusted learning analytics. Paper presented at the international conference on learning analytics & knowledge. Edinburgh, United Kingdom

Dyckhoff AL, Zielke D, Bültmann M, Chatti MA, Schroeder U (2012) Design and implementation of a learning analytics toolkit for teachers. J Educ Technol Soc 15(3):58–76

Ferguson R (2012a) Learning analytics: drivers, developments and challenges. Int J Technol Enhanced Learning 4(5/6):304–317. doi:10.1504/ijtel.2012.051816

Ferguson R (2012b) The state of learning analytics in 2012: A review and future challenges a review and future challenges. Knowledge Media Institute, The Open University, UK

Ferguson R, Buckingham Shum S (2011) Learning analytics to identify exploratory dialogue within synchronous text chat. In G. Conole, D. Gašević (eds) International conference on learning analytics and knowledge. ACM Press, Banff, Canada, p. 99. doi:10.1145/2090116.2090130

Ferguson R, Clow D, Macfadyen L, Essa A, Dawson S, Alexander S (2014) Setting learning analytics in context: overcoming the barriers to large-scale adoption. Paper presented at the international conference on learning analytics and knowledge. Indianapolis, IN, USA

Gašević D, Dawson S, Rogers T, Gasevic D (2016) Learning analytics should not promote one size fits all: the effects of instructional conditions in predicting academic success. Internet High Educ 28:68–84. doi:10.1016/j.iheduc.2015.10.002

Gašević D, Dawson S, Siemens G (2015) Let's not forget: learning analytics are about learning. TechTrends 59(1):64–75

Goh T-T, Seet B-C, Chen N-S (2012) The impact of persuasive SMS on students' self-regulated learning. Br J Educ Technol 43(4):624–640. doi:10.1111/j.1467-8535.2011.01236.x

Goldstein PJ, Katz RN (2005) Academic analytics: the uses of management information and technology in higher education. ECAR Research Study: Educause Center for Applied Research

Graf S, Ives C, Rahman N, Ferri A (2011) AAT: a tool for accessing and analysing students' behaviour data in learning systems. In: Conole G, Gašević D (eds) International conference on learning analytics and knowledge. ACM Press, Banff, Canada, pp 174–179

Jayaprakash SM, Moody EW, Eitel JM, Regan JR, Baron JD (2014) Early alert of academically at-risk students: an open source analytics initiative. J Learn Anal 1:6–47

Jones D, Beer C, Clark D (2013) The IRAC framework: locating the performance zone for learning analytics. In: 30th conference of the Australasian society for computers in learning in tertiary education. Macquarie University, Sydney, pp 446–450

Kahn I, Pardo A (2016) *Data2U: scalable real time student feedback in active learning environments.* Paper presented at the international conference on learning analytics and knowledge. Edinburgh, UK, pp 25–29

Kift SM (2008) The next, great first year challenge: sustaining, coordinating and embedding coherent institution–wide approaches to enact the FYE as "everybody's business". Paper presented at the Pacific Rim First Year in higher education conference. Hobart, Australia

Kift, S. M. (2009). Articulating a transition pedagogy to scaffold and to enhance the first year student learning experience in Australian higher education. (pp. 62): Australian Learning and Teaching Council

Knight S, Littleton K (2015) Discourse-centric learning analytics: Mapping the terrain. J Learn Anal 2(1):185–209

Kobsa A (2007) Privacy-enhanced web personalization. In The adaptive web. Springer, pp 628–670

Krause K (2005) Understanding and promoting student engagement in university learning communities. Paper presented as keynote address: engaged, inert or otherwise occupied, pp 21-22

Krumm AE, Waddington RJ, Teasley SD, Lonn S (2014) A learning management system-based early warning system for academic advising in Undergraduate engineering. In: Larusson JA, White B (eds) Learning analytics: from research to practice. Springer Science + Business Media, New York, USA, pp 103–119

Kruse A, Pongsajapan R (2012) Student-centered learning analytics. In CNDLS thought papers. Georgetown University

Liu DYT, Rogers T, Pardo A (2015) Learning analytics—are we at risk of missing the point? Paper presented at the 32nd conference of the Australasian society for computers in learning in tertiary education. Perth, Australia

Liu DYT, Taylor CE, Bridgeman AJ, Bartimote-Aufflick K, Pardo A (2016) Empowering instructors through customizable collection and analyses of actionable information. Workshop on learning analytics for curriculum and program quality improvement. Edinburgh, UK

Lockyer L, Heathcote E, Dawson S (2013) Informing pedagogical action: aligning learning analytics with learning design. Am Behav Sci 57(10):1439–1459. doi:10.1177/0002764213479367

Long P, Siemens G (2011) Penetrating the fog: analytics in learning and education. EDUCAUSE Rev 48(5):31–40

Lonn S, Aguilar S, Teasley SD (2013) Issues, challenges, and lessons learned when scaling up a learning analytics intervention. Paper presented at the international conference on learning analytics and knowledge. Leuven

Lonn S, Krumm AE, Waddington RJ, Teasley SD (2012) Bridging the gap from knowledge to action: putting analytics in the hands of academic advisors. Paper presented at the international conference on learning analytics and knowledge. Vancouver, Canada, Apr 29–May 2

Macfadyen LP, Dawson S (2012) Numbers are not enough. Why e-learning analytics failed to inform an institutional strategic plan. J Educ Technol Soc 15(3):149–163

Macfadyen LP, Dawson S, Pardo A, Gašević D (2014) Embracing big data in complex educational systems: the learning analytics imperative and the policy challenge. Res Pract Assess 9 (2):17–28

Massingham P, Herrington T (2006) Does attendance matter? An examination of student attitudes, participation, performance and attendance. J Univ Teach Learn Pract 3(2):3

Nelson K, Clarke J (2014) The first year experience: looking back to inform the future. HERDSA Rev High Educ 1:23–45

Norris D, Baer LL, Leonard J, Pugliese L, Lefrere P (2008) Action analytics. Measuring and improving performance that matters in higher education. EDUCAUSE Rev 43:42–67

Pistilli MD, Willis JE, Campbell JP (2014) Analytics through an institutional lens: definition, theory, design and impact. In: Larusson JA, White B (eds) Learning analytics: from research to practice. Springer, New York, pp 79–102

Rodgers JR (2001) A panel-data study of the effect of student attendance on university performance. Aust J Educ 45(3):284–295

Rogers EM (2003) Diffusion of innovations, 5th edn. Free Press

Shacklock X (2016) From bricks to clicks—the potential of data and analytics in higher education. Higher Education Commission

Siemens G, Baker R (2012) Learning analytics and educational data mining: towards communication and collaboration. In 2nd international conference on learning analytics and knowledge, Vancouver. ACM, pp 252–254

Slade S, Prinsloo P (2013) Learning analytics: ethical issues and dilemmas. Am Behav Sci 57 (10):1510–1529. doi:10.1177/0002764213479366

Slade S, Prinsloo P (2014) Student perspectives on the use of their data: between intrusion, surveillance and care. Paper presented at the European distance and E-learning network. Oxford, UK

Superby J-F, Vandamme J, Meskens N (2006) Determination of factors influencing the achievement of the first-year university students using data mining methods. In Workshop on educational data mining. Citeseer, pp 37–44

Tanes Z, Arnold KE, King AS, Remnet MA (2011) Using signals for appropriate feedback: perceptions and practices. Comput Educ 57(4):2414–2422. doi:10.1016/j.compedu.2011.05. 016

Tinto V (2006) Research and practice of student retention: what next? J Coll Stud Retention 8 (1):1–19

Verbert K, Govaerts S, Duval E, Santos JL, Assche F, Parra G et al (2014) Learning dashboards: an overview and future research opportunities. Pers Ubiquit Comput 18(6):1499–1514. doi:10. 1007/s00779-013-0751-2

Verpoorten D, Westera W, Specht M (2011) A first approach to "Learning Dashboards" in formal learning contexts. Paper presented at the 1st international workshop on enhancing learning with ambient displays and visualization techniques. Palermo, Italy

West D, Huijser H, Lizzio A, Toohey D, Miles C, Searle B, et al (2015) Learning analytics: assisting Universities with student retention, Final Report (Part 1). Australian Government Office for Learning and Teaching

Wise AF (2014) Designing pedagogical interventions to support student use of learning analytics. In: Pardo A, Teasley SD (eds) International conference on learning analytics and knowledge. ACM Press, pp 203–211. doi:10.1145/2567574.2567588

Chapter 6
Overcoming the MOOC Data Deluge with Learning Analytic Dashboards

Lorenzo Vigentini, Andrew Clayphan, Xia Zhang and Mahsa Chitsaz

Abstract With the proliferation of MOOCs and the large amount of data collected, a lot of questions have been asked about their value and effectiveness. One of the key issues emerging is the difficulty in the sense—making from the data available. The use of analytic dashboards has been suggested to provide quick insights and distil the large volume of learner interaction data generated. These dashboards hold the promise of providing a contextualized view of data and facilitating useful research exploration. However, little has been done in defining how these dashboards should be created, often resulting in a proliferation of systems for each new research agenda. We present our experience of building MOOC dashboards for two different platforms, namely *Coursera* and *FutureLearn*, motivated by a set of design goals with input from a diverse set of stakeholders. We demonstrate the features of the system and how it has served to make data accessible and useable. We report on problems faced, drawing on analyses of think-aloud sessions conducted with real educators, which have informed our dashboard process.

Keywords Learning and teaching data · Dashboards · MOOCs · Visualization · Sense-making

L. Vigentini (✉)
School of Education, UNSW Australia, Kensington, NSW, Australia
e-mail: l.vigentini@unsw.edu.au

L. Vigentini · A. Clayphan · X. Zhang · M. Chitsaz
Portfolio of the Pro-Vice Chancellor Education, UNSW Australia,
Kensington, NSW, Australia
e-mail: a.clayphan@unsw.edu.au

X. Zhang
e-mail: Xialisa.zhang@gmail.com

M. Chitsaz
e-mail: m.chitsaz@unsw.edu.au

© Springer International Publishing AG 2017
A. Peña-Ayala (ed.), *Learning Analytics: Fundaments, Applications, and Trends*, Studies in Systems, Decision and Control 94,
DOI 10.1007/978-3-319-52977-6_6

171

Abbreviations

HCI   Human computer interation
HTML   Hypertext markup language
JSON   JavaScript object notation
MOOC   Massive open online course
SUS   System usability scale

6.1   Introduction

Over the past few years, MOOCs (Massive Open Online Courses) have become the center of much attention both from the public media (Cormier 2009; Bates 2012; Cooper and Sahami 2013; Dominique 2015) as well as the research community (Daniel 2012; Amo 2013; Baggaley 2013; Davis et al. 2014; Drachsler and Kalz 2016). Despite the hype of big data in education and the potential associated with the ability to collect and analyse large amounts of information about students' learning behaviours (Arnold and Pistilli 2012; Verbert et al. 2013b), one of the biggest limitations is finding how to expose this data in a meaningful and relevant way for different stakeholders, being students, instructors, researchers or developers (Duval 2011; Dernoncourt et al. 2013; Verbert et al. 2013b). Visualization of data and the ability to manipulate visualizations has been demonstrated to provide useful insights (Pauwels et al. 2009; Duval 2011). These are grouped into dashboards, which have been advocated to enable individuals, researchers and policy makers— quick insight into data. Dashboards are often justified by themes such as providing awareness, reflection and sense making (Allio 2012). However, as dashboards have become popular, they have also brought an over-expectation of benefits with them (Davis et al. 2014; Leon Urrutia et al. 2016). In fact, dashboards in and of themselves do not automatically confer learning or awareness gains. It is a combination of effective design, requirements elicitation, and an understanding of stakeholder objectives, which may permit dashboards to be potentially powerful tools for aiding data exploration.

Learning analytics has been suggested to provide a theoretical framework to make sense of learner's interactions with online courses (Verbert et al. 2013a; Corrin et al. 2015; Corrin and de Barba 2015; Drachsler and Kalz 2016). In fact, one of the purposes of learning analytics is to visualise learner activity so that educators can make informed decisions about possible interventions (Clow 2013; Bayne and Ross 2014; Stephens-Martinez et al. 2014). Verbert and colleagues represented the process aided by learning analytics in Fig. 6.1. In order to achieve impact in the improvement of learning and teaching, data is the key entry level in order to articulate appropriate questions.

In this chapter, the authors report about their experience of developing an analytics dashboard for two different MOOC platforms (Coursera and FutureLearn,

Fig. 6.1 Adapted from the learning analytics process model (Verbert et al. 2013a)

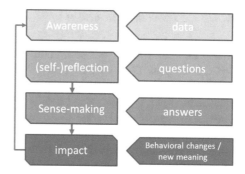

two leading MOOC providers in the USA and the UK respectively), focusing on the challenges of the process and providing useful insights for others embarking on similar initiatives. Firstly the work is placed in the context of existing work in the field of Learning Analytics where the problem is explored in more detail, discussing key aspects in relation to prior works. Then, the development and implementation process will be described focusing on the process and some of the challenges that the team faced, and finally the results from a small case usability study will be reported to bring support to further development of this work.

6.2 How This Work Enhances the LA Field

The sort of dashboard creation process for MOOCs reported here is not new; however it has a number of distinctive features, which provide an opportunity to extend the work carried out in the learning analytics community.

First of all, a number of design goals were established at the outset to cater for multiple stakeholders in a holistic way: this moves away from the specific focus on the teachers (Stephens-Martinez et al. 2014; Corrin et al. 2015), the students (Arnold and Pistilli 2012; Corrin and de Barba 2015; Kia et al.) or institutional research/BI practitioners (Campbell et al. 2007; Mohanty et al. 2013).

Secondly, we had an opportunity to work with multiple platforms which, given the restriction with the data sharing agreements imposed by different providers, makes collaborations between institutions less likely to occur. With exception of a few (Siemens et al. 2011; Cobos et al. 2016), most of the other works published thus far only focus on one platform at the time.

Thirdly, we leveraged on an educational lens to integrate institutional reporting matched with an academic analytics perspective. This allowed focusing on questions at multiple levels of analysis (micro, meso and macro) depending on the relevance to the individual, the course or the institution. This approach is original because the design process focused on generalizable features first, relevant not only in MOOCs, but extending to other learning management systems or learning technology tools as well.

6.2.1 Problem Statement

When the data team at our institution was tasked to report the findings from our MOOC courses (20+ courses now), it quickly became apparent there were a number of different and competing goals. These ranged from broad institutional questions, down to fine-grained details within particular modes of assessment. As such, the process quickly led to an explosion in the number of questions being raised.

These sorts of problems are common and have been chronicled by others (Seaton et al. 2013, 2014; Verbert et al. 2013b; Qu and Chen 2015; Leon Urrutia et al. 2016) leading to the recommendation of standardisation at various levels from the data (Dernoncourt et al. 2013; Veeramachaneni et al. 2014a) to the sets and types of analytics (Duval 2011; Siemens et al. 2011; Verbert et al. 2013b) or the types of visualisations deemed useful (Stephens-Martinez et al. 2014).

In our context, this rapidly drove a nail into initial ad hoc or exploratory approaches to analysis that were meant to provide quick solutions to some of the questions raised by instructors, which didn't necessarily satisfy other stakeholders, nor make the process efficient.

When each of the courses closed, as expected, the analytical team faced high expectations to produce interesting results from the analysis, particularly given that at the time there was not a fully-fledged dashboard available. Team members faced a number of challenges in order to come up with solutions. For example, course academics (instructors) were keenly waiting on analysis and reports on their course data and their research questions; Educational developers wanted answers about their design; and Senior Managers wanted to know whether their investment had significant returns. In addition, the process to actually get the data had a significant delay, adding pressure to the process and the team with a number of key issues:

- Manual data requests required to obtain the data from the educational platforms (Coursera) and bespoke third party systems after course completion, see Fig. 6.2.
- A steep learning curve due to complexities/volume of data from each system.
- Unsustainable processes: Short timeframes to answer a large number of research questions in ad hoc report formats.
- Concurrent priorities with teaching/research focus: institutional research versus course instructors' own research questions.
- Lack of transparency in requests from stakeholders and interested parties.

The need to establish a methodology to build a reproducible workflow became apparent—going from data extraction, to transformation to delivery. We were confronted by a number of problems, which we used as pointers to define our initial needs analysis of the requirements and scope:

- *Time*: A large amount of effort is required to transform data into a useable format. Doing this by hand is not a long-term solution. How can this be automated? What needs to be automated? What should we strive towards?

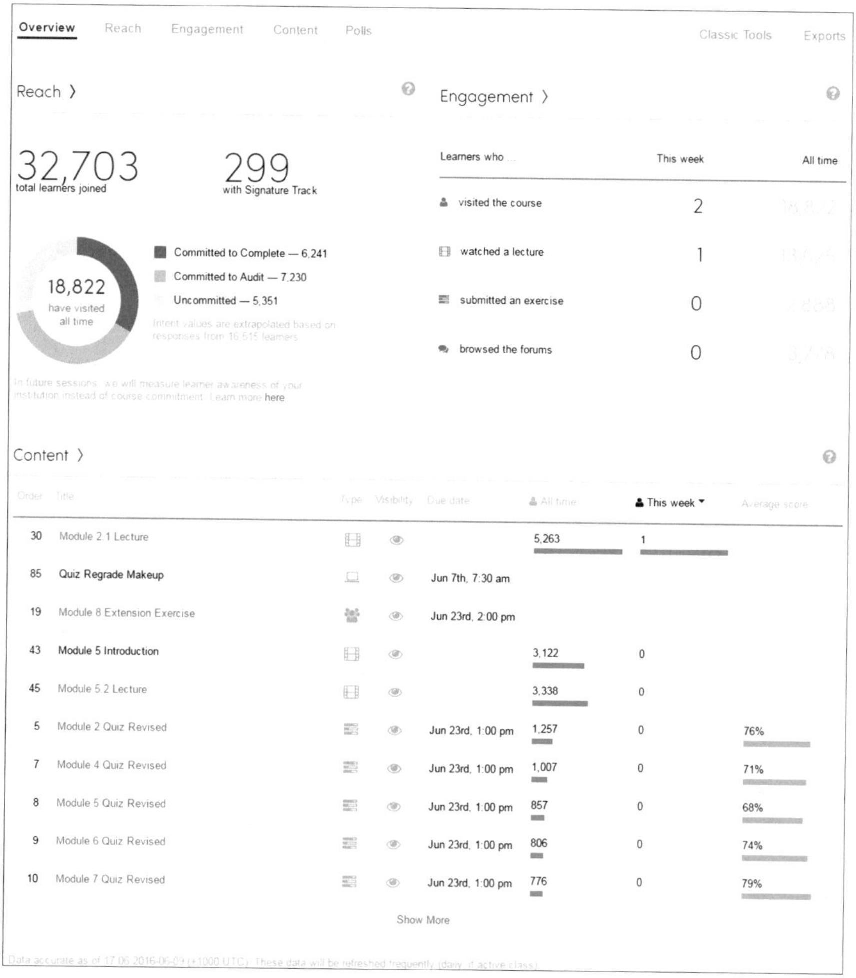

Fig. 6.2 Screen capture of the Coursera analytics dashboard for one of our courses

- *Volume*: There is an ever-growing base of data and variants in the data schemas. What are the common questions being raised, and is this collated someplace? Are these questions transparent between different members of the data team working on them and the stakeholders asking them?
- *Competencies*: How can we best utilize resources, both in terms of systems and people? Are stakeholders able to understand the output?

In order to address the above challenges, the team developed a process grounded on five building blocks:

1. Articulate a reporting framework, which keeps into consideration the needs of different stakeholders.

2. Develop a semi-automated data transformation workflow.
3. Determine which tool is most appropriate for the needs of stakeholders.
4. Design and build a dashboard, which provides flexibility in the exploration and addresses the majority of the questions raised by stakeholders.
5. Provide a framework for the scalable and sustainable reproduction of the process for each MOOC.

6.2.2   Related Work to Dashboards

Although analytics dashboards to support organisations in their decision making processes have been around for some time (Pauwels et al. 2009), those focusing on learning analytics have a shorter history. A useful classification according to their purpose was proposed by Verbert et al. (2013b). The first type of dashboard supports traditional face-to-face lectures; another type of dashboard supports face-to-face group work; further, another type of dashboard is used to support awareness, reflection, sense-making and behavioural changes in online or blended learning. In this chapter we focus specifically on the latter. In fact, in the current MOOC space there are relatively few examples of learning analytics dashboards, with mostly disjointed approaches and ad hoc implementations. For example, Coursera provides a dashboard to instructors with a live view of the data, but the granularity of the information does not necessarily cater for all stakeholders needs or wants (such as the ability for instructor-specific stratification—see Figs. 6.2 and 6.3–*left* for examples of high-level views of course data). Other vendors, such as FutureLearn, take an even further removed option, with no visual dashboard, but rather just a list of key course metrics (Fig. 6.3–*right*).

Although these representations provide a high-level view of MOOCs that may be suitable to get a general sense and may satisfy academic managers; educators are likely to ask more probing and sophisticated questions about contributing factors leading to certain patterns of engagement (Stephens-Martinez et al. 2014).

On the other hand, other platforms like EdX (Seaton et al. 2013; Ruiz et al. 2014; Fredericks et al. 2016; Pijeira Díaz et al. 2016) recently added analytical plug-in modules, that users can install on their own systems, to provide detailed views of how learners engage with the platform. These however may be too detailed and/or complex to cater for the casual user.

There are also examples of external visualizations tools and dashboards (Veeramachaneni et al. 2014b; Qu and Chen 2015; Cobos et al. 2016; Davis et al 2014; Leon-Urrutia et al 2016; Kia et al.), which provide specific representation of behaviours in MOOCs, however the majority of these tools are not open. This often means that teams in different institutions end up replicating similar processes. Furthermore, given the specific reasons why the dashboard is developed, it is natural to ask whether the information represented conveys useful insights to stakeholders outside initial developments.

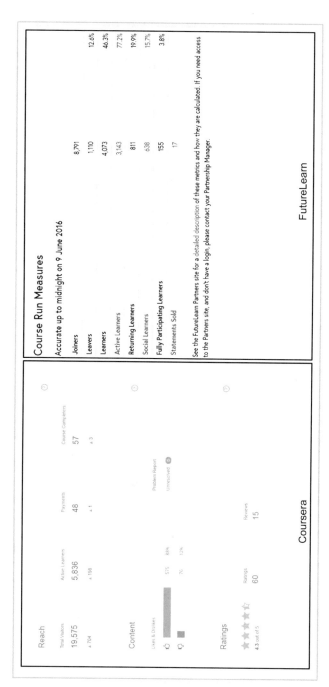

Fig. 6.3 On the *left*, key figures from the Coursera dashboard. On the *right* key metrics from FutureLearn

Table 6.1 Data sources/data tracking in learning analytics from various published works

Data sources	Data tracking
– Artefacts produced[a,b]	– Application logs[a]
– Social interaction (forums)[a,b]	– Camera[a]
– Resource use[a,b]	– Microphone[a]
– Time spent[a,b]	– Depth sensors[a]
– Exercise/Test results[a,b]	– Biofeedback[b,c]
– Assessment/Course grades[b]	– Eye trackers[d]

Sources [a](Verbert et al. 2013a); [b](Romero et al. 2008); [c](Govaerts et al. 2012); [d](Vatrapu et al. 2013)

Nevertheless, as noted in Verbert and colleagues' comparison of different learning analytics dashboards (Romero et al. 2008; Govaerts et al. 2012; Vatrapu et al. 2013; Verbert et al. 2013a) there are common elements that can be used as a starting point. Table 6.1 provides a summary (albeit not comprehensive) of different types of sources and types of tracking from currently published literature.

6.3 Design and Development of the Dashboard

A number of design goals were established from the process of exploring the problems around the analytics space. These were: (1) the need for reproducible processes; (2) flexibility; (3) transparency and (4) extensibility. Based on these, we proposed a framework to enable data exploration efforts to take place that attempted to be platform agnostic. A number of questions emerged which are explored in more detail in the next sections.

Before stepping into the description of the rationale and method for planning, designing and developing the dashboard, it is necessary to quickly describe the type of data available in MOOCs. Table 6.2 provides an overview of four categories of data available in MOOCs: these are commonly valid across MOOCs and Virtual Learning platforms, but the granularity of the details vary. For example while rich demographic data may be available to institutions for credit-bearing courses that they offer, in the MOOC space, information about participants is limited and in general rather sparse. On the contrary, the logs of online activity in MOOCs are more sophisticated than most on-campus blended courses. Notably, as demonstrated in Table 6.1, a large proportion of works published, focused primarily on activity data and performance data, which provide a limited window on the learning experience (Vigentini and Zhao 2016). Coursera and EdX provide a great deal of information about the interaction with videos, however FutureLearn at this point in time does not provide any details about 'in-video' behaviors.

Table 6.2 Types of data available in MOOCs and other learning platforms

Demographic data: general student demographics, including age, gender, language, education level, and location. Demographic data is commonly acquired during the registration process, and additional demographic data can be acquired via feedback surveys	Performance data: student performance based on both formative and summative assessments. This is collected from homework, quizzes, and exams, but it also includes results from pre-course surveys/activities designed to diagnose student knowledge before they take the course
Activity data: how and when students are using the resources, such as watching videos, reading material, submitting homework, taking quizzes, or using the discussion forum. Most platforms break down usage by content and media type (i.e. page views, assignment views, textbook views, video views)	Feedback data: student feedback coming from polls, surveys and comments in forums. Additionally information about student learning goals and motivation and intended use of the material

6.3.1 What Is the Most Appropriate Tool to Build the Dashboard in?

A preliminary analysis into a number of commercial visualization platforms was conducted. This consisted of assessing (1) availability, (2) flexibility, (3) adaptability, (4) export features and (5) data privacy. The following tools were examined: Ubiq; Tableau; SiSENSE; The Dash; Dashzen; Ducksboard; Klipfolio; Leftronic; Qlik; Drillable; Logi and Infocaptor.[1]

We also examined the issue of data privacy. A detailed analysis made clear that not one tool could solve all the issues we had, with most tools coming short with some features or a hefty price tag.

A systematic evaluation and review of the comparison is beyond the scope of this chapter; however it should be noted that different products had fundamental differences in the visualisation capabilities, management of the analytic workflow and varying technical affordances. Therefore we opted to build a lightweight scaffold that allowed multiple tools to be used. This web-based architecture was driven by a theoretical representation of the data as well as the intent to produce a range of visualisation tools (widgets) that helped to explore the data from multiple perspectives. This process was strongly informed by the general framework in Siemens et al. (2011) and the attempt to integrate visualisations at Harvard, Stanford and Berkeley (Dernoncourt et al. 2013; Pardos and Kao 2015).

For what concerns the data format, we considered the MOOCDB data structure (Veeramachaneni et al. 2014a), which seemed to promise a unified approach for data coming from EdX and Coursera, but had to discard the options for two main reasons: (1) the pace at which Coursera changed the format of their data exports made the process unwieldy and difficult to maintain the code up to date; and (2) the

---

[1]Please refer to the note at the end of the chapter with links to the respective websites.

level of transformation required would have led to a substantial loss of information even before we began the analysis of the data. Furthermore, the standard proposed did not actually account for considerable platform differences nor the contextual assumptions determined by the design of the user interface.

6.3.2   How Can We Best Answer the Questions Posed?

After considering the list of questions asked and informed by the data available we developed a framework to allow different ways to organise, display and get to the required information. The framework is based on two top level groups: report categories and functional domains. These are shown in Table 6.3 and then visually in Fig. 6.4 providing the basis for component re-use so that visualizations built for one course, could be easily adapted and re-used in another course. Figures 6.5 and 6.6 provide examples of the form of visualizations we have built to date.

Report categories represent standard reports organised by specific labels. In general the categories refer to the course as a whole. The representations under this label focus on questions that instructors or academic managers would want answers to in a broad way.

Table 6.3 Broad reporting framework based on categories and functional domains

Report categories	Functional domains
'Overview' comprises the funnel of participants to model the student's journey in the course, from when a learner becomes aware of the course to the completion of it. It covers the size of the MOOC, breakdown of the participants by status and achievement level, type and level of activities	*'Videos'* focuses on the daily, weekly, monthly use, broken into actions of downloads and views. Videos are listed by module/week and can be drilled down to individual levels.
'Who are the participants' covers visualisations related to user characteristics: stratification on demographic variables such as age, gender distribution, and education and employment status	*'Content'* explores the daily, weekly, monthly use of course aspects outside of videos.
'What participants do' involves user activities and use of course content: videos, forums, quizzes and peer assessments	*'Forums'* contain daily, weekly, monthly use broken into actions of posts and comments. It also provides insight into participant interactions
'Overview of assessment' looks at the cognitive aspects of student performance including both formative and summative assessments	*'Activities'* focuses on the engagement opportunities, a student has in the course to help them learn. It refers to both formative and summative activities
'Research' provides insights drawn from data mining techniques. For example, this includes cluster analysis based on users' activities	*'Evaluation'* shows the methods used in the course to evaluate the student experience and effectiveness of the MOOC
	'Social Media' contains detail about the reach of the MOOC through certain distribution channels

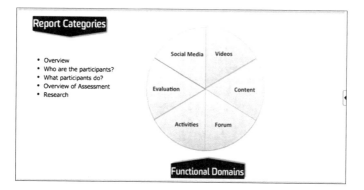

Fig. 6.4 Dashboard frontend for our Coursera courses showing the 3 main entry points for visualizations: report categories, functional domains and the full site map from the right panel activated by pulley button (*arrow*)

Functional domains arrange reports and visualisations according to what their purpose is in the MOOC. The key difference of analysis and representations under this label is the level of granularity of the analysis. Instructors with specific questions about their content or educational developers/learning technologists detailed questions about what worked or not in their learning design. Furthermore, we designed the dashboard as a data distribution point (following in the footsteps of moocRP (Pardos and Kao 2015) allowing individuals wanting to explore further the data to download the data from the visuals provided.

However, unlike moocRP we felt that the workflow to make data available needed to be less sophisticated allowing individuals to extract what they needed after they identified valuable visualisations, so that they could quickly drill-down to answer their questions.

6.3.3   How Can We Make Use of Effective Visual Design?

We largely based our framework and dashboard design on the principles from the work by (Duval 2011; Siemens et al. 2011). In summary, these were: (1) provide viewers with the information they need quickly and clearly, (2) stay away from clichés or gimmicks; (3) focus on what is important; and (4) align to educational objectives and learning goals.

The technical solution took into account fundamental differences in the data provided by the two platforms and also two different workflows.

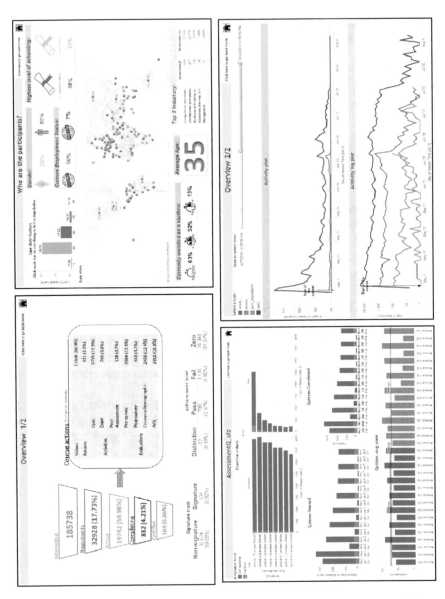

Fig. 6.5 Examples of visualizations from the MOOC Dashboard. *Top*—overview and demographics; *bottom*—assessment performance and timeline of activities

Visual Walkthrough of the main user interface components:

Home page overview:

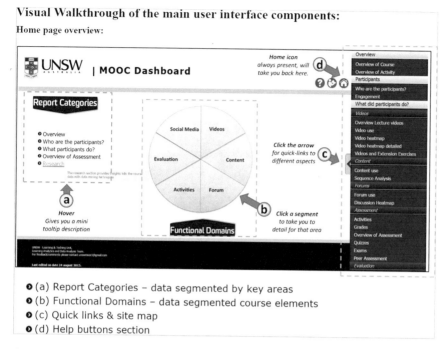

● (a) Report Categories – data segmented by key areas
● (b) Functional Domains – data segmented course elements
● (c) Quick links & site map
● (d) Help buttons section

Fig. 6.6 Visual walkthrough of the dashboard developed for Coursera

6.4 Building the Dashboards

The development of the building blocks (or widgets) and the aggregation into dashboard panels was an iterative process based on four stages:

1. Design the dashboard elements/blocks
2. Prototype the visualizations
3. Test the dashboard with different users to collect feedback
4. Evaluate and iterate over the design.

The dashboard framework for Coursera is relatively simple. Two HTML pages (a home page and a panel page) are the frontend organizing the content and menus. In the backend there is a single JSON configuration file per course that is used to dynamically populate the html elements and the critical component is a set of visualizations created and packaged in Tableau, which are served based on the configuration file. Examples are shown on the next page (Figs. 6.6 and 6.7).

The dashboard for FutureLearn is still at the early stages of development, but it uses a different architecture, relying on Shiny dashboard and R scripts to generate the various visualizations (Fig. 6.8). These are then framed as widgets and organized according to a similar taxonomy as the Coursera dashboard.

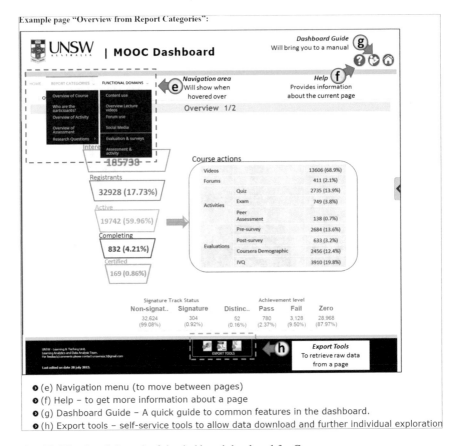

Fig. 6.7 Visual walkthrough of the dashboard developed for Coursera

6.5 Evaluating the User Experience

Initial feedback from instructors has been very positive, however we have also been evaluating the user experience more formally by designing an in-house process for a think-aloud and cognitive walkthrough protocol (Wharton et al. 1992; Fonteyn et al. 1993; Rieman et al. 1995; Azevedo et al. 2013). Think aloud methods are cost-effective, robust, flexible and relatively easy to administer (Nielsen 1994; Conrad et al. 1999).

Based on the review of the literature and the questions raised by our instructors it was evident that instructors need to understand the effectiveness of resources, activities, grading rubrics and support methods in relation to the set learning outcomes and in order to continuously improve the course (Churchill et al. 2013). Academic managers are more interested in the 'bigger picture' and draw comparison between credit-bearing courses offered by the university and the courses offered for free as MOOCs. Another dimension of interest for our courses was the

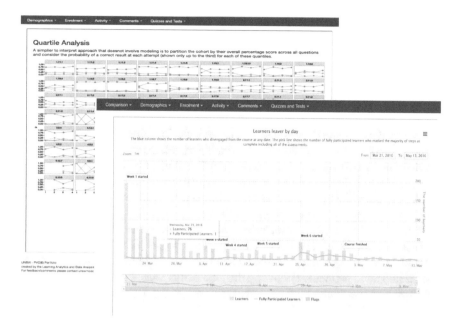

Fig. 6.8 Snapshot of two widgets for the FutureLearn dashboard representing the quartile distribution of quiz answers and the prevalence of high degrees in participants

Table 6.4 Overview of activities, in terms of the number of participants who answered the question successfully, the average time taken to do the activity, and the most common pathway utilized to obtain the answer reported

Activity	Correct response ratio	Mean time	Most common method to arrive at response
(1) How many times on average was the first video from the first week/module viewed by non-signature track students?	7/11	4:45	Functional domains
(2) What percentage of people passed the course?	10/11	2:43	Report categories
(3) How many students finished the peer assessment for <exercise x>?	11/11	2:03	Report categories
(4) How many people made a forum post on <date x>?	9/11	3:42	Functional domains

evaluation of effectiveness for different pedagogical approaches, which led to the design of a set of test activities to determine whether the dashboard fulfilled its intended purpose.

The test activities were designed on real scenarios presented to us by educators and project support staff, yielding an element of authenticity. The four questions (see Table 6.4) were intended to be simple, but non-trivial as they required some level of exploration and integration of information.

6.5.1 Study Design

To gain feedback about the dashboard interface, we developed a protocol to inform a quasi-experimental interaction study bringing in real-users to test the dashboard, and drew on human-computer interaction (HCI) techniques—a think-aloud process combined with a SUS usability questionnaire (Brooke 1996). Each session was conducted individually, accompanied with screen and audio capture (using QuickTime recording on the computer utilized for testing).

The protocols aimed to standardize the testing sessions and allow for some comparability in the observations. The procedure began with a background questionnaire, to collect general information about the participants' computer skills and past dashboard experience. Following this, the participants were each given a sheet that explained the think-aloud process, accompanied by a quick warm-up exercise (not using the dashboard), so as to allow participants to become familiar with the think-aloud procedure. After the warm-up exercise, participants were allowed to browse the dashboard interface for strictly 5 min. After the 5 min had elapsed, the main task began; this consisted of the following four activities:

1. Activity 1. How many times on average was the first video from the first week/module viewed by students?
2. Activity 2. What percentage of people passed the course?
3. Activity 3. How many students finished the peer assessment for <exercise x>?
4. Activity 4. How many people made a forum post on <date x>?

Each activity allowed the participants to get the answer however they liked, and at the end of each activity, they rated on a 6-point scale, the ease with which they were able to answer the question (6 being strongest agreement). Each activity asked how the participant navigated to their answer, with one of the following:

- "I used the Report Categories"
- "I used the Functional Domains"
- "I used the Navigation Sidebar"
- "Other (Please explain)".

Upon the conclusion of the activities, a SUS questionnaire was administered, which sought to reveal usability insights to the interface. Lastly, a semi-structured post-interview questionnaire was administered to tease out what participants thought would be required to fully utilize the dashboard, including thoughts on training, features found most/least useful, and additional elements they would like added to the dashboard to enrich their and others experience.

6.5.2 Apparatus

The machine used for each session was a regular 13-inch MacBook, with a screen resolution of 1280×800 pixels. This apparatus was chosen, given the near ubiquity of laptops used around campus, its portability, as well as to situate the experiment within common hardware, for example compared to a dual or triple screen setup which is less common in the university workplace, and which would have unduly added bias to the interface exploration. Each participant was also provided with a computer mouse, which they could use in-lieu of the trackpad, if they so wished.

6.5.3 Participants

Eleven participants were sourced from UNSW Australia, who had not been exposed to the MOOC dashboard previously. Each session was run individually by one interviewer. For the think-aloud components, minimal help was provided (e.g. if the participant managed to close the web-browser, this was re-opened for them). The age range was 26–54 with a mean age of thirty-seven. Participants had a diverse background set, drawing on a pool of academics, project managers, as well as educational support staff. Four participants had previous use with dashboards.

Overall, participants rated themselves as above average with regard to the use of computers and average with ease of use with new technology. Participants reported that their visibility to teaching data had mainly been used on an ad hoc basis, either for personal research projects or for helping strengthen cases for promotion. Most participants commented that they would like more timely access to data if at all possible, to help in both their research and teaching efforts.

6.6 Results

Three forms of analysis are reported. First is a study of the outputs from the four activities undertaken in each session. This sought to find out features used as related to the pedagogical design of the dashboard, and also the ease of use with finding information for a person first exposed to the dashboard without prior training from analytical/data team members. Next, the usability of the interface is evaluated, as drawn from a SUS questionnaire. Lastly responses from a post-interview questionnaire are analyzed. In addition, free-form comments within each section were used to support interpretation of the analysis and to learn how participants perceived the experience.

6.6.1 Activity Analysis

The analysis of activity was studied in terms of how many participants reported the correct (i.e. expected) response; the time spent toward determining their response; and how the participants arrived at their answer. Table 6.4 summaries these aspects.

For activity 1 (how many times on average was *the first video* from *the first week/module* viewed by *non-signature track students?*), the profile of ease of completing the task was: 2× strongly disagree; 4× disagree; 1× somewhat disagree; 3× somewhat agree; and 1× agree, for an overall score of 2.72/6, see Fig. 6.9. This question took the longest on average to answer by participants, and was also the question, participants least answered accurately (that is, arrived at the expected response). Participants commented on a number of issues, for example:

> 'The expanse of information available made it a slower process than I would have liked when trying to find the information. I assume this would become easier the more I used the tool'

> 'Screen size is small, which makes it difficult to read titles'; 'lots of data on one screen'

> 'There is no way to understand what is meant by first video'.

It appears a number of participants had conceptual issues regarding the use of.

The word 'first video', where review of the think-aloud recordings revealed participants asking themselves, 'first as in any video of the module watched first, first as in the first video on the page'. The particular question required navigation to the video tab, and hovering over the top-most video within the first week/module, which would have revealed the required answer. Half of the participants reported totals rather than averages (for the purpose of the question, these were marked as accurate). The four people who did not arrive at the correct answer, pulled their answer from un-related tabs which they misinterpreted as being videos, or tried to

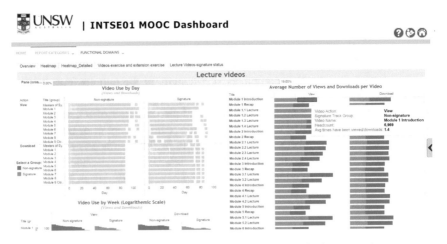

Fig. 6.9 Capture of the panel for Activity 1: the answer is shown in the *pop-up box*

filter to just the first week based on assumptions of when they thought the first week began. Only 1 participant decided to give up in the end (after 7 min). Of the 11 participants, all but one commented that they would have liked more accompanying documentation; however it was not observed that a single participant referred to the help-menu in the top right corner. This highlights that an on-board process, explaining the features and navigation layout may be needed for future deployments. Figure 6.9 shows a screen capture of the task with the solution.

For activity 2 (What percentage of people *passed* the course?), the profile of ease of completing the task was: 1× somewhat disagree; 5× somewhat agree; 3× agree; 2× strongly agree; for an overall score of 4.54/6. From participant's comments, even though 10 out of 11 people reported the correct response, there were still a small number of participants (4 people) who were confused by the usage of the word 'passed', for example:

definition of passed is not clear

what is the difference between certificate and completed.

Otherwise most people were able to report the response, based on the literal description of passed (as had been named in the dashboard), for example:

the question was clear and it asked for a pretty broad/simple answer it was a fairly easy task. obvious to me where to find the answer

the pop-ups were the key to helping me find the information I needed but I ignored them at first.

Half of the comments reported, still asked for further help dialogs or information. A couple of comments mentioned they would have liked 'Report Categories' to be called 'Menu', and 'Functional domains' to be called 'Pie chart', to match their mental models of the dashboard as they perceived it. Figure 6.10 provides the screen capture of the task.

For activity 3 (How many students *finished* the *peer assessment* for <exercise x>?), the profile of ease of completing the task was: 2× strongly disagree; 2× disagree; 2× somewhat agree; 4× agree; 1× strongly agree; for an overall score of 3.63/6. Again participants mentioned that there was:

'Way too much info on screen'/'too much data on one page'.

However, it was mentioned that the pop-up's over graphs were useful, but overall, participants wanted less graphs per page. Figure 6.11 shows the screen capture of the task.

For activity 4 (How many people made a forum post on <date x>?), the profile of ease of completing the task was: 2× strongly disagree; 1× disagree; 2× somewhat disagree; 2× somewhat agree; 3× agree; 1× strongly agree; for an overall score of 3.73/6. For the purposes of this question, the participants needed to direct themselves to the forum section of the dashboard, and click on the forum

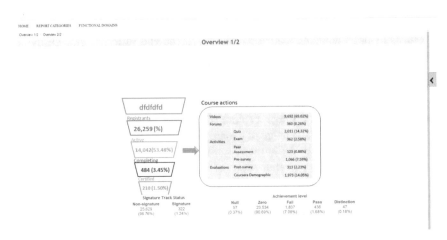

Fig. 6.10 Capture of the panel for Activity 2: the solution is in the *table bottom right*

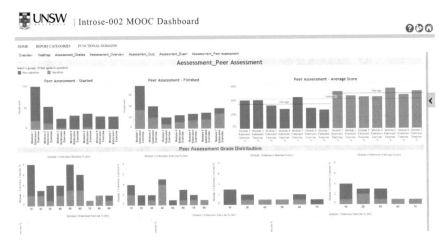

Fig. 6.11 Capture of the panel for Activity 3: answer in the *second chart* in the *first row*

graph to reveal the split between posts and comments. Roughly half were able to do this, however, if they reported the combined posts and comments, this was not deducted from the number reported as correct. This resulted in a few participants wondering if this activity was 'a trick question'. This was again, combined with a similar theme seen in the previous activities of their being a lot of options/ways to navigate to information. Participants also raised concerns about terminology use and jargon they may not be familiar with. Figure 6.12 shows the screen capture.

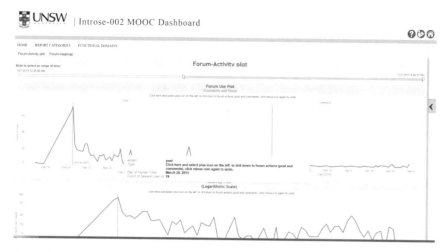

Fig. 6.12 Capture of the panel for Activity 4: answer in the popup box

6.7 System Usability Analysis

The SUS usability questionnaire consists of ten questions, with each odd numbered question posed in a positive frame, and each even numbered question posed in a negative way. Thus, for each odd-numbered question, the closer to 'strongly agree' the better and for each even-numbered question, the closer to 'strongly disagree' is better. The questions in combination can be used to formulate a SUS score, which can help inform an interpretation of the usability of the system under investigation.

The score itself works on a sliding scale, with 85+ as excellent, 70–85 as good, 50–70 as okay, 35–50 as poor, and under 35 meaning a lot of improvement is required in terms of usability. The profiles of each question are given in Table 6.5. The SUS scores are shown in Fig. 6.13.

Overall, what can be glanced from the participants first use of the dashboard (without any aids, tutorials or prior demonstrations), is that scores largely fell between the 'poor' to 'okay' range. A number of the free-form comments mentioned wanting to have an initial walkthrough session, and most mentioned they believe their scores would revise upward the more time they spent with the interface. Thus the fact that only two participants registered in the 'very poor' category (P7/P8), shows that the initial design, whilst suffering cosmetic issues was largely usable.

The individual questions brought forth that, people would likely continue to use the system, they did not find the dashboard overly technically complex, however they would prefer the dashboard to have less graphs/images per tab/webpage. This was a clear indicator that users found the provided information in its current form as partly overwhelming. Users reported that they likely would not need someone technical to help them out, but did ask for an initial primer with the data team to help with explorations. Overall, most questions, were near neutral in response.

Table 6.5 SUS scoring table, from http://usabilitygeek.com/how-to-use-the-system-usability-scale-sus-to-evaluate-the-usability-of-your-website/

Question	SD	D	N	S	SA	Overall
(1) I think that I would like to use this system frequently	1	1	3	4	2	3.45/5
(2) I found the system unnecessarily complex	0	1	2	6	2	3.81/5
(3) I thought the system was easy to use	1	4	5	0	1	2.63/5
(4) I think that I would need the support of a technical person to be able to use the system	2	5	1	2	1	2.54/5
(5) I found the various functions in this system were well integrated	0	3	6	2	0	2.91/5
(6) I thought there was too much inconsistency in this system	3	4	2	2	0	2.27/5
(7) I would imagine that most people would learn to use this system fairly quickly	1	6	1	2	1	2.63/5
(8) I found the system very cumbersome to use.	0	1	3	6	1	3.63/5
(9) I felt confident using the system	1	6	3	0	1	2.45/5
(10) I needed to learn a lot of things before I could get going with this system	0	2	3	5	1	3.45/5

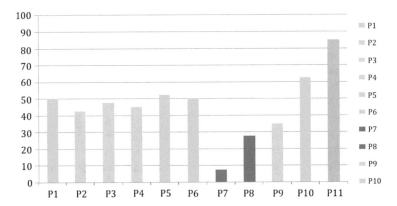

Fig. 6.13 Distribution of scores in the SUS

6.8 Post-interview Questionnaire Analysis

The semi-structured post-interview questionnaire asked six questions to the participant, the first three were about the think-aloud process, and the later three questions about aspects of the dashboard. Overall, no participants had objections or issues with the think-aloud protocol employed. The last three questions are now reported separately.

6.8.1 What Guidance/Training Do You Think Is Necessary to Use the Dashboard?

The participants asked that there be more explanations about terms (interestingly many of these explanations were present on the interface, generally within tooltips or question-mark icons, though most participants did not appear to notice these). Some asked for an annotated page describing the functions of the dashboard. Again, this was present in the help icon in the top right, with none of the participants investigating it. Some asked for a video guide to accompany a FAQ section. Participants reported that:

> if it was exactly as it is, without any modifications, no training or support documents are necessary just more time to understand the different aspects also information built into the system such as the question marks.

An overwhelming theme that emerged was the want for more hands-on demonstrations of possible ways to use the dashboard, for example:

> walked through it at least once to explain what features are where

> If you are not very familiar with technology (like a lot of academics are) I think it would be very frustrating to be honest. If you were very comfortable with technology I think you would work your way around it but I think you would definitely need some sort of small group workshops to get comfortable using it, and some sort of support mechanism to go back to

> someone from the data team to go through the main elements of the dashboard to explain what data is being displayed and how you would go about getting it. Documentation to walk you through the steps to get it. Would be good to know if the data is something that could be used to develop the next courses.

6.8.2 Which Features Were Most Useful and Why?

Participants commented on a range of features, for example, the homepage with the report categories and functional domains as it was 'uncluttered', and 'aesthetically the visual representation of students worldwide was good. At least one participant commented on liking the use of the tooltips, as in:

> pop up info when hovering over graphs. That was very informative when looking at something unfamiliar that you are not sure about.

Within the same question, a number of participants mentioned the confusion with the use of calendar weeks, compared with say course weeks, for example:

> found referring to calendar week number rather than course week number very confusing (i.e. starts at week 31 on forum heatmap page).

Also participants liked that the charts were interactive. For example:

I liked the interactive charts. The plot charts being able to change the dates and that you can hover above things to get info. It just needs to be laid out a bit better because the navigation is a bit clunky and doesn't really do it justice. The more interactive and the more simple it is to get to that stage the better. A pre-set report that you could just click on and it was there would be good. There is a lot of information there, maybe it needs to be split up a bit better and the navigation needs to be sorted but I did like the hovering thing.

6.8.3   Are There Any Additional Features You Would like to See?

The additional features turned into more of a commentary to a request to reduce the amount of information per tab/webpage. For example:

There is a lot of information on each page, maybe split some of the pages, into separate tabs/pages.

too much info on screen at the moment. Need the option to have just one chart on the screen, i.e. click on it to fill the screen. Or, have the option to build your own by screen selecting which charts you want to compare. Just the individual charts on each screen would be best. Think it is just too messy as it is. If you want to find specific piece of info it is too hard to find I think.

'I would like to see less features in lots of pages.' Only one person mentioned an actual feature request, and that was a 'search mechanism'.

6.9   Conclusion and Future Directions

In this chapter we presented an evolving process aimed to create a sustainable and reusable dashboard for MOOCs, which was intended to provide a tool for a variety of stakeholders to make sense of what is happening in the MOOCs developed and delivered by our institution in two different MOOC platforms. The challenges of the process and the choices made in the implementation have been described in Sect. 6.3.

As the development process relies on an active engagement with stakeholders, the prototyping phase of widgets is a responsive process to target stakeholders' needs. These are then included in the main panels of the dashboard following a principled approach, which relies on a framework informed by both data and reporting needs and affordances. User testing, like in the small scale study presented in the second half of the chapter, drives the process of development demonstrating responsiveness to the stakeholders' needs.

In the description of the implementation process we have established workflows, requiring minimal technical skills to generate a visually pleasing layout with our prototypes. Our framework is a step towards removing difficulties that have commonly plagued multi-tool adoption and advances the work carried out by others trying to solve similar problems (Seaton et al. 2013, 2014; Verbert et al. 2013b; Qu and Chen 2015; Leon Urrutia et al. 2016).

Based on the feedback from participants, we believe that the framework proposed goes towards the right direction in alleviating the disorientation typical of users beginning to make sense of MOOC user activity and our preliminary experiences are promising and the lessons learnt, in aiding the wider MOOC-related community in their own data related exploration efforts.

As indicated, the process is evolving and it is our intention to continue the developments, clarifying in more detail the elements of the framework by subdividing the functional domains and reporting categories according to different layers of analysis which target specific stakeholders. For example bursts of activity might be caused by problems with learning materials or inappropriate student behaviour: when these become visible to designers or instructors, this can provide an opportunity to counteract and resolve issues quickly. Activity bursts can also be caused by other factors, such as a particular learning design and as such highlight a particular topic or idea of interest for instructors. Academic managers might also be interested in such bursts because they highlight good (or bad) practice, which others should learn from and re-use (or potentially avoid).

There is no doubt that dashboards offer great opportunities for understanding MOOC activity (Qu and Chen 2015; Leon Urrutia et al. 2016) and the effectiveness of the pedagogies implemented in MOOCs; this chapter provides useful and practical observations in order to further the development of data-driven efforts to represent learning-in-action in online learning environments.

In terms of the development, it is our intention to make the dashboard developed for FutureLearn open. On one hand this will allow other institutions to learn from their MOOCs, on the other, it will open up the possibility for further collaborations with others, improving the quality and effectiveness of the visualisations.

Appendix

Please note that some of the products at the time of publication may not be available as some companies have already been acquired/merged by September 2016.

Coursera	http://coursera.com
Dashzen	https://www.dashzen.com/
Drillable	https://www.salient.com/drillable-dashboards/
Ducksboard	https://ducksboard.com/
FutureLearn	http://futurelearn.com
Infocaptor	http://www.infocaptor.com/
Klipfolio	https://www.klipfolio.com/

(continued)

(continued)

Leftronic	https://www.leftronic.com/
Logi	http://www.logianalytics.com/expertise/dashboards-and-reports/
Qlik	http://www.qlik.com/us/
Shiny Dashboard	https://rstudio.github.io/shinydashboard/
SISENSE	https://www.sisense.com
Tableau	https://tableau.com
The Dash	https://www.domo.com/solution/beyond-business-dashboards
Ubiq	http://ubiq.co/

References

Allio MK (2012) Strategic dashboards: designing and deploying them to improve implementation. Strategy Leadersh 40:24–31. doi:10.1108/10878571211257159

Amo D (2013) MOOCs: experimental approaches for quality in pedagogical and design fundamentals. Proceedings of the first international conference on technological ecosystem for enhancing multiculturality. ACM, New York, pp 219–223

Arnold KE, Pistilli MD (2012) Course signals at Purdue: using learning analytics to increase student success. Proceedings of the 2Nd international conference on learning analytics and knowledge. ACM, New York, pp 267–270

Azevedo R, Harley J, Trevors G, et al (2013) Using trace data to examine the complex roles of cognitive, metacognitive, and emotional self-regulatory processes during learning with multi-agent systems. In: Azevedo R, Aleven V (eds) International handbook of metacognition and learning technologies. Springer, New York, pp 427–449

Baggaley J (2013) MOOC rampant. Distance Educ 34:368–378. doi:10.1080/01587919.2013. 835768

Bates T. (2012) What's right and what's wrong about Coursera-style MOOCs | Tony Bates. http://www.tonybates.ca/2012/08/05/whats-right-and-whats-wrong-about-coursera-style-moocs Retrieved from 23 April 2015

Bayne S, Ross J (2014) The pedagogy of the Massive Open Online Course (MOOC): the UK view. Higher Education Academy, New York

Brooke J (1996) SUS-A quick and dirty usability scale. Usability Eval Ind 189:4–7

Campbell JP, DeBlois PB, Oblinger DG (2007) Academic analytics: a new tool for a new era. Educ Rev 42:40

Churchill D, King M, Fox B (2013) Learning design for science education in the 21st century. J Inst Educ Res 45:404–421

Clow D (2013) MOOCs and the funnel of participation. In: Proceedings of the third international conference on learning analytics and knowledge. ACM, pp 185–189

Cobos R, Gil S, Lareo A, Vargas FA (2016) Open-DLAs: an open dashboard for learning analytics. Proceedings of the third (2016) ACM conference on learning @ scale. ACM, New York, pp 265–268

Conrad F, Blair J, Tracy E (1999) Verbal reports are data! A theoretical approach to cognitive interviews. In: Proceedings of the federal committee on statistical methodology research conference. Tuesday B sessions, pp 11–20

Cooper S, Sahami M (2013) Reflections on Stanford's MOOCs. Commun ACM 56:28–30. doi:10. 1145/2408776.2408787

Cormier D (2009) What is a MOOC. Retrieved from https://www.youtube.com/watch?v=eW3gMGqcZQc

Corrin L, de Barba P (2015) How do students interpret feedback delivered via dashboards? Proceedings of the fifth international conference on learning analytics and knowledge. ACM, New York, pp 430–431

Corrin L, Kennedy G, de Barba P, et al (2015) Loop: a learning analytics tool to provide teachers with useful data visualisations. Perth, Australia, pp 409–413

Daniel J (2012) Making sense of MOOCs: musings in a maze of myth, paradox and possibility. J Interact Media Educ 2012: Art. 18

Davis HC, Dickens K, Leon Urrutia M, Sanchéz Vera M. del M, White S (2014) MOOCs for universities and learners an analysis of motivating factors. Presented at the 6th International Conference on Computer Supported Education. Retrieved from http://eprints.soton.ac.uk/363714/

Leon-Urrutia M, Cobos R, Dickens K, White S, Davis H (2016) Visualising the MOOC experience: a dynamic MOOC dashboard built through institutional collaboration. In Research Track https://www.emoocs2016.eu Graz, Austria, pp 461–471

Dernoncourt F, Taylor C, O'Reilly U-M, et al (2013) MoocViz: a large scale, open access, collaborative, data analytics platform for MOOCs. In: NIPS workshop on data-driven education, Lake Tahoe, Nevada. Retrieved from http://groups.csail.mit.edu/EVO-DesignOpt/groupWebSite/uploads/Site/MoocViz.pdf

Dominique A (2015) MOOCs, SPOCs, and LAPs: the evolving world of education. In: Huffington Post UK. http://www.huffingtonpost.co.uk/anton-dominique/moocs-spocs-and-laps-the-_b_4492046.html. Accessed 12 Apr 2015

Drachsler H, Kalz M (2016) The MOOC and learning analytics innovation cycle (MOLAC): a reflective summary of ongoing research and its challenges. J Comput Assist Learn n/a–n/a. doi:10.1111/jcal.12135

Duval E (2011) Attention Please!: learning analytics for visualization and recommendation. Proceedings of the 1st international conference on learning analytics and knowledge. ACM, New York, pp 9–17

Fonteyn ME, Kuipers B, Grobe SJ (1993) A description of think aloud method and protocol analysis. Qual Health Res 3:430–441. doi:10.1177/104973239300300403

Fredericks C, Lopez G, Shnayder V et al (2016) Instructor dashboards in EdX. Proceedings of the third (2016) ACM conference on learning @ scale. ACM, New York, pp 335–336

Govaerts S, Verbert K, Duval E, Pardo A (2012) The student activity meter for awareness and self-reflection. CHI '12 extended abstracts on human factors in computing systems. ACM, New York, pp 869–884

Kia FS, Pardos ZA, Hatala M Learning dashboard: bringing student background and performance online

Leon Urrutia M, Cobos R, Dickens K, et al (2016) Visualising the MOOC experience: a dynamic MOOC dashboard built through institutional collaboration

Mohanty S, Jagadeesh M, Srivatsa H (2013) Big data imperatives: enterprise "Big Data" warehouse, "BI" implementations and analytics Apress

Nielsen J (1994) Usability engineering. Elsevier

Pardos ZA, Kao K (2015) moocRP: an open-source analytics platform. Proceedings of the second (2015) ACM conference on learning @ scale. ACM, New York, pp 103–110

Pauwels K, Ambler T, Clark BH et al (2009) Dashboards as a service: why, what, how, and what research is needed? J Serv Res. doi:10.1177/1094670509344213

Pijeira Díaz HJ, Santofimia Ruiz J, Ruipérez-Valiente JA et al (2016) A demonstration of ANALYSE: a learning analytics tool for open edX. Proceedings of the third (2016) ACM conference on learning @ scale. ACM, New York, pp 329–330

Qu H, Chen Q (2015) Visual analytics for MOOC data. IEEE Comput Graph Appl 35:69–75. doi:10.1109/MCG.2015.137

Rieman J, Franzke M, Redmiles D (1995) Usability evaluation with the cognitive walkthrough. Conference companion on human factors in computing systems. ACM, New York, pp 387–388

Romero C, Ventura S, García E (2008) Data mining in course management systems: Moodle case study and tutorial. Comput Educ 51:368–384. doi:10.1016/j.compedu.2007.05.016

Ruiz JS, Díaz HJP, Ruipérez-Valiente JA et al (2014) Towards the development of a learning analytics extension in open edX. Proceedings of the second international conference on technological ecosystems for enhancing multiculturality. ACM, New York, pp 299–306

Seaton DT, Bergner Y, Chuang I, et al (2013) Towards real-time analytics in MOOCs. In: IWTA@ LAK

Seaton DT, Bergner Y, Chuang I et al (2014) Who does what in a massive open online course? Commun ACM 57:58–65. doi:10.1145/2500876

Siemens G, Gasevic D, Haythornthwaite C, et al (2011) Open learning analytics: an integrated and modularized platform. Open University Press

Stephens-Martinez K, Hearst MA, Fox A (2014) Monitoring MOOCs: which information sources do instructors value? Proceedings of the first ACM conference on learning @ scale conference. ACM, New York, pp 79–88

Vatrapu R, Reimann P, Bull S, Johnson M (2013) An eye-tracking study of notational, informational, and emotional aspects of learning analytics representations. Proceedings of the third international conference on learning analytics and knowledge. ACM, New York, pp 125–134

Veeramachaneni K, Halawa S, Dernoncourt F, et al (2014a) MOOCdb: developing standards and systems to support MOOC data science

Veeramachaneni K, O'Reilly U-M, Taylor C (2014b) Towards feature engineering at scale for data from massive open online courses

Verbert K, Duval E, Klerkx J et al (2013a) Learning analytics dashboard applications. Am Behav Sci 57:1500–1509. doi:10.1177/0002764213479363

Verbert K, Govaerts S, Duval E et al (2013b) Learning dashboards: an overview and future research opportunities. Pers Ubiquitous Comput 18:1499–1514. doi:10.1007/s00779-013-0751-2

Vigentini L, Zhao C (2016) Evaluating the "student" experience in MOOCs. Proceedings of the third (2016) ACM conference on learning @ scale. ACM, New York, pp 161–164

Wharton C, Bradford J, Jeffries R, Franzke M (1992) Applying cognitive walkthroughs to more complex user interfaces: experiences, issues, and recommendations. Proc SIGCHI Conf Hum Factors Comput Syst, pp 381–388

Chapter 7
A Priori Knowledge in Learning Analytics

Jean Simon

Abstract Learning Analytics (LA) can be data driven: the process is oriented essentially by data and not according to a theoretical background. In this case, results can be, sometimes, not exploitable. This is the reason why some LA processes are theory driven: based on A Priori Knowledge (APK), on a theoretical background. Here, we investigate the relationship between APK and LA. We propose a "2-level framework" that considers LA as a level 2 learning process and includes five components: stakeholders, goals, data, technical approaches and feedbacks. Based on this framework, a sample of LA related works is analyzed to exhibit how such works relate LA with APK. We show that most of the time the APK used for LA is the learning theory sustaining the student's learning. However, it can be otherwise and, according to the goal of LA, it is sometimes fruitful to use another theory.

Keywords Learning analytics · A priori knowledge · Learning theories · 2-level framework · Theory driven · Data driven

Abbreviations

APK   A Priori Knowledge
AT   Activity Theory
BI   Business Intelligence
EDM   Educational Data Mining
ITS   Intelligent Tutoring System
LA   Learning Analytics
LMS   Learning Management System
MOOC   Massive Open Online Course
PLE   Personal Learning Environments
SNA   Social Network Analysis

J. Simon (✉)
Université de la Reunion, ESPE de la Réunion, Allée des Aigues Marines,
97400 Saint-Denis, France
e-mail: jean.simon@univ-reunion.fr

© Springer International Publishing AG 2017
A. Peña-Ayala (ed.), *Learning Analytics: Fundaments, Applications,*
and Trends, Studies in Systems, Decision and Control 94,
DOI 10.1007/978-3-319-52977-6_7

199

TEL   Technology Enhanced Learning
ZPD   Zone of proximal development

7.1 Introduction

According to the philosopher Serres (2007), there were two main revolutions in the history of humanity: First being the invention of writing and second being the invention of information technology or more simply, the computer. Writing has allowed the storing of information on a medium other than the brain and, thus, the outsourcing of memory and knowledge. Computers allow the processing of information by a medium other than the brain and, thus, the outsourcing of many other cognitive processes.

Before the invention of writing, human beings learned not only from their own experiences but also from that of their community and the knowledge embedded within that community. With the invention of writing, two phenomena occurred, first human beings were able to outsource their knowledge and second they could gain access to more knowledge, written in tablets and later in books. At that moment in time, some members in the community became specialized to help others to learn: teachers (see left part of Fig. 7.1). Teachers could transfer their own knowledge, recommend some learning experiences, and suggest books to read. Nowadays human beings can always learn by the previous methods but there is one more tool to help them: the computer (see right part of Fig. 7.1). One important point is that, as computers are able to process information, they can take, in part, the role of the teacher and help students to learn.

Under Technology Enhanced Learning (TEL) (Balacheff et al. 2009) there are many different kinds of digital devices designed to improve learning. For instance, Khalil and Ebner (2015) point out the following: Personal Learning Environments (PLE) (Attwell 2007), Adaptive Hypermedia Educational Systems (Brusilovsky and Millán 2007), Interactive Learning Environments (Scaife et al. 1997), Learning Management Systems (LMS) (Watson and Watson 2007), Virtual Learning Environments (Dillenbourg et al. 2002), Immersive Learning Simulations (Kennedy et al. 2013), Intelligent Tutoring Systems (ITS) (Aleven et al. 2008), Massive Open Online Course (MOOC) (Fournier et al. 2014).

All of these systems interact with students to help them in their learning and these interactions leave trace data, also known as log data (Gašević et al. 2016). These data can be studied to optimize learning. We are then in the field of Learning Analytics (LA). The most used definition of LA (van Harmelen and Workman 2012; Ferguson 2012; Lockyer et al. 2013) is: "Learning analytics is the measurement, collection, analysis and reporting of data about learners and their contexts, for purposes of understanding and optimizing learning and the environments in which it occurs". The goal of LA is to enable institutions or teachers to offer

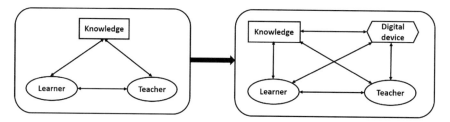

Fig. 7.1 Evolution of learning conditions

students better opportunities to learn (van Harmelen and Workman 2012) but also to invite students to self-reflection (Dyckhoff et al. 2012) and to help them find more personalized ways to learn. LA uses different techniques that convert data into useful information for the stakeholders (Dyckhoff et al. 2012). For instance:

- Educational Data Mining (EDM): Data mining, machine learning and statistics (Romero and Ventura 2010).
- Social Network Analysis (SNA): Where network is seen as nodes and links (Scott 1988).
- Business Intelligence (BI): Transforms raw data into useful information for business analysis purposes (Chen et al. 2012).

However, there are some problems. One of them is that collected data may be too numerous. So, it is difficult to define, in all data, the ones which will be relevant to be analyzed (Chatti et al. 2012). As Wise and Shaffer (2015) explain, in a sample of two million learners, it can be frequent to find some results relevant only for sub-groups and to overgeneralize them. In another way, when data are chosen semi-randomly, the results of LA can be difficult to interpret (Schneider and Pea 2015; Berland et al. 2014). So there is a need for a priori knowledge (APK), for theory, to overcome these difficulties (Lockyer et al. 2013; Wise et al. 2013; Wise and Shaffer 2015).

Another problem is that this APK is rarely explicit (Kelly et al. 2015). Tools are made by teams who have certain pedagogical beliefs (Shum and Ferguson 2012). For example, those beliefs will be used to choose one set of data rather than another, or some algorithm of datamining rather than another. If these assumptions are implicit, some invisible biases can distort the findings (Greller and Drachsler 2012). To overcome this problem, it is necessary that any tool development relies on theory-led design that will make the underlying assumption explicit (Kelly et al. 2015; Lockyer et al. 2013).

The role of APK is well studied in Machine Learning where it is used to design better algorithms, e.g. (Li and Nashashibi 2011). In LA, these kinds of studies are just beginning. Here, we investigate the relationship between LA and APK. For this study, in Sect. 7.2, we explicit what we call APK. In Sect. 7.3, we propose a "2-level framework" which considers LA as a level 2 learning process and which includes five components: stakeholders, goals, data, technical approaches, feedbacks. In Sect. 7.4, based on this framework, a sample of LA related works is analyzed to exhibit how such works relate LA with APK. Finally, in Sect. 7.5, we present the reciprocal benefits of this relationship for LA and APK and also some problems that can appear.

7.2 A Priori Knowledge

APK is the knowledge that a learner already has before beginning a new learning. This knowledge comes from previous learnings. In the following, we present the APK of student, teacher or LA and the main theories used as APK by LA.

7.2.1 APK of Student, Teacher, Learning Analytics

A student is not a tabula rasa. When he learns, he already has an APK upon which he builds his new knowledge. This APK can be constituted of already acquired knowledge in relation with the domain he studies or old experiments in this domain.

When a teacher teaches, he learns too. He learns on his student's learning. A basic example of this kind of learning is the assessment. Through assessments, a teacher learns on which part of their learnings students have succeeded or failed. Without this learning, a teacher cannot help his students. The teacher's learning is thus a learning on learning. For this learning, a teacher has also an APK. This APK is constituted of knowledge about the studied domain in which he is an expert. This APK is also, and especially, constituted of knowledge about the way a human being learns. This APK is then essentially constituted by pedagogy and learning theories.

As for teachers, LA constitutes a learning on learning. From data, LA will produce new knowledge about the learning processes of the students. The "person" who benefits from this learning on learning is the one who will get the outcome of LA: student, teacher, institution or, like in ITS, computer.

When LA is data driven, we could assume that there is no APK. However, as we have seen in the introduction, even in this case there is an APK but this APK is implicit: LA is not "neutral" (Shum and Ferguson 2012).

Most of the time, as we will see, the goal of LA is to help students in their learning or teachers to improve their students' learning. It may be the reason why the APK used, on which relies the LA process, is often constituted of learning theories. However, according to the goal of LA, the APK may be constituted by other theories than learning theories. We give now a brief overview of those different theories: learning theories and others.

7.2.2   Learning Theories

Learning theories aim to understand and to improve learning. They provide a conceptual framework which allows to interpret what is observed. They try to provide also solutions to problems which can occur during the learning. In our case, they are used, explicitly or implicitly, to design the learning system. However, learning theories have emerged in a time when TEL did not yet exist. Among these theories some have a wider spectrum than others (Siemens 2005):

- Behaviorism (Skinner 2011). The brain is considered as a black box. The teacher has access only to the observables of the student's behavior. Learning is designed according to "task based" learning where information is broken in small units.
- Cognitivism (Bruner and Austin 1986; Bloom et al. 1956). Knowledge is organized in a hierarchy and the learner will move from one step to the following one in this hierarchy.
- Constructivism (Piaget 1950). The learner builds his new knowledge on previous knowledge and through his interactions with his environment.
- Social constructivism (Vygotsky 1986). Same principle as constructivism but with the difference being the group helps the learner in building his knowledge.

Most of the others learning theories used by LA as APK, belong to one of those mainstreams. For example, according to Bourdeau and Grandbastien (2010), in the field of ITS, main used learning theories are: Bloom's Mastery Learning, Anderson's Cognitive Theory, Vygotsky's Zone of Proximal Development and Gagne's instructional design theory. Bloom's Mastery Learning (Bloom et al. 1956) consists of verifying that the student has achieved a level of mastery of a first learning before beginning a second learning. Anderson's Cognitive Theory (Anderson 1992) breaks down learning in small and irreducible tasks to be done. Those two theories belong to behaviorism or cognitive mainstreams. The next one, Vygotsky's Zone of Proximal Development (ZPD), belongs to social constructivism. A learner's ZPD is the distance between what the learner can do by himself and what he can do with help. The last one, Gagne's instructional design theory (Gagne 2013) belonged at its origin to behaviorism and cognitivism mainstreams. In this theory, teaching is broken down into five moments: analysis, design, development, implementation, and evaluation.

Nowadays, there are two more learning theories that sustain more and more TEL systems: constructionism and connectivism. In constructionism (Papert 1980), learning is based on project. The student learns by building artefacts. Constructionism belongs to the constructivist mainstream. In connectivism (Siemens 2005), the student learns in the network and through the network. Connectivism incorporates most of the social constructivism ideas. All those learning theories constitute a point of view on how learning appears. As we will see, some of them are more difficult to embed in a LA device.

7.2.3 Other Paradigms of a Priori Knowledge

APK for LA can be constituted by theories other than learning theories. For example, one of those theories, coming from the field of sociology, is SNA (Scott 1988). SNA studies social structures and sees them in terms of nodes and links. It is often used in LA when learning is seen in a social constructivist perspective. In this perspective, the group is used to improve the learning of each of its member. So it is important to understand who is interacting with whom. SNA helps to exhibit these interactions. For example, a network centered on the teacher will not correctly work. It is better to have a lot of interactions between peers. SNA is easy to implement in a LA device.

Another one, Activity Theory (AT) (Engeström 2014), is sometimes considered as a learning theory but its spectrum is larger than learning. In the activity, the subject pursues a goal that results in an outcome. To do this, he uses tools and acts within a community. His relation to this community is defined by rules. To achieve the goal, it may be necessary to establish a division of labor within the community. If we consider the origins of this theory, AT belongs to the social constructivism paradigm. However, AT has a very strong descriptive power. This descriptive power allows it to distinguish and to compare different learning situations obeying to different learning paradigms (behaviorism, constructivism and others). AT is easy to implement too (Simon 2014).

In the following, we will see some other theories coming from different fields used as APK. However, before closing this section, we have to precise that, sometimes, APK is not constituted by theory but by previous learnings. Like for human beings, APK can be built on previous experiments. In this approach, the LA process begins without APK like a data driven process. Then, the results of this first process constitute the APK for the next process.

7.3 The 2-Level Framework

To study the relationship between APK and LA, we propose a 2-level framework. This framework considers LA as a level 2 learning process and uses five characteristics to study it.

7.3.1 Learning Analytics Seen as a Level 2 Learning Process

In Fig. 7.1, we have presented the student's learning. Here, we call it "level 1 learning process" because, on this learning process, another learning process can appear which is, thus, a learning on learning. It can be operated by the student

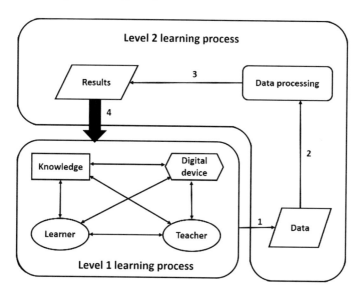

Fig. 7.2 LA is a level 2 learning process

himself, the teacher or the institution. For instance, when a teacher evaluates a
student, he gets information about his student's learning: the teacher learns on his
student's learning. This kind of learning, learning on learning, we call it "level 2
learning". Nowadays, this type of learning can also be operated, partially or totally,
by a computer.

LA is a process which follows an iterative cycle (Chatti et al. 2012). In Fig. 7.2,
we can see this process. (1) Data are collected from the student's learning process
(2) those data are processed, (3) this processing gives results that (4) can be
interpreted and reinjected in the learning process. Most of the research (Chatti et al.
2012; Dyckhoff et al. 2012; Khalil and Ebner 2015; Clow 2012) considers LA
according to this cycle.

The point which makes the difference between our work and the aforementioned
research, is that LA is considered here as a level 2 learning process. As we said, it is
a learning process, in the sense, that this process will produce new knowledge based
on data. It is a level 2 learning process because it is a learning on learning: studied
data are produced by student's learning process. It is a necessity to distinguish those
two levels to study the relationship between LA and APK. The reason is that APK
of LA (level 2 learning) should not be necessarily the learning theory used to shape
the student's learning (level 1 learning). We can imagine, for example, that the
process of level 1 is shaped according to behaviorism and the process of level 2
according to AT. As far as we know, no paper on the subject takes this point into
account.

7.3.2   Five Characteristics of Learning Analytics

Several authors have proposed a framework of characteristics to study LA (Greller and Drachsler 2012; Chatti et al. 2012; Cooper 2012). All those frameworks are built around four questions (Chatti et al. 2012): Who? Why? What? and How? Who are the stakeholders, especially the target: the student? Why: the goal of the analysis? What data to analyze? How the analysis is performed? As those authors, our framework will take into account first four characteristics corresponding to these four questions: stakeholders, goals, data and technical approaches. Then, it will be added a fifth one that summarizes the whole process as a two levels process: feedbacks.

Stakeholders. We begin with the stakeholders because, ultimately, they define the goal and the shape of the entire learning system. To simplify the analysis, we consider four different types of stakeholders: the student (learner), the teacher who helps the learner (educator, instructor), the institution that offers the conditions to learn (administration, university), and the researcher who studies one of the processes: the student's learning or LA. LA is built to answer their questions and allow them to prepare the next actions (Kelly et al. 2015). We can distinguish two kinds of stakeholders: the actors and the observers. The actors are students, teachers and institutions. The observers could be eventually the same but also researchers, software developers, analytics experts (Cooper 2012).

LA is not the same according to whom gets the feedback. Work is done on different scales and different granularities (Ferguson 2012). A teacher is more interested in understanding the learning of his own students, the university, in contrast, looks at all of their students. Moreover, due to the rapid evolution of the technical society and lifelong learning, learning and teaching move from schools and universities to workplaces (Dawson et al. 2015). One example of this phenomenon is the apparition of the MOOCs. Thus, stakeholders are changing too.

Goals. The general goal of LA is to improve student's learning by giving him a better learning environment or by giving his teacher tools to better evaluate and understand his learning (Ferguson 2012). This general goal can be broken down in smaller objectives according to the different stakeholders. According to Chatti et al. (2012), Steiner et al. (2014), for institutions, different objectives can be: Monitoring and analysis to take decisions about the whole process, improving the learning environment, the design of the learning device; Prediction of the future student's possible failure to avoid potential dropouts.

For teachers, objectives can be: Monitoring and analysis to understand the learning process in order to decide the future steps, the future learning activities; Prediction of the future student's failure to be able to offer assistance when needed; Tutoring and mentoring along the whole learning process; Assessment of the learning process; Personalization and recommendation: suggesting to the learner what to do next while leaving the control to him. For students, objectives can be: Formative and summative (self) assessment of the learning process; Reflection: improving student's self-reflection on his learning process, providing comparison with past experiences, with other learners. For researchers, objectives can be: Allowing the evaluation of a learning theory (Greller and Drachsler 2012). Improving learning digital devices. As we can see, some of these objectives can overlap. According to the goal which, itself, depends on the stakeholders, LA and its APK will be different.

Data. Data studied by LA, belong to the educational field (Romero and Ventura 2010). Dillenbourg (2016) call these data "learners' behavioral particles". By themselves they do not support a lot of semantics but when there are thousands of millions of them, it is possible to draw rules and predictions about the students' behavior.

These data have been produced by the different actors during their interaction with the learning digital device. Dyckhoff et al. (2012) distinguish data according to what they represent: data about users, data about content, data about assessment, data about activity, data about events. Sometimes data come from "special" devices: cameras, sensors, like skin conductivity, heartbeat, EEG, gesture sensing, infrared imaging, and eye tracking (Blikstein 2013). One challenge for LA is to collect relevant data. This is one of the main reasons invoked by researchers for using APK.

Technical Approaches. Here we speak about the technical approaches of LA (level 2 process) and not the ones concerning the learning digital devices (level 1 process). LA can use different technologies: statistical analysis techniques, Machine Learning, EDM, SNA, Natural Language Processing, BI and others (Dyckhoff et al. 2012; Greller and Drachsler 2012; Shum and Ferguson 2012).

All of these techniques follow the cycle: data collection, data preprocessing, data processing, results. For Dyckhoff et al. (2012), LA techniques should obey to the following requirements. *Usability*: Users (teachers, students) must be able to use the system without difficulty; *Usefulness*: LA is supposed to allow the improvement of learning; *Interoperability*: LA can be used on data from different learning devices; *Extensibility*: LA can be used on different scales; *Reusability*: LA should use simple blocks that can be used in more complex functions; *Real-time operation*: to allow user getting information just-in-time; *Data Privacy*: to protect the identities of the users. For the moment each one of those requirements is a challenge and at the end, interpreting the results is also a challenge (Wise 2014; Wise and Shaffer 2015). As we will see, extensibility and reusability of LA have an impact on the choice of the APK.

Feedback. The object of the feedback is to adjust the situation for a better learning experience. The feedback can be in the direction of the institution to decide the future of the course. It can be towards the teacher to improve the efficiency of his course (Dyckhoff et al. 2012). It can be towards the student, to allow him to eventually change his learning strategy by self-reflection. It can be also towards the learning digital device, like in ITS.

If the feedback is for the teacher or student, it will be given during the process by visualization tools, and not at its end. By this way, it allows formative assessment. When the feedback is towards the learning digital device, the object will be to optimize the intelligent tutor (Aristizabal 2016).

As feedback depends on stakeholders and goals of LA and as it is the relation between the two levels (LA and student's learning), feedback summarizes the whole process. It is a key to understanding it.

7.3.3 Comparison with Other Frameworks

There are some similarities and some differences between the 2-level framework and frameworks presented in other researches (Cooper 2012; Khalil and Ebner 2015; Greller and Drachsler 2012; Chatti et al. 2012). Among similarities, all frameworks consider the whole process like a cycle. All of them, also, take into account the four questions: Who? Why? What? How? Consequently, all of them integrate the four characteristics: stakeholders, goals, data and techniques.

Most of the differences will come from the objectives pursued in these papers. All of these frameworks have the goal to help design LA devices. Opposed to our work, none of them has for an objective to study precisely the relationship between LA and APK. It could be the reason why, in two of these papers, (Khalil and Ebner 2015; Chatti et al. 2012), there is no mention of any type of APK. In (Cooper 2012) and in (Greller and Drachsler 2012), APK is briefly mentioned as a characteristic to take into account. Cooper speaks of "embedded theory" and Greller and Drachsler of "pedagogies".

Thus, actually, the main difference between the 2-level framework and those frameworks is the way to see the whole process. If all papers consider the whole process like a cycle, no one considers that this cycle is going from one level to an upper level and reciprocally. In the same way, no one considers LA as a level 2 learning process. We have explained above the necessity to adopt this point of view for our goal: We have to be able to distinguish the learning theory sustaining the student's learning from the APK of LA. As other frameworks do not take into account the 2 levels, they do not focus on the feedback. For them, feedback is just a part of the technique.

7.4 Relationship Between APK and LA According to the 2-Level Framework

Using the 2-level framework, here, we analyze the papers taking into account LA and APK. First, we explain how we have selected them. Then, we observe the relationship between APK and LA according to each of the five characteristic we have presented. These five characteristics are intertwined in a LA system, thus, some points exhibited in one section could also appear in another. It will be especially true for the "stakeholders" and the "goals" characteristics, because the goals depend on the stakeholders.

7.4.1 Papers Identification

To identify the papers, firstly, we look at all papers in the proceedings of the two mains conferences concerned by LA: EDM conferences and LAK conferences. Most of those papers do not speak about APK. When we look at the main keywords (used more than 10 times in the proceedings keywords) we obtain for EDM conferences: "Intelligent Tutoring Systems" (32 times), "Educational data mining" (30), "Knowledge tracing" (20), "Student modeling" (18), "Natural language processing" (12), "Datamining" (12), "Machine learning" (11), "Learning" (11), "MOOC" (10), "Learning analytics" (10).

For LAK conferences we obtain "Learning analytics" (146), "Datamining" (17), "MOOCs" (17), "Social Network Analysis" (16), "Educational Data Mining" (13), "Visualization" (16), "Higher Education" (14). Among those keywords there is no one about learning theory or APK. If we look at keywords about learning theories ["Behaviorism" (or "behaviorist"), "Cognitivism(t)", "Constructivism(t)", "Social constructivism(t)", "Constructionism(t)", "Connectivism(t)", "Learning theory (ies)", "Learning science(s)]" we obtain 6 occurrences in all EDM conferences papers and 9 occurrences in LAK conferences. So references to a possible APK or learning theory appeared rarely in keywords.

We wanted to go deeper and looked at references to learning theories in the body of the papers in LAK proceedings. We obtain: "Behaviorism(t)" (5 occurrences), "Cognitivism(t)" (4), "Constructivism(t)" (83), "Social(socio-)constructivism(t)" (14), "Constructionism(t)" (18), "Connectivism(t)" (163). We observe that if the learning theories are not in the focus of the papers, they stay, for a part of them, in their background but very rarely. However, this should be moderated by the fact that we only looked at the mainstreams of learning theories and we used their precise names. For example, "social constructivism (t)" cannot appear while "Vygotsky" can appear more than once.

Secondly, we review all the papers of the volume 2, No 2 (2015) of the Journal of Learning Analytics "Special section: Learning analytics and learning theory".[1] Finally we search in databases or search engines and chain from known research papers (Liyanagunawardena et al. 2013).

7.4.2 A Priori Knowledge and Stakeholders

Outcomes of LA are given back to stakeholders for interpretation and action. When stakeholders are educational institutions, like Universities, we are more in the field of academic analytics. In that field, Shum and Ferguson (2012) consider there is not a lot of consideration about learning theory and pedagogy. For Fournier et al.

---

[1]https://epress.lib.uts.edu.au/journals/index.php/JLA/issue/view/358.

(2014), most of those institutions support LMS which are in the field of classical learning paradigms and, as such, easy to integrate in institutional learning.

Stakeholders can also be teachers or students. In constructivism, LA is often used as a formative assessment: the teacher will interpret the outcome to choose the next activity for the learner. To get a "just-in-time" adaptation, feedback has to be given during the learning process and must be easily understandable. In constructivism, LA can also help the student in his self-reflection. To succeed, the student has to be able to understand the purpose of the learning activity and what are the most productive actions to do (Wise 2014). Here also, feedback has to be just-in-time and easy to understand to avoid cognitive overload. As the teacher and student have not the same reference frame, feedbacks have to be different (Wise 2014).

In social constructivism, the focus will move from the student to the group of students and teacher. LA has to give information to the teacher, student and group of students, on the way the group is working. SNA (Bakharia and Dawson 2011) and automated discourse analysis (Kelly et al. 2015) are often used to give this information. Here also, it is important to get the information just-in-time to adjust the collaborative process between learners (van Leeuwen 2015). For Wise (2014), teacher and students should have access to analytics. The interpretation of the analytics can thus be a space of negotiation between them. Nowadays, researchers try to propose learning environments more open and networked like PLEs or MOOCs. By doing that, the trend is to move from social constructivism towards connectivism.

In summary, from what precedes, we see that, when learning theories are used as APK for LA, then, LA is often used as formative assessment for the level 1 learning. It can be sometimes otherwise according to the goal.

7.4.3 A Priori Knowledge and Goals

Most of the authors in the field agree to say that the main goal of LA is to improve learning (van Leeuwen 2015; Shum and Ferguson 2012; Chatti et al. 2012; Cooper 2012), (Greller and Drachsler 2012). However, LA can have also other goals.

Improving Learning. Tempelaar et al. (2013) consider that LA can help the teacher to predict learner success or failure, to suggest learning resources, to increase reflection, to improve social learning environments, to detect undesirable learner behaviors, and to detect the emotions of the learners. LA can help the teacher also to get more confidence in what he is proposing to the learner (van Leeuwen 2015). It can help the group to better understand how it works (Perera et al. 2009). It can help the learner in his self-reflection (Shum and Ferguson 2012).

With recent learning theories, the assessment is more difficult (Anderson and Dron 2012). In constructionism, one difficulty is to measure the learning achievements (Berland et al. 2014). This pedagogy is centered on the "21st century skills": creativity, innovation, critical thinking, problem solving, communication, and collaboration. The acquisition of those skills by the student cannot be assessed by traditional current assessment techniques. EDM and others LA techniques are essential tools which allow the assessment to occur (Blikstein 2011; Berland et al. 2014). For Swenson (2014), more radical, the goal of LA is to allow the disruption and the modification of education. In this way, we see the emergence of connectivism (Siemens 2005) where LA constitutes a key tool. As with constructionism, the assessment of the skills cannot be done by traditional tools. EDM and LA are essential to evaluate the connectedness of the learner's knowledge (Knight et al. 2014).

Greller and Drachsler (2012) say that the type of pedagogy used should appear explicitly in the goals of LA, but it is rarely the case. All goals are defined to improve learning but the reference to a learning theory is rarely explicit. However, it is a necessity because, as Gašević et al. (2016) note, all LAs do not work on all learning systems. They show, for example in the case of predictive analytics that some predictive models successful with one type of course, designed according to behaviorism, fail with another, designed according to constructivism.

Other Goals. For Brooks et al. (2012) too, the goal of LA is to understand learning but they indicate that the way its outcomes are used is up to the administrators, instructors, and instructional designers. The same information can have different effects according to the stakeholder who receives it. For example, the goal of LA might be to evaluate the dropout rate in a course. An important dropout rate might result in the modification of the course by the teacher or the closure of the course by the administrator. This last case recalls that learning theories are not the only ones that sustain LA. We must not forget that BI is one of the founding theories of LA (Dyckhoff et al. 2012). BI is designed to help a company to make better business decisions (Chatti et al. 2012). At the time where institutions must educate *more* students at *lower* costs (Mehaffy 2012), the goals might be not only pedagogical. The ethical aspects of these goals have to be considered (Atkisson and Wiley 2011).

Other possible goals for LA are to sustain researchers in their studies. LA can help to confirm and improve well established learning theories (Miyamoto et al. 2015). The strength of LA to improve learning theories is based on two essential points. LA works with millions of data which facilitates the generalization of the results (Dillenbourg 2016). These data come from real life and can avoid some biases which possibly appear in experiences in laboratories. This is the reason why Fournier et al. (2011) suggest that analytics becomes a part of social science and consider there is a need for the apparition of a more data-driven computational social science.

When outcomes of LA are oriented towards administration or researchers, it could happen that there is no feedback towards teachers or students. If there is one, it will follow a circuitous path, for example by changing the whole design of the course.

In brief, the relationship between the goal of LA and APK is rarely explicit. Most of the time, this goal can be summarized in "improving learning" which agrees with all learning theories. In this case, LA is used as a tool for assessments, and for some new learning theories (constructionism or connectivism), it is an unavoidable tool. The APK of LA is often defined by the learning theory sustaining the learning device (level 1 learning process).

However, the goal of LA can also be defined independently of this level 1 process. For example, this can appear when the stakeholder who decides to launch LA is out of this level 1 process: an administrator who wants to know if a course is profitable, or a researcher who wants to confirm his theory. In this case, the APK can be another theory than a learning theory. These different goals of LA will generate different relationships between data (collect and preprocessing) and APK. In some cases, it could go as far as using "special" devices to get these data.

7.4.4 A Priori Knowledge and Data

During LA process, the focus will be put on data during two main steps: data collection and data preprocessing (Romero and Ventura 2010). For these authors data preprocessing will consist in data selection; data cleaning to detect and correct errors; data integration when data come from different sources; data transformation which can consist of calculating new attributes from the existing ones. These are important steps because the results of LA process can be biased if they are not properly done (Fournier et al. 2014).

Log Data. When data are log data, one difficulty for LA is to select the "good" data because there could be hundreds of thousands of them (Wise and Shaffer 2015). With numerous data, it is always possible to find some rules which are true on a subset of all the dataset. Moreover, with a totally data-driven LA, it can happen that some data can be ignored while there are theoretically important (Sao Pedro et al. 2012). For Berland et al. (2014), selecting relevant data is one of the greatest challenges of LA. To win this challenge, it is necessary to have recourse to theory which will allow to define the context and what are the most relevant data in this context (Wise et al. 2013).

In the papers, theories used for selecting data are, most of the time, implicit and often rely on experts of the domain, e.g. (Sao Pedro et al. 2012). For example, in the "evidence model" (Rupp et al. 2012), there are continual exchanges between subject-matter experts and statisticians to get features of the student's work that provide "evidence" about student proficiencies. For Li et al. (2015) finding deep functional features is both time-consuming and error-prone. For Kennedy et al. (2013) in some domains, like in surgery, it is difficult even for experts to explain and find "right" features. So they propose datamining approaches of behavior and interactions of both expert practitioners and novice students to extract them.

When the theory is explicit another problem is data semantic. A clear data semantic is important for the LA process (Demchenko et al. 2014). With some learning theories which have been developed when computers did not exist yet, features used for the LA process are necessarily "proxy indicators". For example, "time online" will be considered as "time on task", "opening a web page" as "document reading" (Hewitt 2015).

In connectivism or constructionism, data come from different sources. In PLE for example, data will be generated by different tools allowing filtering and sorting information, creating, aggregating and publishing new information (Anderson and Dron 2012). Kop and Carroll (2011) give some examples of tools that can be used in a connectivist approach: blog, del.icio.us, Moodle discussion, Flickr, Second Life, Google Groups, Facebook, YouTube, NetVibes. Collecting data from all these tools is a real challenge because of their highly divergent and distributed nature (Anderson and Dron 2012).

Data from "Special" Digital Devices. Most of the time, data are produced by different stakeholders using a digital learning device and they consist in logs of computer activities. But sometimes, often when LA is done for research, researchers will add some other data coming from different "special" devices.

In constructionism, Worsley and Blikstein (2013), Blikstein (2013), Berland et al. (2014) propose to adopt a multimodal LA. Multimodal LA allows supporting more process-oriented assessment (Worsley and Blikstein 2013). The idea is to get a more "holistic picture of the learner" (Dyckhoff et al. 2012). In those analytics, data come from cameras, wearable sensors, biosensors (e.g., skin conductivity, heartbeat, and EEG), gesture sensing, infrared imaging, and eye tracking (Blikstein 2013).

For example, in Computer Supported Collaborative Learning, learning theories suggest to put the focus on joint visual attention (Schneider and Pea 2015). Researchers use high-frequency sensors (such as eye-trackers) which allow students to see the gaze of their partner displayed on the screen in real-time. Dyads of students getting access to this device perform better than others. Research about affective state of the student, has recourse to biosensors (Pinkwart 2016). One of the goals of this research is to underpin affective tutors in ITS. The idea is to improve the student model so that the tutor can diagnose frustration or satisfaction and can also uncover the reason for that state (Rajendran et al. 2013).

In embodied learning environments (Lindgren and Johnson-Glenberg 2013), data are collected by Mixed-Reality devices. In these immersive learning environments, learning is operated by the student for a part through interactions between his body and the mixed-reality. An important part of data comes from gesture sensing device.

In brief, APK helps to select more relevant data. This APK can be implicit when, for example, this is the one of experts who indicate the "good" features. When APK is explicit, it is often the learning theory sustaining the level 1 learning process. If this learning theory is connectivism, collecting data is a real challenge because of their distributed nature. In LA more oriented for research, data can come from "special" devices like eye-tracking and others.

7.4.5 A Priori Knowledge and Technical Approaches

APK is implicitly or explicitly embedded in the technical approaches of LA. Most of the time, the APK will be the learning theory used for the learning design (Lockyer et al. 2013) of the level 1 learning process but it is not always the case.

Focus of the Technical Approach According to the A Priori Knowledge.
According to the learning theory used for the level 1 learning process, the technical
approach will not have the same focus. In a cognitive-behaviorist approach, the
focus is put more on the cognitive skills of the students. For example, Bayesian
Knowledge Tracing is used in cognitive tutors to determine if and when the learning
of a skill occurs (Desmarais and Baker 2012). Other techniques study personality
and performance of the students. In Magnisalis et al. (2011), stereotype theory is
used to categorize students and then build an individual learner model which
embeds those skills. Some emotional states can also be integrated in this learner
model (Bourdeau and Grandbastien 2010).

In a constructivist perspective, Lockyer et al. (2013) consider that there are two
key issues in the learning design: checkpoint and process analytics. Checkpoint
analytics are data indicating that the learner has met the prerequisite for learning.
Process analytics provide information on the way the learner processes information
and knowledge. Shum and Ferguson (2012) suggest to use the following analytics:
Content Analytics which analyses data produced by users; Disposition Analytics
which analyses motivation to learn; Context Analytics which analyses context of
learning (e.g. mobile learning).

In social constructivism, there is a shift of the focus of LA from the individual
towards the group because the collaborative activities in the group promote students
understanding (Bakharia and Dawson 2011). This shift has appeared as well as in
ITS learning devices than in other less automated learning devices. The idea in ITS,
is to provide tutor with a social intelligence (Bourdeau and Grandbastien 2010) to
create a Vygotskyan intelligent tutor (Desmarais and Baker 2012) which is able to
identify skills necessary to collaborate successfully. Shum and Ferguson (2012)
suggest to use SNA which analyzes interpersonal relationships, and Discourse
Analytics which analyzes data produced by language interactions. For (Bakharia
and Dawson 2011), SNA provides tools for instructors or students to interpret the
activity in the group. The goal is to discover relevant structures in social networks
(Anaya et al. 2015). Bakharia and Dawson (2011) note that there is a correlation
between an individual's connectivity and his academic success. Discourse analytics
is often used with SNA (De Liddo et al. 2011; Joksimović et al. 2015; Swenson
2014). In social constructivism, dialogue between learners is a key point for
learning. It is important to understand how each learner engages with other learner's ideas to build his own knowledge (De Liddo et al. 2011).

It is interesting to note, here, that SNA and Discourse analytics are techniques
but also theories. Moreover, these theories are not learning theories and they are not
used to shape the level 1 learning process but to shape the LA process. As we have
said SNA comes from the field of sociology (Scott 1988). Discourse analytics
comes from areas such as exploratory dialogue, latent semantic analysis and
computer-supported argumentation (Ferguson 2012).

More Complex Technical Approaches According to A Priori Knowledge.
Cognitive or behaviorist learning digital devices, as in some ITS, often present very
structured learning environments (Berland et al. 2014). In those kind of environments, it is easy to infer structure from data. As we have seen in the previous

section, in learning devices shaped by other learning theories, like constructionism or connectivism, data may come from different sources. The problem to extend EDM or LA techniques to those more opened environments is not trivial (Berland et al. 2014). To face this problem multimodal LA can be one solution and offers new insights to understand students' learning (Blikstein 2013).

For (Blikstein 2013): Multimodal LA is "*a set of techniques that can be used to collect multiple sources of data in high frequency (video, logs, audio, gestures, biosensors), synchronize and code the data, and examine learning in realistic, ecologically valid, social, mixed-media learning environments.*" Each source corresponds to a point of view and multimodal approach coalesces these different points of view (Blikstein 2013). By doing this, multimodal LA allows a finer grain analysis of the student's behavior (Worsley and Blikstein 2013). Some other researchers claim for a mixed approach, e.g. (Fournier et al. 2011): quantitative and qualitative. If the quantitative approach will be mainly done by machine, they suggest that the qualitative approach would be done by human experts who will give more meaning to the experiences.

In brief, we see that there is a dialectic process between APK and LA techniques which allows more and more complex learning systems. From the well-defined learning environments shaped according to cognitivist, behaviorist theories, we go towards ill-defined environments shaped according to socio-constructivism, constructionism or connectivism. During this move, the focus of LA is shifting from the individual to the group, and takes into account more diversified data. Two interesting remarks to point out are, first, some APKs used by LA, like SNA, do not come from the field of learning theories and, second, the general trend is to diversify the techniques to diversify the points of view and therefore acquire more information on the learning process.

7.4.6 A Priori Knowledge and Feedback

We devote a special section to APK and feedback because it is through the feedback that the full circle is closed (Aristizabal 2016). The feedback leads to interpretation, interpretation leads to actions, actions lead to modifications of student's behavior, modifications give new data and new LA process (Greller and Drachsler 2012). Thanks to feedback, level 1 learning process and level 2 learning process constitute one and the same process. As we have said, the feedback summarizes the whole process. One key point about feedback is its level of automation.

Level of Automation. Greller and Drachsler (2012) distinguish only two levels of automation: outcome is used either by a human or by a system to reflect and to act. Brooks et al. (2012), for their part, distinguish three ways to give this feedback: automated as in ITS, semi-automated as in "nudge" analytics that prompt individuals to take action (Ferguson 2012), and non-automated through teacher or peer help interaction. For them, there is no better choice between those three levels of automation. All the different techniques can improve learning. For Lockyer et al.

(2013), the question is more to bridge the technical–educational divide to provide the right feedback on how well the learning has attained its goal.

The level of automation is related to the actions that will follow. In case of an automated system, computer agents determine specific interventions. These can be simple as sending a notification or recommendation to one of the stakeholders. They can be more complex and the outcome of the LA is interpreted according to some system of rules leading to automatic changes in the learning system (Wise et al. 2013). This appears essentially in ITS, e.g. (Roll et al. 2010; Olney et al. 2012). In (Olney et al. 2012), the ITS named GURU presents different strategies: scaffolding, information-elicitation, direct instruction. When the process is totally automated, the APK is often implicit. For example, learning theories embedded in GURU come from 50-h of expert human tutor observations (Olney et al. 2012). When the APK is explicit, it refers most of the time to behaviorism or cognitivism.

The level can be also semi-automated as in "nudge" analytics (Ferguson 2012). In Course Signals at Purdue, for example, (Arnold and Pistilli 2012), the student is warned when his data indicate that he is in danger to fail. By this way, he knows that he has to take action. To help him, "actionable intelligence" is set up which can guide him to appropriate help resources and explain how to use them (Arnold and Pistilli 2012). We are more in constructivism.

Non-Automated Process. When the process is non-automated, the questions are: Who get the feedback, teachers, and students, both? Do they get same or different analytics? (Wise 2014). Most of the times, students or teacher will be informed of the outcome through a visual interface device. Those visual interfaces will depend of the learning theory explicitly or implicitly embedded in the learning device (Wise 2014). Lockyer et al. (2013) give some examples of these visualization tools with pedagogical background: Reports (Individual and cohort monitoring), SNA (Social-constructivist models of learning), Student dashboards and monitoring (Self-regulated learning), Individual and group monitoring (Individual and cohort monitoring), Learning content interaction (Individual and cohort monitoring), Discourse analysis (Social learning and argumentation theory). All of these pedagogical backgrounds belong to constructivism or social-constructivism.

The difficulty in designing such visualization tools is between giving an overview or a work overload to user (van Leeuwen 2015). Both, teacher and student, have to understand rapidly the information and not to be overwhelmed by it (Wise 2014). This is the reason why, most of the time, the feedback will take the shapes of diagrams like social network diagrams. These diagrams are designed in such way that user can take the information at a glance.

Once the user has the information, he has to be able to interpret it. This interpretation depends on his knowledge of the context: user has to correctly understand relationship among technology functionality, observed interactions, and educational theory (Lockyer et al. 2013). It is important for interpretation that the user has a model of learning for the particular environment to define what will be a productive activity in this context (Wise 2014).

One crucial factor is time: when should analytics be consulted during the teaching and learning process (Wise 2014)? Van Leeuwen (2015) distinguishes two levels to answer. On a macro level, LA can detect patterns from completed courses. For example, it is used to detect students who might fail. In this case, we are more in an evaluative assessment that can serve administrators to decide the future of a course, help researchers in seeing the merits of a learning theory and aid teachers with the modification of the course. On a micro level, LA can be used for real-time assessment to support teachers and allow them to give help just-in-time to their students. In this case we are more in a formative assessment.

In brief, the feedback is a key moment of the process. We can consider that its main characteristics are its level of automation, its main user, its periodicity in the learning process. According to the level of automation, the APK is not the same. No paper takes into account the type of feedback when it is for the institution or for researcher. It is the reason why, most of the time, the feedback is designed according to the learning theory supporting the level 1 learning process. If the feedback is for teacher or student, it has to be given to them during the process to provide just-in-time, useful, and context-sensitive information (Lockyer et al. 2013) and not after. One more time, LA constitutes a formative assessment.

One remark before closing this section: Greller and Drachsler (2012) consider LA as a thermometer that helps to regulate learning. But, when the full process is totally automated like in ITS, LA is not just a thermometer, it is a part of a thermostat which regulates automatically the student's learning through the feedback.

7.5   An Insight of APK, Learning Theories and LA Ties

The foregoing leads to the following observations. In most of the papers, the APK on which relies LA is rarely explicit. When the APK is explicit, most of the time, it is a learning theory. In this case, most of the time, this learning theory is the one used to shape the student's learning (level 1 learning process). In this case, most of the time, LA will be used as an assessment for the student's learning. In this case, most of the time, the assessment will be formative. All those points need to be questioned here. To do this, we go back from the last point to the first one. We begin by looking at LA used as assessment, then at the relationship between learning theories and LA and finally at the relationship between APK and (learning) theories.

7.5.1   Learning Analytics and Assessment

Most of the time, LA is used for assessment. Assessment depends from the underlying learning theory. Knight et al. (2014) question "classical" assessment; for them "'Success' can no longer be defined as a matter of regurgitating, unaided, the correct information". In one way, the shift in learning theories from cognitive/behaviorist theories towards connectivism through social constructivism corresponds to a shift from simple environments to more complex ones (Pinkwart 2016). The corresponding assessments has to shift from assessments for well-defined domains, where there is a strong domain theory to structure it, towards ill-defined domains where assessments are much more difficult (Aleven et al. 2008).

Well-defined environments elicit students' response and data which can be easily evaluated. This cannot be done in ill-defined environments (Gobert et al. 2013). For (Mislevy et al. 2012), in constructivism, assessment is really difficult due to the fact that there is not only assessment of knowledge but also of skills. The difficulty is to know which behaviors or performances can be studied to assess. This is why they prone the use of "evidence-centered design" which put the focus on inferential elements of educational assessment. In the same way, Gobert et al. (2013) note that classical test theory and item response theory cannot really assess inquiry skill. The reason is that this skill is multidimensional and multi-stepped in nature.

Knight et al. (2014) distinguish LA designed for summative assessment and LA designed for formative assessment. They advocate the second one. Using assessment in a summative or in a formative way is an indication of the pedagogy used in the learning system. There cannot be a constructivist learning system where assessment is used only for summative evaluation. Formative assessment is consubstantial of constructivist theory and its followings: socio-constructivism, constructionism, connectivism. For most of the papers reviewed here, this is the case. They indicate that LA is used to improve learning by giving feedback to teacher to orient further activities of students, e.g. (Brooks et al. 2012; Bakharia and Dawson 2011), or directly to student for self-reflection, e.g. (Dyckhoff et al. 2012; Chatti et al. 2012).

Evidently, LA can also offer summative assessment even in constructivism learning systems. Thus, we are more in evaluation and the goal will be different. Academic institutions need evaluation to deliver certificates or researchers will use it to confirm learning theory. Moreover, with the development of other learning systems (like MOOCs) LA is more and more needed for summative assessment. With the apparition of LA and those news learning systems, Shum and Ferguson (2012) wonder if educational institutions will stay the only ones to do summative assessments and certify advance learning.

7.5.2 Learning Theories and Learning Analytics

Here we present the reciprocal contributions of learning theories and LA and show that there is a dialectical process between them. This dialectical process benefits both to learning theories and to LA.

Contribution of Learning Theories to Learning Analytics. We have seen that APK helps researchers to select relevant data, to choose different techniques according to the goal of LA, to define the type of feedback. Wise and Shaffer (2015) note that one difficulty in LA, is the researchers' degrees for freedom which can be too large and do not allow to take decisions. By providing a coherent framework APK can help to limit it. One of the limits will be to define the context in which learning is operated (Knight et al. 2014; Gašević et al. 2016).

Lockyer et al. (2013) consider that the learning design is a form of documentation that provides the informative context. This is a necessity because without a description of the context, it is difficult to improve learning (Dyckhoff et al. 2012). Learning is always situated in context (Kelly et al. 2015). Moreover, the description of the contextual environment allows for comparison (Greller and Drachsler 2012). Wise and Shaffer (2015) follow this trend when they say that theory would help the researcher to generalize results to other contexts. For Dawson et al. (2015), the question of context is to know if the LA used in one context will be suitable for another one. Moreover, if we refer to the definition of LA (in the Introduction) we see that context is part of it.

Contribution of Learning Analytics to Learning Theories. Among the papers, it appears that one of the main contributions of LA to learning theories, is to confirm (or refute) them on some points. For example, Worsley and Blikstein (2011) confirm Beck's research, according to which increasing student's expertise tends to increase student's self-confidence. For Fournier et al. (2011), human sciences could be reinforced by analytics because analytics can generate other methods to support qualitative research. Those methods can increase rigor and replicability, especially in constructionism (Berland et al. 2014). They can do that because their data are extracted from real learning situations and not from ad hoc situations built in laboratory (Desmarais and Baker 2012; Lee et al. 2015; Martinez-Maldonado et al. 2015).

They can also increase rigor and replicability because these real life situations generate numerous data which are statistically relevant (Demchenko et al. 2014). This appears particularly with MOOCs (Anderson and Dron 2012) where numerous data give a lot of information about how learners behave and interact with content and others students. These numerous data need new techniques and cannot be processed by classical standard database management systems. These new techniques, by automatically extracting interesting rules, allow to avoid painstakingly handmade classification and to speed research. For example, Schneider and Pea (2015) propose techniques which automatically annotate hours of videos and hundreds of transcript pages and allow to graph the evolution of particular behaviors.

Another trend to improve research is to generate data with special devices which bring more information on the learning process. As we have seen, different devices, like eye-tracker, sensors, allow to multiply points of view on a same phenomenon and by this way to acquire a deeper understanding of this phenomenon (Schneider and Pea 2015). Researchers, like Blikstein (2013), suggest to use multimodal analytics. Multimodal analytics allows a fine-grained data collection and analysis that help researchers to examine student's learning in an unprecedented scale.

Dialectic Between Learning Theories and Learning Analytics. As note Merceron et al. (2015), theories will not emerge from an ocean of data. To be interpreted correctly data need theory. On the opposite, theories must rely on data to be confirmed. There is a dialectic between learning theory and LA. For Mislevy et al. (2012), this dialectic improves understanding and practices. Wise and Shaffer (2015) ask for a dialogue in the field to understand how LA can draw on and contribute to theory.

Speaking about constructionism (Berland et al. 2014), note that collaboration between researchers in the EDM and constructionist traditions will allow constructionist researchers to make rich inferences about learning, while providing EDM researchers with many new research questions. The challenge is to bring together those two research communities which have different conceptions of what learning is and how it can be measured.

Some researchers like (Aristizabal 2016) suggest getting a theory of education from datamining. The idea is that discovery of theory is possible by acquisition of data. According to them, theory could be more a bottom-up process in contrast to "theory generated by logical deduction from a priori assumptions" (Aristizabal 2016). However, even in their framework they indicate that EDM researchers have to work with psychologists, teachers and professionals of neurosciences for collecting relevant data and improving mining techniques. Once again we see that there should be a dialectical process between LA and learning theories.

Kelly et al. (2015) claim that the connection between learning design and learning theory has to be explicit. For Gašević et al. (2016), studies in LA should: (1) Indicate the connection between LA and decades of research in learning theories; (2) Explain in what way the outcome of LA will improve and then adjust this research by confirming (or disconfirming) some findings.

7.5.3 A Priori Knowledge and Learning Theories

When the APK of LA is explicit, most of the time, it is the learning theory sustaining the student's learning. And otherwise? We show here that sometimes it can be difficult to make APK explicit, we show also that the implementation of learning theory in LA can be partial and finally we point out that the APK used by LA should not be necessarily the learning theory sustaining the level 1 learning process.

A Priori Knowledge Not Mentioned or Not Explicit. Even if the APK is not mentioned, there is always one embedded in a LA device. As we wrote, a LA device is designed by a team and the members of this team have pedagogical beliefs (Shum and Ferguson 2012). Selection of data and technique will be done according to these beliefs.

The APK can also not be a theory. The APK can be the one of the expert who chooses relevant data or the relevant technique like in (Olney et al. 2012; Rupp et al. 2012). The APK can also be built through datamining approaches, grounded in the behavior of expert practitioners. In both case, the APK is the one of experts. This APK can come from theories these experts have learned but also from their own experiences. Thus, it is difficult to make this APK explicit.

Interpretation and Partial Implementation of Learning Theory in LA. When LA uses a learning theory as APK, it often interprets this theory as we have seen, in Sect. 7.4.4, with the "proxy-indicators". It can also refer just to a part of this theory.

For example, papers about ITS, often referred to the "Two Sigma" effect of Bloom (Cen et al. 2007; Olney et al. 2012; D'Mello et al. 2010; Mazoue 2014) and not to the whole theory of Bloom. According to the "Two Sigma" effect, if the student is tutored one to one, he will perform two standard deviations better than students in "classical" environments like classrooms. Researchers use this effect to justify ITS. However, when they analyze the reasons of the "Two Sigma" effect, they fall implicitly in broader theory. For example, according to Chi et al. (2008), the efficiency of tutoring may be explained by a student-centered hypothesis according to which students construct their own knowledge (constructivism paradigm), the tutor-centered hypothesis which relies on pedagogical strategies (behaviorist or cognitivist or constructivist paradigm) and the interaction hypothesis which is a blending of both and which promotes collaboration (social constructivism paradigm).

The "spacing effect" is another part of theory which is used in LA (Miyamoto et al. 2015; Svihla et al. 2015). The "spacing effect" is drawn from psychology literature. The idea is that dividing study time in multiple sessions improve learning. Multiple study sessions are better than one massed session (Miyamoto et al. 2015). Studying learners' behavior in MOOCs, Miyamoto et al. (2015) show that "the number of sessions students initiate is correlated with certification rate".

The fact that theories are interpreted or partially implemented in LA can present some biases. For Pardos (2015), LA researchers must engage deeper with adequate learning theories to embed assumptions from the theoretical frameworks in LA. For example, in his commentary of the paper of Miyamoto et al. (2015), he notes that the "spacing effect" underlying their paper concerns more rote memorization than learning. This is partly taken into account by Svihla et al. (2015) who distinguish retention from learning.

APK Used by LA Should Not Be Necessarily the Learning Theory Sustaining the Level 1 Learning Process. Which theory to use as an APK for LA? There is not a single answer to this question because the answer depends on the precise goal of LA, and the goal depends on the stakeholder launching the LA process.

When the stakeholder is not involved directly in this level 1 learning process, his goal can be something else than improving the learning process. In this case, the APK used for LA can be not a learning theory. For example, Clow (2013) uses the funnel of participation theory to explain the very big drop-off in MOOCs. This theory is borrowed from "marketing funnel". It models a customer going from knowing that a product exists to buying this product. Between those two steps there is a big drop-off. Another example, will be when the institution wants to know if a course is profit-making and will use BI to look at data.

When the goal will be to improve learning, then the goal will depend from the learning theory sustaining the level 1 learning process. But to achieve this goal the theory used for APK by LA can be another one. This is what happens, when the level 1 learning process is designed according to social constructivism and, the level 2 according to SNA. Thus, SNA is used as APK by LA to understand how the groups behave (Bakharia and Dawson 2011; Kop and Carroll 2011; Fournier et al. 2011; Ferguson 2012). Another example is given by (Worsley and Blikstein 2013) who propose a constructionist environment and to study this environment they use Knowledge In Pieces theory. This theory considers that students build their knowledge by dynamically articulating and reorganizing atomistic intuitions rather relying on theoretical systems.

Moreover, the use of the learning theory sustaining the level 1 learning as APK for LA may be a problem when we want to generalize. Gašević et al. (2016) show that a LA designed for one course will not correctly work on another. So, one challenge is to develop scalable LA which can support different courses designed according to different learning theories. The APK used by LA must have a large enough scope to take into account those different learning theories. This is what Wise (2014) calls "extensibility" and "reusability". AT, by its strong descriptive power, is one possible candidate for that. In Simon (2014), we show that AT used as APK allows to take into account different types of learning systems.

7.6 Conclusion

In this article, we consider that LA is a learning on learning process. All learning processes rely on APK. So, all LAs implicitly or explicitly rely also on some APKs. To study the relationship between APK and LA, we use the 2-level framework. This framework considers LA as a level 2 learning process and is constituted of five characteristics: stakeholders, goals, data, technical approaches and feedbacks. For each of those characteristics, we look at the relationship between APK and LA in selected papers. We show that most of the time the APK used for LA is the learning theory shaping the student's learning. We present then an insight of APK and LA ties in which we show that it can be otherwise. According to the goal of LA, it can be fruitful to use other theories than learning theories.

In the near future, LA will be unavoidable for at least three reasons:

- Skills of the 21st century are not the ones of the previous centuries,
- Data will come also from learning in classrooms,
- There is an economical pressure to automatize, at least partially, education.

First point, according to (Knight et al. 2014) and others, the 21st century skills are creativity, innovation, critical thinking, problem solving, communication, and collaboration. These skills, as we have seen, are hardly evaluated by traditional assessments (Blikstein 2011). They can be assessed nowadays because interactions between student and other students, student and teacher and student and knowledge are more and more mediated by computers. In so doing, information generated by these interactions are digitized and can be processed by computers. This process is a LA process.

Second point, until now, LA has been essentially concerned with distance or blended learning. It does not concern classrooms because, in classroom, information is rarely digitized. This should change with the apparition of digital tablets which can easily be integrated in classrooms. With tablets, information will be digitized and automatically processed (Simon 2015). For example, digital school-books on tablets will be more and more interactive, closer to ITS than to traditional schoolbooks. Data provided by these interactions could be mined by researchers (and schoolbooks editors) to improve learning. This is the field of LA.

Third point, there is an economical pressure to automatize, at least partially, education (Anderson and Dron 2012; Mehaffy 2012). As we have seen the goal is to educate more students, with greater learning outcomes, at lower costs (Mehaffy 2012). This cannot be done without LA. In this case, LA will be used as well as for formative assessments than summative assessments. The automation of teaching is already done in ITS. It can happen in classroom with tablets as we have just seen. It can be useful also in MOOCs (Kay et al. 2013).

If LA becomes unavoidable, its results should be reliable. In those different contexts, with different goals, the choice of the "right" APK for LA is going to be more and more crucial.

In conclusion, we want to point out two problems that can appear with LA. The first one is related to the necessity to guarantee an ethical use of data (Fournier et al. 2011; Shum and Ferguson 2012). We can wonder if each digitized data is not a loss of freedom. To avoid that, the use of these data has to be controlled. The second is related to the fact that, to learn, the student will rely more and more on the network (Shum and Ferguson 2012). Here we can wonder if the human being is not becoming only a node in a network. Both questions go beyond LA field and concern many aspects of the society of the 21st century.

References

Aleven V, Ashley K, Lynch C et al (2008) Intelligent tutoring systems for Ill-defined domains: assessment and feedback in Ill-defined domains. In: Proceedings of the 9th international conference on intelligent tutoring systems, Montreal, pp 23–27

Anaya AR, Boticario JG, Letón E (2015) An approach of collaboration analytics in MOOCs using social network analysis and influence diagrams. In: Proceedings of the 8th international conference on educational data mining, Madrid, pp 492–496

Anderson M (1992) Intelligence and development: a cognitive theory. Blackwell, London

Anderson T, Dron J (2012) Learning technology through three generations of technology enhanced distance education pedagogy. Eur J Open Distance E-Learn 15

Aristizabal A (2016) An epistemological approach for research in educational data mining. Int J Educ Res 4:131–138

Arnold KE, Pistilli MD (2012) Course signals at Purdue: using learning analytics to increase student success. In: Proceedings of the 2nd international conference on learning analytics and knowledge, ACM, New York, pp 267–270

Atkisson M, Wiley D (2011) Learning analytics as interpretive practice: applying Westerman to educational intervention. In: Proceedings of the 1st international conference on learning analytics and knowledge, ACM, New York, pp 117–121

Attwell G (2007) Personal learning environments-the future of eLearning? Elearning Papers 2:1–8

Bakharia A, Dawson S (2011) SNAPP: a bird's-eye view of temporal participant interaction. In: Proceedings of the 1st international conference on learning analytics and knowledge, ACM, New York, pp 168–173

Balacheff N, Ludvigsen S, De Jong T, Lazonder A, Barnes S-A, Montandon L (2009) Technology-enhanced learning. Springer

Berland M, Baker RS, Blikstein P (2014) Educational data mining and learning analytics: applications to constructionist research. Technol Knowl Learn 19:205–220

Blikstein P (2011) Using learning analytics to assess students' behavior in open-ended programming tasks. In: Proceedings of the 1st international conference on learning analytics and knowledge, ACM, New York, pp 110–116

Blikstein P (2013) Multimodal learning analytics. In: Proceedings of the third international conference on learning analytics and knowledge, ACM, New York, pp 102–106

Bloom BS, Engelhart MD, Furst EJ et al (1956) Taxonomy of educational objectives: the classification of educational goals. David McKay Company, New York

Bourdeau J, Grandbastien M (2010) Modeling tutoring knowledge. In: Advances in intelligent tutoring systems. Springer, Heidelberg, pp 123–143

Brooks C, Greer J, Gutwin C (2012) Using an instructional expert to mediate the locus of control in adaptive e-learning systems. In: Proceedings of the 2nd international conference on learning analytics and knowledge, ACM, New York, pp 84–87

Bruner JS, Austin GA (1986) A study of thinking. Transaction Publishers

Brusilovsky P, Millán E (2007) User models for adaptive hypermedia and adaptive educational systems. In: The adaptive web. Springer, pp 3–53

Cen H, Koedinger KR, Junker B (2007) Is over practice necessary?-Improving learning efficiency with the cognitive tutor through educational data mining. Front Artif Intell Appl 158, 511

Chatti MA, Dyckhoff AL, Schroeder U et al (2012) A reference model for learning analytics. Int J Technol Enhanced Learn 4:318–331

Chen H, Chiang RH, Storey VC (2012) Business intelligence and analytics: from big data to big impact. MIS Q 36:1165–1188

Chi MT, Roy M, Hausmann RG (2008) Observing tutorial dialogues collaboratively: insights about human tutoring effectiveness from vicarious learning. Cogn Sci 32:301–341

Clow D (2012) The learning analytics cycle: closing the loop effectively. In: Proceedings of the 2nd international conference on learning analytics and knowledge, ACM, New York, pp 134–138

Clow D (2013) MOOCs and the funnel of participation. In: Proceedings of the third international conference on learning analytics and knowledge, ACM, New York, pp 185–189

Cooper A (2012) A framework of characteristics for analytics. CETIS Analytics Ser 1

Dawson S, Mirriahi N, Gasevic D (2015) Importance of theory in learning analytics in formal and workplace settings. J Learn Analytics 2:1–4

De Liddo A, Shum SB, Quinto I et al (2011) Discourse-centric learning analytics. In: Proceedings of the 1st international conference on learning analytics and knowledge, ACM, New York, pp 23–33

Demchenko Y, De Laat C, Membrey P (2014) Defining architecture components of the big data ecosystem. In: Proceeding of international conference on collaboration technologies and systems (CTS), 2014, IEEE, pp 104–112

Desmarais MC, Baker RS (2012) A review of recent advances in learner and skill modeling in intelligent learning environments. User Model User-Adap Inter 22:9–38

Dillenbourg P (2016) The evolution of research on digital education. Int J Artif Intell Educ, 1–17

Dillenbourg P, Schneider D, Synteta P (2002) Virtual learning environments. In: 3rd Hellenic conference "Information & Communication Technologies in Education," Kastaniotis editions, Greece, pp 3–18

D'Mello S, Olney A, Person N (2010) Mining collaborative patterns in tutorial dialogues. JEDM-J Educ Data Mining 2:2–37

Dyckhoff AL, Zielke D, Bültmann M et al (2012) Design and implementation of a learning analytics toolkit for teachers. Educ Technol Soc 15:58–76

Engeström Y (2014) Learning by expanding. Cambridge University Press

Ferguson R (2012) Learning analytics: drivers, developments and challenges. Int J Technol Enhanced Learn 4:304–317

Fournier H, Kop R, Sitlia H (2011) The value of learning analytics to networked learning on a personal learning environment. In: Proceedings of the 1st international conference on learning analytics and knowledge, ACM, New York, pp 104–109

Fournier H, Kop R, Durand G (2014) Challenges to research in MOOCs. J Online Learn Teach 10:1

Gagne RM (2013) Instructional technology: foundations. Routledge, London

Gašević D, Dawson S, Rogers T et al (2016) Learning analytics should not promote one size fits all: the effects of instructional conditions in predicting academic success. Internet High Educ 28:68–84

Gobert JD, Sao Pedro M, Raziuddin J et al (2013) From log files to assessment metrics: measuring students' science inquiry skills using educational data mining. J Learn Sci 22:521–563

Greller W, Drachsler H (2012) Translating learning into numbers: a generic framework for learning analytics. Educ Technol Soc 15:42–57

Hewitt J (2015) Commentary on "distributed revisiting: an analytic for retention of coherent science learning". J Learn Analytics 2:102–106

Joksimović S, Dowell N, Skrypnyk O et al (2015) How do you connect?: analysis of social capital accumulation in connectivist MOOCs. In: Proceedings of the fifth international conference on learning analytics and knowledge, ACM, New York, pp 64–68

Kay J, Reimann P, Diebold E et al (2013) MOOCs: so many learners, so much potential... IEEE Intell Syst, 70–77

Kelly N, Thompson K, Yeoman P (2015) Theory-led design of instruments and representations in learning analytics: developing a novel tool for orchestration of online collaborative learning. J Learn Analytics 2:14–43

Kennedy G, Ioannou I, Zhou Y et al (2013) Mining interactions in immersive learning environments for real-time student feedback. Australas J Educ Technol 29:172–183

Khalil M, Ebner M (2015) Learning analytics: principles and constraints. In: Proceedings of world conference on educational multimedia, hypermedia and telecommunications, pp 1326–1336

Knight S, Shum SB, Littleton K (2014) Epistemology, assessment, pedagogy: where learning meets analytics in the middle space. J Learn Analytics 1:23–47

Kop R, Carroll F (2011) Cloud computing and creativity: learning on a massive open online course. Eur J Open Distance E-Learn 14

Lee HS, Gweon GH, Dorsey C et al (2015) How does Bayesian knowledge tracing model emergence of knowledge about a mechanical system? In: Proceedings of the fifth international conference on learning analytics and knowledge, ACM, New York, pp 171–175

Li N, Matsuda N, Cohen WW et al (2015) Integrating representation learning and skill learning in a human-like intelligent agent. Artif Intell 219:67–91

Li H, Nashashibi F (2011) Robust real-time lane detection based on lane mark segment features and general a priori knowledge. In: IEEE international conference on robotics and biomimetics (ROBIO). Springer, Heidelberg, pp 812–817

Lindgren R, Johnson-Glenberg M (2013) Emboldened by embodiment six precepts for research on embodied learning and mixed reality. Educ Researcher 42:445–452

Liyanagunawardena TR, Adams AA, Williams SA (2013) MOOCs: a systematic study of the published literature 2008–2012. Int Rev Res Open Distrib Learn 14:202–227

Lockyer L, Heathcote E, Dawson S (2013) Informing pedagogical action: aligning learning analytics with learning design. Am Behav Sci 57(10):1439–1459

Magnisalis I, Demetriadis S, Karakostas A (2011) Adaptive and intelligent systems for collaborative learning support: a review of the field. IEEE Trans Learn Technol 4:5–20

Martinez-Maldonado R, Pardo A, Mirriahi N et al (2015) The LATUX workflow: designing and deploying awareness tools in technology-enabled learning settings. In: Proceedings of the fifth international conference on learning analytics and knowledge, ACM, New York, pp 1–10

Mazoue JG (2014) The MOOC model: challenging traditional education. Educause Rev 47:25–42

Mehaffy GL (2012) Challenge and change. Educause Rev 47:25–42

Merceron A, Blikstein P, Siemens G (2015) Learning analytics: from big data to meaningful data. J Learn Analytics 4–8

Mislevy RJ, Behrens JT, Dicerbo KE et al (2012) Design and discovery in educational assessment: evidence-centered design, psychometrics, and educational data mining. JEDM-J Educ Data Min 4:11–48

Miyamoto YR, Coleman CA, Williams JJ et al (2015) Beyond time-on-task: the relationship between spaced study and certification in MOOCs. J Learn Analytics 2:47–69

Olney AM, D'Mello S, Person N et al (2012) Guru: a computer tutor that models expert human tutors. In: Intelligent tutoring systems. Springer, Heidelberg, pp 256–261

Papert S (1980) Mindstorms: children, computers, and powerful ideas. Basic Books

Pardos ZA (2015) Commentary on "Beyond time-on-task: the relationship between spaced study and certification in MOOCs". J Learn Analytics 2:70–74

Perera D, Kay J, Koprinska I et al (2009) Clustering and sequential pattern mining of online collaborative learning data. IEEE Trans Knowledge Data Eng 21:759–772.

Piaget J (1950) Introduction à l'épistémologie génétique. Presses Universitaires de France, Paris

Pinkwart N (2016) Another 25 years of AIED? Challenges and opportunities for intelligent educational technologies of the future. Int J Artificial Intell Educ 1–13

Rajendran R, Iyer S, Murthy S et al (2013) A theory-driven approach to predict frustration in an ITS. IEEE Trans Learn Technol 6:378–388

Roll I, Aleven V, Koedinger KR (2010) The invention lab: using a hybrid of model tracing and constraint-based modeling to offer intelligent support in inquiry environments. In: Intelligent tutoring systems. Springer, Heidelberg, pp 115–124

Romero C, Ventura S (2010) Educational data mining: a review of the state of the art. IEEE Trans Syst Man Cybern Part C Appl Rev 40:601–618

Rupp A, Levy R, Dicerbo KE et al (2012) Putting ECD into practice: the interplay of theory and data in evidence models within a digital learning environment. JEDM-J Educ Data Min 4:49–110

Sao Pedro MA, Baker RS, Gobert JD (2012) Improving construct validity yields better models of systematic inquiry, even with less information. In: User modeling, adaptation, and personalization. Springer, Heidelberg, pp 249–260

Scaife M, Rogers Y, Aldrich F et al (1997) Designing for or designing with? Informant design for interactive learning environments. In: Proceedings of the ACM SIGCHI conference on human factors in computing systems, ACM, pp 343–350

Schneider B, Pea R (2015) Does seeing one another's gaze affect group dialogue? A computational approach. J Learn Analytics 2:107–133

Scott J (1988) Social network analysis. Sociology 22:109–127

Serres M (2007) Les nouvelles technologies: La révolution culturelle et cognitive. In: Proceeding of 40 ans de l'INRIA, Interstices

Shum SB, Ferguson R (2012) Social learning analytics. Educ Technol Soc 15:3–26

Siemens G (2005) Connectivism: a learning theory for the digital age. Int J Instr Technol Distance Learn 2

Simon J (2014) Data preprocessing according to activity theory. In: European conference on data mining proceedings, Lisbon, Portugal, pp 245–249

Simon J (2015) Digital tablets: a Trojan virus for ICT in education? In: Proceedings of the Asian conference on technology in the classroom 2015, Kobe, Japan, pp 111–122

Skinner BF (2011) About behaviorism. Vintage

Steiner CM, Kickmeier-Rust MD, Albert D (2014) Learning analytics and educational data mining: an overview of recent techniques. Learn Analytics Serious Game 6

Svihla V, Wester MJ, Linn MC (2015) Distributed revisiting: an analytic for retention of coherent science learning. J Learn Analytics 2:75–101

Swenson J (2014) Establishing an ethical literacy for learning analytics. In: Proceedings of the fourth international conference on learning analytics and knowledge, ACM, New York, pp 246–250

Tempelaar DT, Heck A, Cuypers H et al (2013) Formative assessment and learning analytics. In: Proceedings of the third international conference on learning analytics and knowledge, ACM, New York, pp 205–209

van Harmelen M, Workman D (2012) Analytics for learning and teaching. CETIS Analytics Ser 1:1–40

van Leeuwen A (2015) Learning analytics to support teachers during synchronous CSCL: balancing between overview and overload. J Learn Analytics 2:138–162

Vygotsky LS (1986) Thought and language, Rev ed. MIT Press, Cambridge

Watson WR, Watson SL (2007) What are learning management systems, what are they not, and what should they become. TechTrends 51(2):28–34

Wise AF (2014) Designing pedagogical interventions to support student use of learning analytics. In: Proceedings of the fourth international conference on learning analytics and knowledge, ACM, New York, pp 203–211

Wise AF, Shaffer DW (2015) Why theory matters more than ever in the age of big data. J Learn Analytics 2:5–13

Wise AF, Zha Y, Hausknecht SN (2013) Learning analytics for online discussions: a pedagogical model for intervention with embedded and extracted analytics. In: Proceedings of the third international conference on learning analytics and knowledge, ACM, New York, pp 48–56

Worsley M, Blikstein P (2011) What's an expert? Using learning analytics to identify emergent markers of expertise through automated speech, sentiment and sketch analysis. In: EDM, pp 235–240

Worsley M, Blikstein P (2013) Towards the development of multimodal action based assessment. In: Proceedings of the third international conference on learning analytics and knowledge, ACM, New York, pp 94–101

Chapter 8
Knowledge Discovery from the Programme for International Student Assessment

Mirka Saarela and Tommi Kärkkäinen

Abstract The Programme for International Student Assessment (PISA) is a worldwide study that assesses the proficiencies of 15-year-old students in reading, mathematics, and science every three years. Despite the high quality and open availability of the PISA data sets, which call for big data learning analytics, academic research using this rich and carefully collected data is surprisingly sparse. Our research contributes to reducing this deficit by discovering novel knowledge from the PISA through the development and use of appropriate methods. Since Finland has been the country of most international interest in the PISA assessment, a relevant review of the Finnish educational system is provided. This chapter also gives a background on learning analytics and presents findings from a novel case study. Similar to the existing literature on learning analytics, the empirical part is based on a student model; however, unlike in the previous literature, our model represents a profile of a national student population. We compare Finland to other countries by hierarchically clustering these student profiles from all the countries that participated in the latest assessment and validating the results through statistical testing. Finally, an evaluation and interpretation of the variables that explain the differences between the students in Finland and those of the remaining PISA countries is presented. Based on our analysis, we conclude that, in global terms, learning time and good student-teacher relations are not as important as collaborative skills and humility to explain students' success in the PISA test.

Keywords PISA · Learning analytics · Big data · Knowledge discovery · Hierarchical clustering

M. Saarela (✉) · T. Kärkkäinen
Department of Mathematical Information Technology,
University of Jyväskylä, 40014 Jyväskylä, Finland
e-mail: mirka.saarela@jyu.fi

T. Kärkkäinen
e-mail: tommi.karkkainen@jyu.fi

© Springer International Publishing AG 2017
A. Peña-Ayala (ed.), *Learning Analytics: Fundaments, Applications, and Trends*, Studies in Systems, Decision and Control 94,
DOI 10.1007/978-3-319-52977-6_8

Abbreviations

ESCS   Economic, social, and cultural status
LA   Learning Analytics
MOOC   Massive open online course
OECD   Organisation for Economic Cooperation and Development
PISA   Programme for International Student Assessment

8.1   Introduction

The original purpose of *Learning Analytics* (LA), as stated by researchers such as Siemens (2013, p. 1383) and Ferguson (2012, p. 306), was to "measure, collect, analyze, and report data about learners and their contexts, for the purposes of understanding and optimizing learning and the environments in which it occurs." Slightly different variants were later offered to characterize the discipline (Pardo and Teasley 2014; Gray et al. 2014; Siemens and Baker 2012). Increased attention to Massive Open Online Courses (MOOCs) (e.g., Wang et al. 2014; Ye and Biswas 2014; Reich et al. 2014; Coffrin et al. 2014; Hickey et al. 2014; Santos et al. 2014; Vogelsang and Ruppertz 2015; Ferguson and Clow 2015; Hansen and Reich 2015; Wise et al. 2016; Hecking et al. 2016) has intensified the need for data-based learning support from the perspective of *big data*. This is evidenced by several articles (e.g., Picciano 2012; Chatti et al. 2012; Siemens and Baker 2012; Chatti et al. 2014; Dawson et al. 2014; Wise and Shaffer 2015; Merceron et al. 2016) as well as by the theme of the 2015 Learning Analytics and Knowledge conference "Scaling Up: Big Data to Big Impact" (see Dawson et al. 2015).

The PISA is a worldwide triennial survey conducted by the Organisation for Economic Cooperation and Development (OECD), resulting in publicly available educational data on a large scale. In addition to assessing the proficiency of 15-year-old students from different countries and economies in reading, mathematics, and science, the PISA provides "data about learners and their contexts" as one of the largest public databases[1] of students' demographic and contextual data, such as their attitudes and behaviors toward various aspects of education. More than seventy countries and economies have already participated in the PISA, and the assessment is referred to as the "world's premier yardstick for evaluating the quality, equity, and efficiency of school systems" (OECD 2013a).

In the PISA studies, data collection is of very high quality, including the development of the appropriate instruments, the procedures, and the storage of the data in public databases. This is evidenced by the large amount of money spent on ensuring quality related to these issues. However, much less money has been invested in the analysis of the collected data, and only a few PISA analysis studies

---

[1]The PISA data can be downloaded from http://www.oecd.org/pisa/pisaproducts/.

have resulted in publications in the scientific field (Olsen 2005a). Rutkowski et al. (2010) argue that the size of the PISA data sets as well as the technical complexities within them may be the reason why more researchers do not work with these freely available and high-quality data.

Our research is motivated by the lack of secondary analysis of the PISA data, which calls for the development and utilization of big data LA methods for making discoveries within the international domain of the PISA. Such methods can then be used to summarize the PISA data sets in novel ways in order to better understand students from diverse countries and the settings in which they learn (Siemens and Baker 2012). Hence, in relation to big data LA, we focus on the international context in an effort to understand national education systems as learning environments. Such a scope for LA was also emphasized by Long and Siemens (2011), who pointed out that LA should occur on the national and international levels, primarily targeting national governments and education authorities. As a classroom is in a school is in a city is in a region is in a country is in a continent, thorough use of educational data and empirical evidence should be linked to those principles and practices of educational systems that are known to have an effect on learning. This is the primary concern in the PISA.

Chatti et al. (2014) introduced a reference model for LA based on four dimensions (stakeholders, objectives, data, and methods) that resembles the critical LA dimensions suggested by Greller and Drachsler (2012). Figure 8.1 illustrates how large-scale educational assessments, such as the PISA, can leverage big data LA according to these dimensions. Specifically, national bodies introduce the objectives (i.e., the factors that constitute good national education systems) for

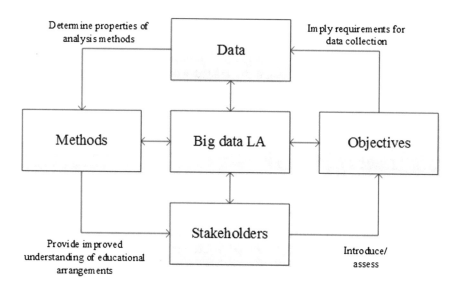

Fig. 8.1 Conducting big data LA for large-scale educational system assessments (cf. Chatti et al. 2014; Greller and Drachsler 2012)

assessing the international student population. Then, large amounts of data representing student background and proficiency are sampled and transformed into derived representations, whose characteristics (the sample to population alignment introducing weights and the rotated test design introducing missing values) must be handled by applied LA methods. When meaningful patterns are found, these are reported back to the educational decision makers.

Ferguson et al. (2014) emphasize the large-scale institutional adoption of appropriate educational patterns. In the best case, the institutional meso-level approaches are aggregated from the upscale local micro-level patterns and from the downscale macro-level characteristics of a good educational system. Thus, meaningful patterns at the macro level (e.g., within a large educational organization) originate from characteristics of a large student population in relation to the rigorously measured learning outcomes.

The structure of this paper is as follows. In Sect. 8.2, we provide necessary background on big data LA and educational knowledge discovery from the PISA. In Sect. 8.3, a relevant review on methodologically related studies is provided, and the forms and complexities of PISA data are described. Next, the overall analysis method is depicted in Sect. 8.4. In Sect. 8.5, the results and interpretations of the hierarchical clustering of the aggregated country profiles are presented and statistically validated. In Sect. 8.6, the PISA results are visualized in a dashboard. Finally, in Sect. 8.7, the empirical work is summarized, and in Sect. 8.8, the overall conclusions are presented.

8.2 Background and Related Work

We next provide the necessary theoretical background for the empirical part of the chapter. First, we explain big data LA and summarize LA methods. Then, we characterize a pool of methodologically related work on the use of clustering in educational data analysis. We observe that methodologically related studies are typically conducted on the micro level of individual courses or tutoring systems.

8.2.1 Toward Big Data LA

As emphasized in the introduction, LA studies are increasingly leveraging big data. The term "big" in "big data" does not solely refer to the amount of data but actually references four "V"s (the first three according to Laney (2001) and the last one as described by Gupta et al. (2014)): (i) *Volume* refers to the size of data sets caused by the number of data points, their dimensionality, or both; (ii) *Velocity* is linked to the speed of data accumulation; (iii) *Variety* stands for heterogeneous data formats, which are caused by distributed data sources, highly variable data gathering, etc.;

and (iv) *Veracity* refers to the fact that (secondary) data quality can vary significantly, and manual curation is typically impossible.

In relation to big data LA, PISA data are characterized by a high volume and low veracity due to missing values, but there is no velocity and small, well-managed variety due to the meticulous design. Moreover, unlike the existing LA studies, the collected student sample is aligned to the whole worldwide student population under study using weights (see the last paragraphs in Sect. 8.2.3). For example, the sample data of the PISA 2012 consists of approximately half a million students, representing 24 million 15-year-old students from 68 different countries and territories.

Chatti et al. (2012) state that different LA techniques for detecting interesting educational patterns originate from four analysis categories: *statistics*; *information visualization*; *data mining* (identifying this with knowledge discovery in databases) in the form of classification, clustering, and association rule mining; and *social network analysis*. Other LA researchers support this notion that data mining and knowledge discovery techniques are one category of the broader set of LA methods. Rogers (2015), for example, lists data mining as one of the more sophisticated quantitative methods in LA, and Siemens (2013) states that knowledge discovery from databases is an LA technique that has become increasingly important.

Generally, with the advent of big data in education, LA methods have shifted from the more traditional data analysis techniques, such as statistics, to more scalable data mining methods (Hershkovitz et al. 2016; Joksimović et al. 2016). In fact, Ferguson (2012) points out that the two main differences between general educational research and the specific research field of LA (according to the LA definition given in the beginning of this chapter) is that LA "make[s] use of pre-existing, machine-readable data, and that its techniques can be used to handle 'big data.'"

Application of data mining and knowledge discovery methods in an educational context typically realizes an *educational knowledge discovery process* that, especially when using an open educational data set like that of the PISA, supports learning and knowledge analytics (Verbert et al. 2012). Several case studies (e.g., Hu et al. 2016; Brown et al. 2016; Grawemeyer et al. 2016; Allen et al. 2016; Chandra and Nandhini 2010) have proven the need for and the success of specific knowledge discovery processes and data analysis methods within the educational domain. However, data from many of the existing educational case studies are specific to certain educational environments or institutions, which complicates the comparison of the techniques and the results provided.

In contrast, the PISA tests are standardized, and the resulting data sets are comparable between different nations and their educational arrangements. Hence, the PISA provides an interesting and novel case for big data LA techniques (Saarela and Kärkkäinen 2014, 2015a, b, c; Kärkkäinen and Saarela 2015), combining the methodological requirements that are due to the above-mentioned technical complexities of the data with comparative educational knowledge discovery.

8.2.2 On Educational Data Analysis Using Clustering

As has been pointed out above, *clustering* is one of the key techniques in the data mining category of the LA methods. Next, we describe a pool of work related to the clustering of educational data as well as the empirical work in Sects. 8.4 and 8.5. This set of papers was primarily identified by scanning through the most relevant publication forums (see Saarela et al. 2016a) in the field, especially the *Journal of Learning Analytics*[2] and the *Conference on Learning Analytics and Knowledge*,[3] restricting the topic to clustering with real educational data sets. The description of the work is organized according to the clustering method used and the size of the clustered educational data set.

Hierarchical Clustering. Logs of 454 online mathematics practice sessions by 69 students were clustered by Desmarais and Lemieux (2013). In that study, pre-processing first transformed the logs into temporal sequences (time series) reflecting the state of interaction between the student and the learning environment. These representations were then clustered using an agglomerative hierarchical method, and the interpretation of the result was based on visualizing the clusters as state sequence diagrams. Three characteristic forms of using the system were identified: (i) exploratory browsing, (ii) short practice sessions, and (iii) exercise-intensive sessions.

Self-regulatory strategies of undergraduate students, especially their characteristics in accessing online learning material, were studied by Colthorpe et al. (2015). Hierarchical clustering of 97 students was able to separate high- and low-performing students, and the low-performing students were characterized by extensive use of lecturing recordings. This could, however, be explained by the form of engagement with the learning material.

Segedy et al. (2015) provided a more in-depth analysis of students' self-regulated interaction with the learning material in an open-ended computer-based learning environment. Student assessment was based on the coherence analysis, whose descriptive metrics for 99 sixth grade students were separated into five clusters using complete-link hierarchical clustering as part of the versatile analysis process. In addition to two very small clusters of (i) confused guessers and (ii) students disengaged from the task, the main clusters characterized the self-regulated interaction patterns of (iii) frequent researchers and careful editors, (iv) strategic experimenters, and (v) engaged and efficient students.

Hu et al. (2016) used hierarchical clustering to analyze the responses of 523 English and Chinese primary school students to a questionnaire about their reading behaviors, reading preferences, and attitudes toward reading. Three main reading profiles were identified, and they were fully characterized by good, moderate, and bad reading habits.

---

[2]See http://learning-analytics.info/.

[3]See http://lakXX.solaresearch.org/, where XX stands for year in which the conference took place. For example, http://lak16.solaresearch.org/ contains a link to the proceedings of the 2016 conference.

Hecking et al. (2016) combined social similarity (i.e., distances in the communication graph of the students) and semantic similarity (i.e., distances between the content-based roles by the students) to construct a socio-semantic block modeling approach for analyzing a MOOC discussion forum. Hierarchical clustering was used in the actual construction of the block model from the derived similarity measure. The analysis of the communication graph of 647 students in 502 threads on 27 forums verified the presence of different roles, with a moderate correlation between the social and the semantic role by a student. Discovery of the three main socio-semantic roles suggested that online discussion forums need better recognition and adaptation to the different user roles.

K-Means. A collaboration of 31 participants in a math discussion board was addressed by Xing et al. (2014) through the lens of activity theory, which links individual and social behavior, using the prototype-based k-means clustering method. In this study, the important phases of the educational clustering process, preprocessing, and interpretation of the clustering result were strongly present. The result consisted of three clusters characterizing (i) learners who were personally participative but less communicative on the group level, (ii) collaboratively participating but shallow learners, and (iii) less participative poor learners.

An automated approach using the k-means clustering algorithm was described by Li et al. (2013) for constructing a student model from the content features of algebra problems. Methodologically versatile preprocessing (feature extraction, min-max scaling, and principal component analysis) and tenfold cross-validation characterized the approach. The experiment with data from 71 students concluded that the clustering-based model was at least as good as the prior manually constructed model, as it was able to reveal previously unidentified and valuable knowledge components of mathematical problem solving. An innovative assessment of the physical learning environment that also used the k-means clustering method was reported by Almeda et al. (2014). The result consisted of four different clusters characterizing the similar content profiles of 30 classroom walls, as decorated by the teachers.

Multiple clustering methods (including k-means and hierarchical clustering) at various stages of the data analysis were applied by Blikstein et al. (2014) to reveal the different patterns and trends of the development of programming behavior in an introductory undergraduate programming course. The overall analysis of 370 participants and 154,000 code snapshots was concluded in multiple ways. First, for different tasks within LA, different kinds of tools are needed, ranging from fast and simple wrap-ups of data to advanced machine-learning methods running on high-performance computing platforms. Secondly, concerning the clustering methods, it is necessary to have either better support to interpret the result of a clustering method or the application of more advanced methods to improve the potential insights and knowledge discovery from data. Thirdly, concerning the domain of the study, the changes in the code update patterns by the students were more strongly correlated with the course performance compared to the size of code updates.

A subset of methods used by Blikstein et al. (2014) were also utilized by Worsley and Blikstein (2014) to analyze the problem-solving patterns of 13 students for open-ended engineering tasks. This LA method was based on the segmentation and extraction of action features from the hand-coded video data. The k-means algorithm produced four clusters whose interpretation could be summarized into two principal dimensions of idea quality and design process, which were both related to students' level of experience.

Expectation Maximization. Bouchet et al. (2013) clustered the derived variables of multiple thematic groups from the log data of 106 college students using an intelligent tutoring system fostering self-regulated learning. They used the expectation-maximization algorithm from Weka, resulting in three clusters as suggested by the knee point (see Saarela and Kärkkäinen 2015a), after careful cross-validation with multiple restarts. The three clusters were generally characterized by varying levels of performance but also reflected (through metadata) differences in the number of self-regulated learning processes in which the students were engaged. Bogarin et al. (2014) also used the expectation-maximization algorithm from Weka and discovered three clusters from the log data of 84 Psychology students training to learn online with Moodle. In particular, a cluster of the most passive online students was detected, of which two-thirds failed the course.

Activity in online discussion forums as a predictor of study success was also studied by López et al. (2012). Methodologically, it was shown that the prototypes obtained from the expectation-maximization clustering algorithm with tenfold cross-validation with Weka software were able to distinguish 114 different and informative cases of university student behavior. Similar to Bogarin et al. (2014), it was concluded that active participation in the course forum was a good predictor of the final grade for the course.

Summary. To summarize this small survey of educational clustering methods, hierarchical clustering, k-means, and expectation maximization were the most common approaches. This was also the conclusion in the review by Peña-Ayala (2014). Similarly, student modeling, including behavior and performance models, was the dominant educational data analysis approach, covering all the assessed research except Almeda et al. (2014) (see Table 11 in the work published by Peña-Ayala 2014). Note that a set of older references concerning the use of clustering in educational settings, as briefly introduced by Bouchet et al. (2013) in Sect. 6, also emphasized the student model as an important part of intelligent online tutoring systems.

8.2.3 LA Approaches Oriented to Analyze PISA Repositories

As concluded in the previous section, clustering is one of the key techniques for analyzing educational data, especially in LA. However, most of the educational clustering studies use small data sets of tens or at most hundreds of students at the micro and meso levels of educational systems. By comparison, the PISA 2012 data set is comprised of around half a million students and represents a population of 24 million people worldwide (see the last paragraph in Sect. 8.3.2).

A considerable amount of literature has been published on the PISA. However, as observed by Olsen (2005a), these publications are mainly national or international reports that have not undergone the peer-review process. Furthermore, many of the peer-reviewed publications dealing with the PISA (e.g., Deng and Gopinathan 2016; Auld and Morris 2016; Rasmussen and Bayer 2014; Yates 2013; Bank 2012; Bulle 2011; Waldow et al. 2014; Grek 2009; Simola 2005; Sahlberg 2011; Kumpulainen and Lankinen 2012) do not present the researchers' own empirical analysis but only refer to the reports or statistics published by the OECD. In the papers where the researchers' own empirical models are being derived and analyzed (e.g., Skryabin et al. 2015; Kriegbaum et al. 2015; Erdogdu and Erdogdu 2015; Tømte and Hatlevik 2011; Zhong 2011; Fonseca et al. 2011), the missing data is most often completely removed, and the sample is analyzed by ignoring the weights and, hence, the population level. Moreover, typically students from only a few countries are being compared in the existing literature, although a very scarce pool exists of comparisons at the level of the whole PISA sample (e.g., Drabowicz 2014; Zhong 2011).

We have also carefully assessed the use of clustering with the PISA data sets and have only been able to identify our own recent publications for the PISA 2012 (Saarela and Kärkkäinen 2014, 2015b, c) and two older publications for the PISA 2003 (Olsen 2005b) and for the PISA 2000 (Kjærnsli and Lie 2004). Thus, our main contributions here are that we augment the traditional PISA analysis by utilizing big data LA methods and work with the data set on the macro level of the whole student population, as conforms to the recommendations given by the OECD (2014b). This population-level scope is a novel setting in big data LA.

8.3 The PISA Profile

In this section, we outline the contextually related work of the chapter. More precisely, since Finland is the primary interest in our clustering application, we introduce the main characteristics of the Finnish educational system, which has performed so well in the PISA assessments, as well as related research. The last part of this section is devoted to a description of the collection and overall processing of the PISA assessment, yielding to multiple forms of publicly available educational data sets on a macro level.

8.3.1 The Finnish Educational System and the PISA

In this paper, our main focus is on Finland in comparison to the other countries that participated in the latest PISA assessment. Traditionally, Finnish students have performed exceptionally well in the PISA tests. The reasons for Finland's success on the PISA, particularly in the 2003 and 2006 assessment cycles, have been analyzed in several studies, and educational stakeholders from all over the world have visited Finland to find explanations for the high-performing students.

Consequently, education became an important asset in Finland's image and identity. In fact, Finland has invested considerably in the international educational export sector (Schatz et al. 2016), and, although Finland's place in the international ranking dropped in the latest PISA assessment, it is still placed the highest in Europe. Here, our goal is to assess the variables that most distinguish Finland from the other countries participating in the PISA.

Finland's high performance in the PISA assessments has been analyzed in several articles. Many of these articles have linked the well-performing students to the highly qualified teachers, who need to have a Master's degree for a permanent position. In particular, it has been argued that, in Finland, being a teacher is one of the most prestigious occupations, as evidenced by the fact that only the best and most motivated students are admitted to the teacher training programs as well as the observation that Finnish teachers enjoy a very high status in society (Morgan 2014; Sahlberg 2011; Linnakylä et al. 2011; OECD 2011; Andere 2015).

A second reason that has been identified as contributing to Finland's high results in the PISA relates to the organization of the national school system. Instead of (i) market-oriented schooling, (ii) standardization of schools and tests, concentrating on measurable performance, and (iii) competition between students and schools, the focus in Finland's schools is more on cooperation, collaboration, and the belief that teachers will support each student's individual learning (Simola 2005; Sahlberg 2011). National curricula as well as explicit learning objectives and standards do exist, but schools and teachers in Finland enjoy great autonomy and decision-making authority (i.e., they can decide on individualized learning strategies and pedagogical methods in order to reach the common educational goals) (Kumpulainen and Lankinen 2012; Linnakylä et al. 2011; OECD 2011).

The fact that schools in Finland are neither competing nor evaluated by standardized tests is one of the reasons why the variance between the Finnish schools is so small[4] (Simola 2005). Additionally, there is a no division of students into different school types or tracks based on their performance. Indeed, all students in Finland attend common, untracked, comprehensive schools of equally good quality

---

[4]According to the 2012 assessment, the between-school variation in Finland is only 6% of the overall math performance, which is the second-lowest figure in comparison with all PISA countries.

Table 8.1 Interaction between culture and education in Finland

Culture	Education
Strong mutual trust	Parents and government trust teachers (indicated by the strong autonomy and authority of the teachers)
Equity and equality (care for others instead of wanting to be the best)	Common untracked comprehensive school systems; free lunch, health care, and school transport; children with special needs study in the same classroom
Indulgent country	Minimal time allocated to studying, broad rich curriculum

from grades 1–9, typically those nearest to their homes. These schools are publicly funded and offer free lunches, health care, and school transport for all pupils (OECD 2011; Linnakylä et al. 2011).

These mutually interdependent and interconnected factors that are associated with Finland's high achievements in the PISA have also been emphasized by Välijärvi et al. (2007), who have concluded that Finland's success can be explained by a combination of "comprehensive pedagogy, students' own interests and leisure activities, the structure of the education system, teacher education, school practices and, in the end, Finnish culture" (see Table 8.1).

Research has shown that culture tends to affect both people's goals and their actions to reach these goals (Hitlin and Piliavin 2004). As has been pointed out above, Finnish people put great emphasis on equity and equality. Several studies have also highlighted the trust that seems to exist in Finnish culture in general and between the educators and the community in particular (Sahlberg 2011; OECD 2011).

The *Hofstede model* (Hofstede 2011) acknowledges the idea that Finland is more of a collaborative than a competitive country. According to the model, Finland's society can be characterized as being highly "feminine," meaning that the most important driving factors in life are to live a good life and to care for others instead of to focus on one's own success and want to be the best. This is interesting when linked to the recent study by French et al. (2015), who found a negative causal relationship between education expenditure and power distance and masculinity. According to this study, the less masculine a country is, the more it invests in education.

8.3.2 Characteristics and Forms of the PISA Data

The OECD states that the PISA results have a high degree of validity and reliability (for example, OECD 2012, 2014b), so they can be used to assess and compare the educational systems of the participating countries. To ensure the validity and reliability of the PISA data, large amounts of money are spent. For example, in Germany alone, the aggregate costs of the PISA assessment have reached 21.5

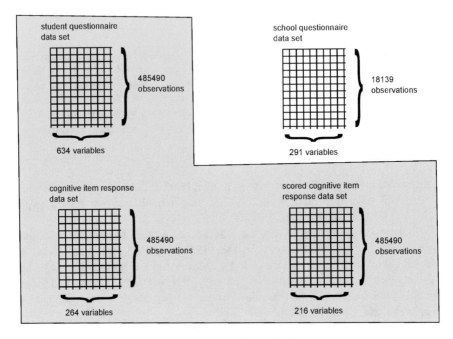

Fig. 8.2 Overview of the 2012 data sets available from OECD

million euros (Musik 2016). However, as was pointed out in the introduction of this chapter, the PISA assessments as well as the resulting PISA data are method-ologically very complex.

As highlighted by the OECD (2012), "the successful implementation of PISA depends on the use, and sometimes further development, of state-of-the-art methodologies and technologies." Since a mixture of different methods is used in this large study, and many variables are derived, it is not obvious how certain values in the publicly available database[5] (see Fig. 8.2) were collected, obtained, and reported. The fact that the PISA data are voluminous and complex can also be concluded based on the time that is needed to publish the PISA data and results: Usually around 1.5 years passes between data collection and when the first PISA results and data are published. For example, the 2012 PISA data collection took place in spring 2012, and its results were published in December 2013.

An overview of the 2012 PISA data is provided in Fig. 8.2. In all three data sets with pink backgrounds in Fig. 8.2, the observations are the assessed students. The basic information about the student (student's ID, country, test language, and school ID) and which test he or she was administered (booklet ID) is provided in all three of these student data sets. The *student cognitive items* and *scored cognitive item response* data sets document the students' responses to the cognitive items and

---

[5]Can be downloaded from http://pisa2012.acer.edu.au/downloads.php.

how these were scored. Altogether, there were 206 different cognitive items in the PISA 2012 data. An example of a cognitive item variable label is "SCIE—P2006 Wild Oat Grass Q4." As can be seen, it includes the domain (in this case, science), the PISA cycle in which the question was first used (the PISA 2006), the name for the particular task unit[6] (*Wild Oat Grass*), and the question number (*4*).

The most informative and meaningful part of the PISA data is the *student questionnaire data set* (see Fig. 8.2). However, as previously mentioned, one of the biggest challenges when working with the PISA data is that many variables in this data set are not direct measurements but rather variables that have already been transformed and preprocessed. For example, the students' abilities/performances in the cognitive tests are summarized in the form of *plausible values*. Plausible values are, as Wu (2005) puts it, "multiple imputations of the unobservable latent achievement for each student." This is explained more thoroughly at the end of this section.

Certain scale indices in the data—indicating, for example, students' attitudes toward school and learning—are also derived variables. This means that in order to be able to work with the PISA data, it is necessary to understand how the many derived variables have been created and how they can be used for further analysis. In the PISA, the *Rasch model*, which is a special case of item response theory, is used for this purpose.

Gray et al. (2014) emphasize the importance of integrating item response theory factors and methods, such as the Rasch model, into the existing LA models. Item response theory models can improve existing models, because they can model latent (i.e., not directly measurable) traits, such as intelligence, ability, or motivation. Moreover, they can be applied even with a large number of missing values. The potential of using item response theory in LA has been shown, for example, by Bergner et al. (2015), who estimated student abilities based on homework scores from an MOOC in which a large number of scores were missing.

The second challenge when working with the PISA data is the high sparsity. Since the assessment material developed for the PISA exceeds the time that is allocated for the test, each student is administered only a fraction of the whole cognitive testing material and only one of the three different background questionnaires. Because of this rotated design, very few variables in the PISA data sets have values for all observations. For example, in the PISA 2012, each student was assigned a test booklet of cognitive items that should be solvable in two hours. However, the comprehensive PISA 2012 cognitive item battery consisted of test items to be solved in six hours.

The *scored item set* (see Fig. 8.2) incorporates 206 scored items for 485,490 students. Nevertheless, because of the different booklets, which always contain only a fraction of the total items, 74% (that is, 73,860,420) of the different item variables have missing values. Similarly, because of the three different background

---

[6]PISA items are organized into units. Each unit consists of a stimulus (consisting of a piece of text or related texts, pictures, or graphs) followed by one or more questions.

questionnaires administered, the majority of the variables in the *student questionnaire data set* are missing approximately one-third of their values. We have discussed sparsity in educational data, particularly in the PISA data, and algorithms to cope with this issue in many of our recent studies (Saarela and Kärkkäinen 2014, 2015a, b, c; Kärkkäinen and Saarela 2015; Saarela et al. 2016b).

Finally, the PISA data are an important example of a large data set that includes weights. Only a fraction of the 15-year-old students from each country takes part in the assessment, but the gathered sample depicts the whole student population by multiplying the students' results by their respective weights, which simply measure how many similar students are represented by one student in the sample. For example, the sample data of the latest assessment consist of 485,490 students, which, when taking the weights into account, are representative of more than 24 million 15-year-old students in the 68 different countries and territories that participated in the PISA 2012.

Both over- and under-sampling has taken place in the PISA for different student groups. As a consequence, in order to state findings that are valid for the whole population, it is important to utilize these weights at each stage of the analysis. The way in which we incorporated the weights into a robust clustering algorithm for sparse data is illustrated and applied in our prior works (respectively, Saarela and Kärkkäinen 2015c, b).

8.3.3 Rasch Model

As described above, because of the different PISA test booklets administered, the actual scored student test data is extremely sparse with a great deal of missing values (74%). The easiest approach for measuring each student's ability would be to average the percentage of the correct answers over the three domains. However, since not all students were presented with the same test items, and the test items varied in their difficulty, this approach is considered unreliable. With the Rasch model, however, the probability of success on a given item can be modeled as a logistic function of the difference between the student and item parameters (Rasch 1960). Hence, the Rasch model enables a comparison of student abilities/test results/characteristics, even if not all students were tested on the same test items.

In the PISA, the Rasch model is employed to estimate both student abilities—depending on their item responses and the item difficulties in the cognitive test—and general student characteristics—depending on their responses on the background questionnaire. Mathematically, in the simplest case of the Rasch model when the test item is dichotomous, the probability that a student i with ability denoted by β_i provides a correct answer to an item j of difficulty δ_j can be stated as follows (8.1):

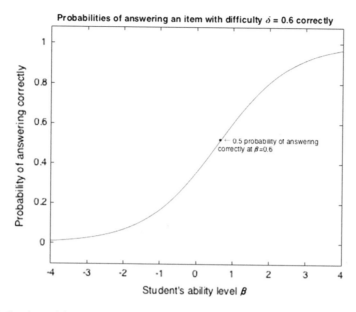

Fig. 8.3 Rasch model example. Probabilities that a correct answer is given to an item with difficulty δ = 0.6 for different student abilities. The probability that a student with ability β = 0.6 will provide a correct answer to this item is 0.5

$$P\left(X_{ij}|\beta_i, \delta_j\right) = \frac{\exp(\beta_i - \delta_j)}{1 + \exp(\beta_i - \delta_j)}. \tag{8.1}$$

When the Rasch model is employed, it iteratively creates a continuum/scale on which both a student's ability and item difficulty are located and where a probabilistic function links these two components. Usually, the item difficulties are estimated first, and this is referred to as the item calibration. The overall objective is to obtain data that will fit the model.

There should be a higher probability that a student should give a correct answer to an easy item than to a difficult item. Similarly, there should be a higher probability that a student with high ability should give correct answers to items than a student with low ability. This is shown in Fig. 8.3, where the probability that a correct answer is given to an item with difficulty δ = 0.6 is plotted for different student abilities. Moreover, as also illustrated in Fig. 8.3, when a student's ability is equal to the difficulty of the item, there is by definition a 50% chance of a correct response in the Rasch model.

To estimate the item difficulty, only the probability of being correct on that item and the ability of the students who completed the item must be known. Likewise, to estimate the student's ability, only the probability of being correct on a set of items and the difficulty of those items must be known (Embretson and Reise 2013). Every item and every student will be located in the scale created with the Rasch model.

Therefore, comparable student ability estimates can be obtained, even if the students were assessed with a different subset of items (OECD 2014b). The only requirement is that some link items exist (i.e., some items in the different test booklets must be the same).

In the PISA, a generalization of the original Rasch model is employed that can score not only dichotomous but also polytomous items (e.g., cognitive items can be scaled as *incorrect, partially correct,* and *correct* and questionnaire Likert-scale data can be scaled as *completely agree, agree, neutral, disagree,* and *completely disagree*). This model is called the one-parameter logistic model for polytomous items.

8.3.4 Plausible Values

There exist many other international large-scale educational assessment studies such as the PISA, including the National Assessment of Educational Progress,[7] the European Survey on Language Competences,[8] the Trends in International Mathematics and Science Study, and the Progress in International Reading Literacy Study.[9] The idea behind the PISA and these other assessments is not to measure and report the proficiencies of individual students. Instead, the primary goal is to provide a reliable overview of the proficiencies and national characteristics of the whole population (OECD 2014b; Marsman 2014). This is the main difference between typical micro- or meso-level LA and big data LA for the PISA.

Plausible values are used to estimate the proficiencies of the population, which, in the PISA, comprises all 15-year-old pupils within the participating countries. Some studies (Monseur and Adams 2008; Wu and Adams 2002; OECD 2014b) have shown that plausible values—in comparison to Weighted Likelihood Estimates, which overestimate, and Expected A Posteriori estimators, which underestimate population variances—produce unbiased estimates for population statistics.

In short, plausible values are random draws from the posterior distribution of a student's ability. These posterior distributions are estimated with a Bayesian approach in combination with the Rasch model. The posterior distribution of a student's ability β_i, given his or her vector of item responses x_i and certain additional variables about the student from the background questionnaire (e.g., gender and many others) that are encoded in a vector y_i, is defined as (8.2):

$$f(\beta|x_i, y_i) \propto P(x_i|\beta, \delta) f(\beta|\lambda, y_i), \qquad (8.2)$$

---

[7]nces.ed.gov/nationsreportcard/.

[8]www.surveylang.org/.

[9]http://timssandpirls.bc.edu/.

where $P(x_i|\beta,\delta)$ denotes a Rasch model given the student's ability β and the difficulties of the items δ in the test, and $f(\beta|\lambda,y_i)$ denotes a population model. This population model for a student i is usually estimated with the latent (called latent because the predictor is unobserved) regression model $\beta_i = y_i^T\lambda + \varepsilon_i$, where $\varepsilon_i = \mathcal{N}(0,\sigma^2)$ (Marsman 2014; OECD 2014b).

In other words, in each country, the student's abilities are assumed to follow a conditional Gaussian distribution, given y_i (i.e., the variables from the background questionnaire). This is the prior distribution. Then, the student takes the PISA test. The statistical model ("likelihood") of the success in the test is a Rasch model, where the probability of success is a logistic function of the unknown but estimated latent ability and the difficulties of the test items (see Eqs. 8.1 and 8.2).

The estimated posterior distribution of the ability of the student is specific for each student, as each student has different values for background variables and test results. This means that success in the PISA test "corrects" our prior beliefs regarding the student's ability. If a student successfully solves a difficult item, this indicates higher ability than success on an easy item. However, the student's exact ability is not known, and it is represented on the population level with five plausible values that are random realizations based on his or her posterior distribution. For this reason, the official PISA protocol (OECD 2012) requires that the same analysis be repeated five times when analyzing student performance, with one analysis for each plausible value.

8.4 Comparison of Students in PISA 2012 Countries Using Aggregated Hierarchical Clustering

The empirical part of this work is focused on comparing the student characteristics of Finland to those of the other countries that participated in the PISA assessment 2012. This comparison is conducted by utilizing three of the four LA techniques described by Chatti et al. (2012) (see Sect. 8.2.1): clustering as one of the core *data mining* techniques, *visualization* of the clustering result to illustrate Finland's position in comparison to the other countries, and, finally, *statistical* testing to verify the findings.

8.4.1 Variables for the Clustering

Our overall analysis method is to apply hierarchical clustering on all PISA 2012 countries/economies, to visualize the similarities between the participating countries through a dendrogram, and to conduct different statistical tests on two distinct levels. For this, we first aggregated the entire sample of half a million students in the PISA 2012 into the population level of each country by computing the weighted

Table 8.2 Overview and identification of the derived PISA variables utilized in this study

PISA variable	ID	PISA variable	ID
Economic, social, and cultural status	1		
Sense of belonging	2	Attitude toward school: learning outcome	3
Attitude toward school: learning activities	4	Perseverance	5
Openness to problem solving	6		
Self-responsibility for failing in math	7	Interest in mathematics	8
Instrumental motivation to learn math	9	Self-efficacy in mathematics	10
Anxiety toward mathematics	11	Self-concept in mathematics	12
Behavior in mathematics	13	Intentions to use mathematics	14
Subjective norms in mathematics	15	Mathematics work ethic	16
Out-of-school study time	17	Learning time (min. per week)—test language	18
Learning time (min. per week)—Mathematics	19	Learning time (min. per week)—science	20
Age at <ISCED 1>	21		
Teacher-student relations	22	Mathematics teacher's support	23
Teacher behavior: formative assessment	24	Teacher behavior: Student orientation	25
Teacher behavior: Teacher-directed instruction	26	Experience with applied math tasks at school	27

means of the available data in a country-wise manner. We used all observations in the PISA 2012 data set. All variables in the PISA student data set (and their possible values) can be found in the codebook.[10] In Saarela and Kärkkäinen (2014, 2015c, b) and Kärkkäinen and Saarela (2015), we utilized the individual variables on a student level that are known to explain performance in mathematics. Here, we used an extended set of variables, including those that are more on the scale of a classroom (e.g., teacher behavior) or a country (e.g., time of formal instruction in certain school subjects) than on an individual student level.

In Table 8.2, all variables used in this study are listed. All are derived variables constructed with the Rasch model using students' answers to the background questionnaire or other already-derived variables. For example, the first variable, the *index of economic, social, and cultural status*, is constructed using the *highest parental occupation*, the student's *home possessions*, and the *highest parental education*, which themselves are derived variables constructed with the Rasch model (OECD 2014b).

The following five variables (i.e., those with the IDs 2–6 in Table 8.2) are generally associated with performance on a student level, while the next ten variables (IDs 7–16) are all related to attitudes toward mathematics. Since mathematics

---

[10]Available at http://pisa2012.acer.edu.au/downloads/M_stu_codebook.pdf.

was the major domain in 2012, attitudes toward this subject received considerable attention in the background questionnaire. Here, we use all ten mathematics indices that together summarize 67 items in the student background questionnaire.

The next five variables in the table (IDs 17–21) are related to how much time students spend studying. Both formal learning time in different subject areas as well as out-of-school study hours are detailed. The last variable, *Age at ISCED 1*, reports the beginning of the systematic education in reading, writing, and mathematics. The last six variables (IDs 22–27) are all on the level of the teacher or teaching method.

8.4.2 Hierarchical Clustering

An issue with the PSA data is the aforementioned absence of a large number of values. Moreover, each student in the PISA data sets has a weight expressing how representative he or she is of the population of all 15-year-old students within his or her country. Therefore, we computed the weighted means of the available data for each variable for each country/economy as inputs for the clustering algorithm. We then normalized our data set using z-scoring and applied hierarchical clustering with Matlab's default settings (i.e., agglomerative single-linkage clustering with the Euclidean distance).

Agglomerative clustering techniques operate in a bottom-up fashion (Zaki and Meira 2014). Hence, we started with each PISA country as a separate cluster. Then, the most similar country clusters C_m and C_n were repeatedly merged so that they formed a new and bigger cluster. The most similar clusters were defined as the ones with the smallest Euclidean distance between a point in C_m and a point in C_n (8.3):

$$\delta(C_m, C_n) = \min\{\delta(\boldsymbol{u}, \boldsymbol{v}) | \boldsymbol{u} \in C_m, \boldsymbol{v} \in C_n\}, \tag{8.3}$$

where $\delta(u, v) = \left(\sum_{i=1}^{d} (u_i - v_i)^2\right)^{\frac{1}{2}}$ (see Zaki and Meira 2014).

To decide the number of clusters in the PISA 2012, the Davies-Bouldin cluster index (Davies and Bouldin 1979) was applied on the z-scored data. As can be seen from Fig. 8.4, the Davies-Bouldin index suggested that there are ten clusters in the data. Therefore, the merging of closest clusters was terminated after ten clusters were formed.

8.5 Results

In this section, we first visualize the hierarchical clustering result of the aggregated PISA countries in the form of a dendrogram. Then, we profile the country clusters according to their geographic and cultural similarities. Finally, we analyze the

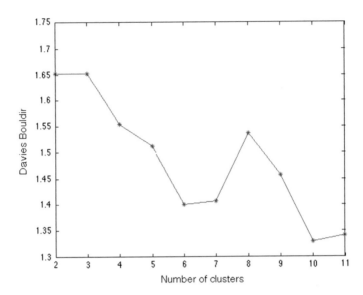

Fig. 8.4 The Davies-Bouldin index suggests that there are ten clusters in the data

clustering results more deeply using statistical tests on two different levels. Since Finland is our primary interest, we first evaluate the differences between all clusters, and then we analyze Finland's cluster and its position within its own cluster.

8.5.1 Visualization and Profiling of the Clusters

Figure 8.5 shows the hierarchical clustering result. Based on the similarities of countries in particular groups, we suggest the following labels for the ten clusters, as documented in Table 8.3.

It is a surprise that Finland is not part of the Nordic/English-speaking cluster to which all other Nordic countries belong. This finding is interesting compared to the classification of Bulle (2011), who introduces "the Northern model: Denmark, Finland, Iceland, Norway, Sweden" as one of the five main OECD educational systems. This indicates that even if the educational systems are similar, it does not necessarily follow that the student characteristics are also similar.

The dendrogram implies that Finland belongs to the Europe cluster and is actually closest to the Netherlands. In the PISA 2012 results summary (OECD 2014a, p. 7), the performances of these two countries in mathematics were found to not be statistically significantly different among many other pairs of countries. In addition, both the Netherlands and Finland are highly feminine cultures according to the *Hofstede model* (Hofstede 2011).

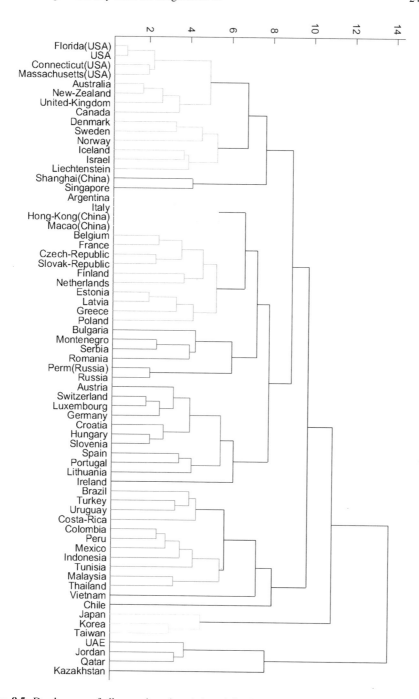

Fig. 8.5 Dendrogram of all countries when their weighted mean is clustered

Table 8.3 Clustering results

ID	Label	Countries/economies
C1	"Nordic/english-speaking"	Australia, Canada, United Kingdom, New Zealand, Florida (USA), Connecticut (USA), Massachusetts (USA), USA, Denmark, Iceland, Norway, Sweden
C2	—	Costa Rica, Israel, Uruguay
C3	"Eastern countries"	Bulgaria, Lithuania, Montenegro, Perm (Russia), Romania, Russia, Serbia
C4	"South America/Africa"	Argentina, Chile, Tunisia
C5	"Developing countries"	Brazil, Colombia, Indonesia, Mexico, Malaysia, Peru, Thailand, Turkey, Vietnam
C6	"High-performing Asian"	Shanghai (China), Singapore
C7	"Kazakhstan"	Kazakhstan
C8	"Arabic"	United Arab Emirates, Jordan, Qatar
C9	"Asian"	Hong Kong (China), Japan, Korea, (Macao) China, Taiwan
C10	"Europe"	Austria, Belgium, Switzerland, Czech Republic, Germany, Spain, Estonia, Finland, France, Greece, Croatia, Hungary, Ireland, Italy, Liechtenstein, Luxembourg, Latvia, Netherlands, Poland, Slovak Republic, Slovenia

As has been explained above, it was unexpected that Finland belonged to the Europe cluster and not to the Nordic/English-speaking cluster. We utilized statistical tests to assess the significance of the single variables and to explain why a particular country was allocated to a certain cluster. Since not all of our variables were normally distributed, we had to use non-parametric tests.

To specifically address the finding of Finland's position, we will first report the differences between all the clusters. Second, we will summarize the differences between Finland and its own Europe cluster; third, we will describe the variables that separate the Europe cluster from the Nordic/English-speaking cluster.

8.5.2 Differences Between All the Global Clusters

A Kruskal-Wallis H test (Kruskal and Wallis 1952) showed that there was a highly statistically significant difference in 20 of the 27 variables between the different clusters. The test statistics of all highly statistically significant variables are provided in Table 8.4. With reference to Table 8.4, variable 25, *teacher behavior: student orientation* (i.e., how much attention that teachers pay to individual students), was the most important in terms of accounting for variance in the cluster membership ($\chi^2(9) = 51,227$, $p < 0.001$).

Subsequently, pairwise comparisons were performed using Dunn's (1964) procedure with a Bonferroni correction for multiple comparisons. This post hoc analysis revealed highly statistically significant differences in the *ESCS* between the

Table 8.4 Kruskal-Wallis H test statistics (all clusters) with a post hoc test

Variable	χ^2 (9)	p	Post hoc test	Variable	χ^2 (9)	p	Post hoc test
1	48,676	★★★	C10–C5, C1–C5	4	38,499	★★★	C9–C1
5	33,306	★★★	–	7	37,399	★★★	–
8	48,701	★★★	C10–C5	9	49,857	★★★	C9–C5, C10–C5
10	30,765	★★★	–	11	42,170	★★★	C1–C5
12	35,298	★★★	–	13	49,549	★★★	C1–C5
14	34,029	★★★	–	15	49,082	★★★	C10–C5
16	39,863	★★★	–	18	40,457	★★★	–
19	36,542	★★★	–	22	42,940	★★★	C10–C5
23	46,378	★★★	C10–C5	24	45,203	★★★	–
25	51,227	★★★	C10–C5	26	42,610	★★★	–

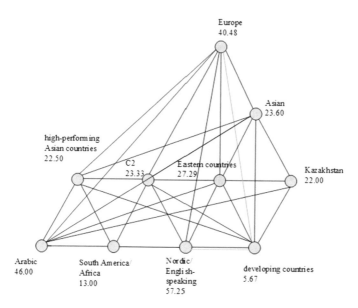

Fig. 8.6 Pairwise comparisons of clusters for ESCS. Statistically significant differences (between developing countries and the Nordic/english-speaking cluster and between developing countries and the Europe cluster) are marked in *yellow*

developing (mean rank = 5.67) and the Nordic/English-speaking clusters (mean rank = 57.25) as well as between the developing and the Europe (mean rank = 40.47) clusters, but not between any other group combination for this variable. This is also illustrated in Fig. 8.6, in which all pairwise comparisons of the different clusters for their ESCS are shown. In the figure, black lines reflect a pairwise comparison that is not statistically significant, while orange lines reflect a statistically significant pairwise comparison.

The last column in Table 8.4 summarizes the post hoc analysis for all the variables. As can be seen from the table, highly statistically significant differences were found in the *attitude toward school: learning activities* (i.e., the degree to which a student sees hard work in school pay off later) between the Asian (mean rank = 5.00) and the Nordic/English-speaking clusters (mean rank = 51.08), in the *interest in* and *enjoyment of mathematics* between the developing countries (mean rank = 56.89) and Europe (mean rank = 14.90) clusters, in the *instrumental motivation to learn mathematics* (i.e., the degree to which a student's hard work in mathematics pays off later) between the developing (mean rank = 57.89) and the Asian (mean rank = 7.80) countries, and between the developing countries and the Europe (mean rank = 19.10) clusters.

Highly statistically significant differences were found for the developing countries cluster when compared with the Nordic/English-speaking cluster with regard to *anxiety toward mathematics* (mean rank C5 = 55.00 vs. C1 = 14.92) and *behavior in mathematics* (i.e., the role of mathematics inside and outside school) (mean rank C5 = 54.11 vs. C1 = 12.17). In addition, highly statistically significant differences were found for the developing countries cluster when compared with the Europe cluster with regard to *subjective norms in mathematics* (mean rank C5 = 51.11 vs. C10 = 15.81) (i.e. how much attention to mathematics is given by friends and family), *teacher-student relations* (mean rank C5 = 51.44 vs. C10 = 14.90), *mathematics teacher's support* (mean rank C5 = 52.22 vs. C10 = 14.43), and *teacher behavior: student orientation* (mean rank C5 = 54.33 vs. C10 = 15.14), respectively. No highly statistically significant differences were found for any other group combination.

Hence, the statistical test on a global level suggests that, overall, the Europe cluster and the developing countries cluster are the most dissimilar to each other. Students in the Europe cluster have a higher economic, social, and cultural status— but students in the developing countries cluster have higher interests, more motivation to learn, and higher subjective norms in mathematics from their friends and family. Furthermore, students in the developing countries tend to report better relations with their teachers.

When comparing Finland to other countries, the rather negative attitudes toward mathematics were already observed in the 2003 assessment cycle. In both *interest in* and *enjoyment of* mathematics, Finland was ranked 37th out of the 40 participating countries (Linnakylä et al. 2011).

Moreover, in a longitudinal study of Finnish students in grade 1 to grade 12 by Metsämuuronen et al. (2012), it was concluded that student contentment in regard to school in Finland decreases significantly from the second to the eighth grade, while it then very slightly increases starting in the ninth grade. The majority (82%)[11] of the Finnish students participating in the PISA are in the ninth grade, and almost all the rest are in the eighth grade (16%). Hence, Finnish students are at the

---

[11]Our own calculation from the PISA 2012 data.

Table 8.5 Wilcoxon signed-rank statistics (Europe—Finland clusters)

Variable	1	7	10	11	16	17	18	24
Z	−3.920	3.920	3.920	3.771	3.808	3.920	3.845	3.920
P	★★★	★★★	★★★★	★★★	★★★	★★★	★★★	★★★

stage in their basic education where their self-reported attitudes toward school are very poor.

Metsämuuronen et al. (2012) suggest that these generally negative attitudes of Finnish students toward education are due to their modesty and honesty: "Part of the explanation in Finland [...] can be the appreciation of honesty and speaking frankly [...] pupils in Finland [...] are relatively humble when they describe their knowledge. This 'humbleness' may also be reflected in attitude measurements."

8.5.3 Differences Between Finland and the Other Countries Within the Europe Cluster

According to the clustering result, Finland is most similar to the countries in the Europe cluster. Table 8.5 summarizes the highly statistically significant variables according to which Finland differs from the remaining countries within its own cluster, as determined by the Wilcoxon signed-rank tests.

As can be seen in Table 8.5, the majority of the Europe cluster has a significantly lower *ESCS* than Finland ($z = -3.92, p < 0.001$). Nevertheless, the Europe cluster majority has a significantly higher *self-responsibility for failing in mathematics* ($z = 3.92, p < 0.001$), *anxiety toward mathematics* ($z = 3.771, p < 0.001$), and *self-efficacy in mathematics* ($z = 3.92, p < 0.001$) than Finland. Furthermore, the Europe cluster in general shows higher scores in many variables that measure emphasis of formal assessment and how much time students spend studying.

In particular, there is a significantly higher *work ethic in mathematics* ($z = 3.808, p < 0.001$) and more *out-of-school study hours* in the Europe cluster than in Finland ($z = 3.920, p < 0.001$). The latter is illustrated in Fig. 8.7, where the weighted average out-of-school study hours for students in all participating PISA countries are plotted. As can be seen from the figure, Finnish students not only study the least outside of school within their own Europe cluster but also compared to all other countries participating in the PISA.

In addition, *learning time (min. per week)—test language* in Europe is significantly greater than in Finland ($z = 3.845, p < 0.001$, see Table 8.5), and Europe has a significantly higher score in *teacher behavior: formative assessment* than Finland ($z = 3.920, p < 0.001$). In summary, these results support the observations by Sahlberg (2011), who writes that educational decision makers in Finland "do not seem to believe that doing more of the same in education would necessarily make any significant difference for improvement."

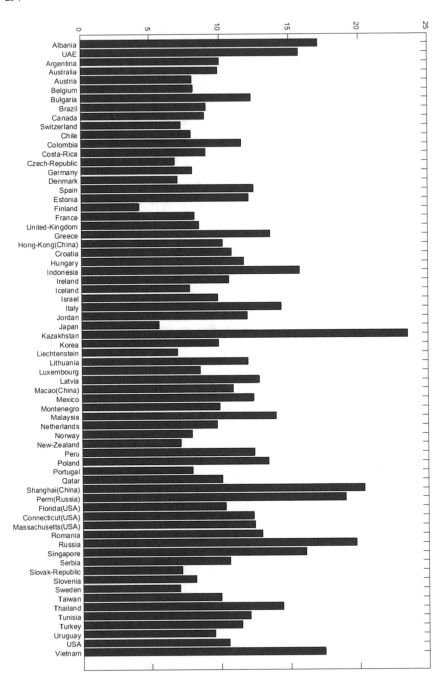

Fig. 8.7 Weighted averages of the out-of-school study hours for all in PISA-participating countries. In comparison to all the other countries, Finnish students study the least after school

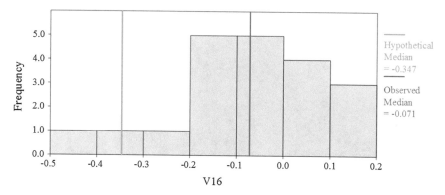

Fig. 8.8 One-Sample Wilcoxon rank test for work ethic: the work ethic of students in Finland is significantly *lower* than the work ethic of students in the Europe cluster

As can be seen from the Wilcoxon signed-rank test result and Fig. 8.8, 15-year-old students in Finland seem to already have a rather relaxed attitude toward formal assessment and investing time in their studies. This is particularly evident in the highly statistically significantly lower work ethic[12] of Finnish students.

It must also be kept in mind that the systematic teaching of reading, writing, and mathematics begins later in Finland than in Europe ($z = -3.435, p < 0.001$). This is illustrated in Fig. 8.9. In Finland, children are seven years old when they start school. Combined with the finding that the hours of formal instruction of certain subjects are, as described in the above paragraph, significantly lower in Finland, this means that Finnish students spend less time at school than students in other countries. This finding has also been emphasized by Kumpulainen and Lankinen (2012).

8.5.4 Europe Cluster in Comparison to the Nordic/English-Speaking Cluster

A Mann-Whitney U test was run to determine if there were differences in the 27 variables between the Europe and the Nordic/English-speaking clusters.

---

[12]The *work ethics* scale index is computed with the Rasch model and by using the extent to which students agree or disagree with the following statements: *I finish my homework in time for mathematics class; I work hard on my mathematics homework; I am prepared for my mathematics exams; I study hard for mathematics quizzes; I keep studying until I understand mathematics material; I pay attention in mathematics class; I listen in mathematics class; I avoid distractions when I am studying mathematics;* and *I keep my mathematics work well organized.*

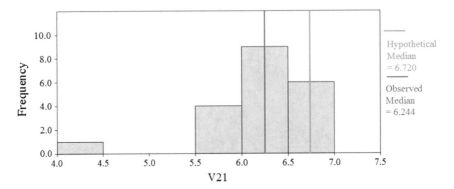

Fig. 8.9 One-sample Wilcoxon rank test for *age at <ISCED 1>*: systematic teaching of reading, writing, and mathematics begins significantly later in Finland than in Europe

Table 8.6 Mann-Whitney U test results comparing the Europe cluster to the Nordic/english-speaking cluster

PISA variable ID	4	8	9	12	15	16	18	22	23	25
U	19	27	5	30	1	38	22	20	28	20
Z	−4.004	−3.705	−4.528	−3.593	−4.678	−3.293	−3.892	−3.967	−3.668	−3.967
p	★★★	★★★	★★★	★★★	★★★	★★★	★★★	★★★	★★★	★★★

Distributions of the 27 variables for the two groups were not similar, as assessed by visual inspection. The test statistics can be found in Table 8.6.

When we combine the test results of the Mann-Whitney U test of the Nordic/English-speaking versus Europe and the Wilcoxon signed-rank test of Europe versus Finland, we find that two variables (16 and 18) augment Finland's special characteristics: *work ethic* and *study time (test language)* are statistically significantly lower in Europe and even lower in Finland. As described above, these variables measure how much time students spend studying and how much they strive for high grades in mathematics.

According to the Mann-Whitney U test, there was a significant ($p < 0.001$) difference in *attitude toward school: learning activities, interest in and enjoyment of mathematics, instrumental motivation to learn mathematics, self-concept in mathematics, subjective norms in mathematics, mathematics work ethic, test language learning time, teacher-student relations, mathematics teacher's support*, and *teacher behavior: student orientation* between the two clusters. In all these variables, the Nordic/English-speaking cluster showed higher values than the Europe cluster. With reference to Table 8.6, *subjective norms in mathematics* seems to be the most important variable that separates the Nordic/English-speaking cluster from the Europe cluster.

The comparisons of the Nordic/English-speaking cluster to the Europe cluster mostly revealed variables that estimate the students' own perception of their merits and importance. It is especially interesting that the self-reported self-concept is significantly lower in Finland, because this PISA 2012 variable actually explains the performance of Finnish students in the PISA mathematics test fairly well, and it is the mathematics scale index that correlates the most with their plausible values in mathematics (Saarela and Kärkkäinen 2014). However, it seems that even if Finnish students evaluate their own skills realistically, they are more modest about them. Generally, students in the Nordic/English-speaking cluster tend to have higher opinions about themselves, are more motivated, and report better relations with their teachers.

The average mathematics performance based on the plausible values of the countries in the Nordic/English-speaking cluster is 495.3, while the mean mathematics performance of the countries in the Europe cluster is higher (500.5). We conclude that learning time and positive student-teacher relations seem to be less important features than collaborative skills or being free from arrogance for explaining students' success in the PISA test.

8.6 Visual LA of the PISA Results

The macro-level LA of the Finnish basic educational system is visualized in the dashboard of Figs. 8.10, 8.11, 8.12 and 8.13 through the lens of the cultural background, the PISA, and our empirical analysis. This dashboard consists of four figures, and its composition was inspired by Ferguson and Shum (2012).

Finland has been a top-performing PISA country in the last five assessment cycles (Fig. 8.10), although the ranking clearly decreased in 2012, especially in mathematics. A certain interesting success factor of the educational system is the cultural deviation from the world's midlevel as a feminine culture with a low power distance (Fig. 8.11). The system is based on the strong autonomy and authority of highly educated teachers, with a small amount of formal assessment and, in particular, a complete lack of national comparative assessments of the learning results (Fig. 8.12). In addition, a rich common curriculum is present for untracked groups

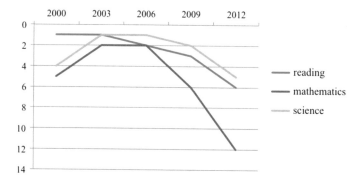

Fig. 8.10 Finland's ranking in the PISA cycles

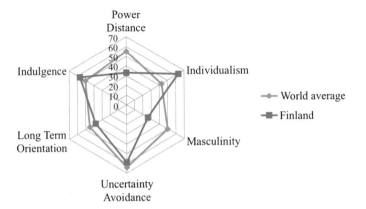

Fig. 8.11 Finland's scores in the Hofstede model dimensions

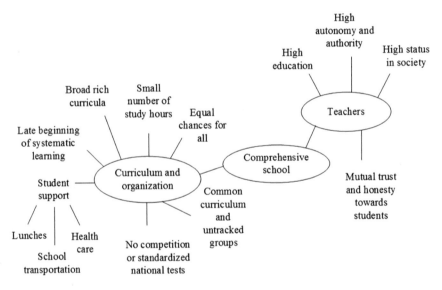

Fig. 8.12 Characteristics of the Finnish educational system

of students, who start late in their systematic learning of reading, mathematics, and science. As a whole, equity and equality characterize the system, which provides strong student support (e.g., in the form of free lunches, health care, and school transportation) (Fig. 8.12).

However, many contradictory factors about the Finnish students in relation to their high PISA results emerged in the empirical LA analysis (Fig. 8.13): they have a low motivation to learn and excel in school, a low interest in school topics, a low work ethic, and an exceptionally small number of extra-school study hours. The importance of their studies, and specifically mathematics, is considered low for their

Fig. 8.13 Finland's student characteristics from clustering. The *red bubbles* indicate alarming characteristics and the *yellow bubble* indicates the characteristic that could be improved

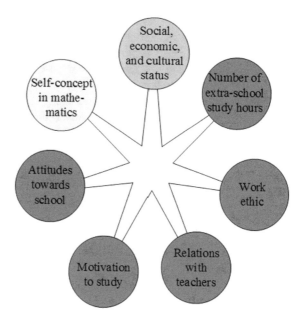

future career. The overall evaluation of the different facets of the dashboard indicates that the lowering trend of the PISA, and particularly the mathematics performance of Finnish students, may continue. To improve the system, so as to perhaps be ranked once again as number one in the PISA, students need to be more motivated and oriented toward schoolwork, extra-school study hours, and mathematics, and to keep their future career orientation clearly in mind. We also hypothesize that the complete common, joint, and untracked subject orientations demotivate the most talented students by requiring minimal effort from them. All these factors provide further challenges to subsequent upper secondary and higher education.

8.7 Discussion

We briefly summarize the empirical findings from the previous sections. These were obtained by utilizing one of the illuminated educational clustering techniques, hierarchical clustering, and by taking into account all the specific demands of the PISA data discussed above. As suggested by the Davies-Bouldin cluster validation index, we first divided the students of all the PISA-participating countries into ten separate groups. The clusters that were found generally could be explained by the culture and geographical location of the countries in them. Nevertheless, Finland surprisingly belonged to the Europe cluster (see Fig. 8.5), while all the other Scandinavian countries belonged to the cluster of Nordic/English-speaking countries. This illustrates how similar educational systems (see Bulle 2011) can be reflected by different student characterizations.

Statistical significance tests of the clustering result revealed why particular countries were allocated to a certain cluster. At first, it seemed that the results of the statistical test were somehow contradictory, as students in better-performing countries had worse student-teacher relations and generally showed less confidence in their own achievements and skills. Moreover, the work ethic of the students in the better-performing Europe cluster was significantly lower than that of the students in the Nordic/English-speaking countries cluster—and the better-performing Finnish students showed a work ethic that was significantly worse than the remaining students in the Europe cluster. However, these findings seem to be connected to and explicable by the existing research related to Finnish culture in general.

As was explained in the literature review about the Finnish educational system and culture, Finnish citizens are modest about their own achievements, and they place great emphasis on equity and equality. The most important driving factors in the life of this highly feminine country are to live a good life and to care for others rather than to focus on one's own success and desire to be the best. This is interesting because, as emphasized in our literature review, French et al. (2015) found a negative causal relationship between education expenditures and power distance and masculinity. Furthermore, Finnish students seem to have an extremely relaxed attitude toward formal assessment and investing time in studies, as can be expected in a feminine country.

Finally, the main success of Finnish students in the PISA seems to a great extent to be related to the relatively better scores of the lowest-scoring Finnish students in comparison with other countries (Andersen 2010), which in turn is supported by the collaborative and ostentation-free thinking in the country. However, as illustrated in Fig. 8.10, Finland's ranking significantly dropped in the latest PISA 2012 assessment (OECD 2013b), and according to the overall characterization of the Finnish students as given and visualized in Fig. 8.13, the negative trend in performance might have continued in the PISA 2015.[13]

8.8   Conclusions

LA is a growing and expanding research field. Traditionally, many studies have concentrated on analyzing educational data originating from a macro or (at the most) meso level. The publicly available and high-quality PISA data sets, on the other hand, provide the opportunity to conduct big data LA research on the macro level, because they comprise data on a whole population of international students.

---

[13]Data from the PISA 2015 will be published by the OECD in December 2016 (National Center for Education Statistics 2016).

In this chapter, we have introduced the background for conducting large-scale LA research on the PISA. We have described the main data sets as well as the complexities within them and discussed how to work with these data. Moreover, we have provided a review of relevant clustering studies within the educational domain. Our empirical work, as discussed in the previous section, provided novel findings and strengthened earlier knowledge on the particularities of the Finnish educational system, which has received a great deal of attention during the 21st century due to the exceptionally good performance of the Finnish students in the PISA tests.

We used quantitative LA methods to identify the main attributes of individual learners that affect their learning experience in the environment where the learning occurs (Fournier et al. 2011). Similar to the reviewed educational clustering studies in Sect. 8.2.2, we analyzed the student model; however, in contrast to these previously reviewed studies, our model represented a prototype of a national student population obtained by weighted aggregation. Concerning Finland, the high-achieving country inside the PISA assessments, it was concluded that an educational system promoting student collaboration, humility, and equity can successfully cope with the challenges of negative attitudes toward mathematics, low work ethic, and little study time outside school. This summarizes the evidence-based knowledge discovered about the long-term impact of educational policies and practices on the achievement targets (Piety et al. 2014). Such a conclusion also provides an example of a national education system assessment using big data LA as illustrated in Fig. 8.1. The international–objectives driven data collection and transformation improves understanding of educational arrangements via proper analysis methods that are able to cope with the specialties of the sampled large-scale data.

Big data LA, as described in Sect. 8.1 and depicted in Fig. 8.1, linking together the four dimensions of LA proposed by Chatti et al. (2014) (see also Greller and Drachsler 2012), encapsulated and supported the overall management of the large-scale educational system assessment based on the PISA data. Our empirical work exemplifies the multiple facets of LA: hierarchical clustering as a data mining technique, visualization of the dendrogram to illustrate the clustering result, and statistical testing to verify the findings. Thus, our work increased the body of knowledge for the macro level of educational systems. We promoted reflection of the main characteristics that differentiate the students in various educational environments, according to the objectives of LA by Chatti et al. (2014) (see Sect. 8.3.3). Our reflections of the PISA results were emphasized in the dashboard in Figs. 8.10, 8.11, 8.12 and 8.13 using different LA visualization tools. This dashboard facilitates awareness and monitoring of critical educational aspects for the Finnish 15-year-old student population (Beheshitha et al. 2016).

As a whole, the PISA—as well as the other large-scale-assessments, such as those mentioned in Sect. 8.3.4—provides a very rich and interesting source for macro-level LA studies. We think that the methods and the framework developed for the publicly available large-scale assessment data sets can and will advance the open architecture of educational applications, which Peña-Ayala (2014) has identified as one of the shortcomings of the current educational data analysis research area.

As part of our future research, we intend to repeat our study using the individual students instead of the country-level aggregation as data for clustering. Furthermore, one of the recent trends in LA focuses on educational process mining (Sedrakyan et al. 2016; Mukala et al. 2015; Trčka et al. 2010). For the traditional pen-and-paper PISA tests, this is not an option. However, for the future PISA cycles, where the tests will be increasingly conducted electronically and log event data will therefore be available (compare the PISA 2012 problem-solving test, which was conducted electronically and log files can be downloaded from the above-cited OECD webpage), this would provide an interesting and promising direction for future research.

Acknowledgements The authors would like to thank Dr. Jouni Välijärvi, Dr. Kari Nissinen, and Dr. Antero Malin from the Finnish Institute for Educational Research for the valuable discussions and support concerning the PISA data and assessment. We also acknowledge B.Sc Susanne Jauhiainen for technical assistance in preparing the final version of the chapter.

References

Allen LK, Mills C, Jacovina ME, Crossley S, D'Mello S, McNamara DS (2016) Investigating boredom and engagement during writing using multiple sources of information: the essay, the writer, and keystrokes. In: Proceedings of the sixth international conference on learning analytics & knowledge, ACM, pp 114–123

Almeda MV, Scupelli P, Baker RS, Weber M, Fisher A (2014) Clustering of design decisions in classroom visual displays. In: Proceedings of the fourth international conference on learning analytics & knowledge, ACM, pp 44–48

Andere E (2015) Are teachers crucial for academic achievement? Finland educational success in a comparative perspective. Educ Policy Anal Arch 23(39):1–27

Andersen FØ (2010) Danish and Finnish PISA results in a comparative, qualitative perspective: how can the stable and distinct differences between the Danish and Finnish PISA results be explained? Educ Assess Eval Accountability 22(2):159–175

Auld E, Morris P (2016) PISA, policy and persuasion: translating complex conditions into education 'best practice'. Comp Educ 52(2):202–229

Bank V (2012) On OECD policies and the pitfalls in economy-driven education: the case of Germany. J Curriculum Stud 44(2):193–210

Beheshitha SS, Hatala M, Gašević D, Joksimović S (2016) The role of achievement goal orientations when studying effect of learning analytics visualizations. In: Proceedings of the sixth international conference on learning analytics & knowledge, ACM, pp 54–63

Bergner Y, Colvin K, Pritchard DE (2015) Estimation of ability from homework items when there are missing and/or multiple attempts. In: Proceedings of the fifth international conference on learning analytics & knowledge, ACM, pp 118–125

Blikstein P, Worsley M, Piech C, Sahami M, Cooper S, Koller D (2014) Programming pluralism: using learning analytics to detect patterns in the learning of computer programming. J Learn Sci 23(4):561–599

Bogarin A, Romero C, Cerezo R, Sanchez-Santillan M (2014) Clustering for improving educational process mining. In: Proceedings of the fourth international conference on learning analytics & knowledge, ACM, pp 11–15

Bouchet F, Harley JM, Trevors GJ, Azevedo R (2013) Clustering and profiling students according to their interactions with an intelligent tutoring system fostering self-regulated learning. J Educ Data Min 5(1):104–146

Brown MG, DeMonbrun RM, Lonn S, Aguilar SJ, Teasley SD (2016) What and when: the role of course type and timing in students' academic performance. In: Proceedings of the sixth international conference on learning analytics & knowledge, ACM, pp 459–468

Bulle N (2011) Comparing OECD educational models through the prism of PISA. Comp Educ 47(4):503–521

Chandra E, Nandhini K (2010) Knowledge mining from student data. Eur J Sci Res 47(1):156–163

Chatti MA, Dyckhoff AL, Schroeder U, Thüs H (2012) A reference model for learning analytics. Int J Technol Enhanced Learn 4(5–6):318–331

Chatti MA, Lukarov V, Thues H, Muslim A, Yousef AMF, Wahid U, Greven C, Chakrabarti A, Schroeder U (2014) Learning analytics: challenges and future research directions. E-learn Educ (Eleed) J 10:1–16

Coffrin C, Corrin L, de Barba P, Kennedy G (2014) Visualizing patterns of student engagement and performance in MOOCs. In: Proceedings of the fourth international conference on learning analytics and knowledge, ACM, pp 83–92

Colthorpe K, Zimbardi K, Ainscough L, Anderson S (2015) Know thy student! combining learning analytics and critical reflections to increase understanding of students self-regulated learning in an authentic setting. J Learn Anal 2(1):134–155

Davies DL, Bouldin DW (1979) A cluster separation measure. IEEE Trans Pattern Anal Mach Intell PAMI 1(2):224–227

Dawson S, Gašević D, Siemens G, Joksimović S (2014) Current state and future trends: a citation network analysis of the learning analytics field. In: Proceedings of the fourth international conference on learning analytics & knowledge, ACM, pp 231–240

Dawson S, Gašević D, Mirriahi N (2015) Challenging assumptions in learning analytics. J Learn Anal 2(3):1–3

Deng Z, Gopinathan S (2016) PISA and high-performing education systems: explaining Singapores education success. Comp Educ 52(4):449–472

Desmarais MC, Lemieux F (2013) Clustering and visualizing study state sequences. In: Proceedings of the 6th international conference on educational data mining, pp 224–227

Drabowicz T (2014) Gender and digital usage inequality among adolescents: A comparative study of 39 countries. Comput Educ 74:98–111

Dunn OJ (1964) Multiple comparisons using rank sums. Technometrics 6(3):241–252

Embretson SE, Reise SP (2013) Item response theory. Psychology Press

Erdogdu F, Erdogdu E (2015) The impact of access to ICT, student background and school/home environment on academic success of students in Turkey: an international comparative analysis. Comput Educ 82:26–49

Ferguson R (2012) Learning analytics: drivers, developments and challenges. Int J Technol Enhanced Learn 4(5–6):304–317

Ferguson R, Shum SB (2012) Social learning analytics: five approaches. In: Proceedings of the second international conference on learning analytics & knowledge, ACM, pp 23–33

Ferguson R, Macfadyen L, Clow D, Tynan B, Alexander S, Dawson S (2014) Setting learning analytics in context: Overcoming the barriers to large-scale adoption. J Learn Anal 1(3): 120–144

Ferguson R, Clow D (2015) Examining engagement: analysing learner subpopulations in massive open online courses (MOOCs). In: Proceedings of the fifth international conference on learning analytics & knowledge, ACM, pp 51–58

Fonseca J, Valente MO, Conboy J (2011) Student characteristics and PISA science performance: Portugal in cross-national comparison. Proc Soc Behav Sci 12:322–329

Fournier H, Kop R, Sitlia H (2011) The value of learning analytics to networked learning on a personal learning environment. In: Proceedings of the first international conference on learning analytics & knowledge, pp 104–109

French JJ, French A, Li WX (2015) The relationship among cultural dimensions, education expenditure, and PISA performance. Int J Educ Dev 42:25–34

Grawemeyer B, Mavrikis M, Holmes W, Gutierrez-Santos S, Wiedmann M, Rummel N (2016) Affecting off-task behaviour: how affect-aware feedback can improve student learning. In: Proceedings of the sixth international conference on learning analytics & knowledge, ACM, pp 104–113

Gray G, McGuinness C, Owende P, Carthy A (2014) A review of psychometric data analysis and applications in modelling of academic achievement in tertiary education. J Learn Anal 1(1): 75–106

Grek S (2009) Governing by numbers: the PISA effect in Europe. J Educ Policy 24(1):23–37

Greller W, Drachsler H (2012) Translating learning into numbers: a generic framework for learning analytics. Educ Technol Soc 15(3):42–57

Gupta D, Sharma A, Unny N, Manjunath G (2014) Graphical analysis and visualization of big data in business domains. In: Big data analytics, lecture notes in computer science (8883). Springer, Berlin, pp 53–56

Hansen JD, Reich J (2015) Socioeconomic status and MOOC enrollment: enriching demographic information with external datasets. In: Proceedings of the fifth international conference on learning analytics & knowledge, ACM, pp 59–63

Hecking T, Chounta IA, Hoppe HU (2016) Investigating social and semantic user roles in MOOC discussion forums. In: Proceedings of the sixth international conference on learning analytics & knowledge, ACM, pp 198–207

Hershkovitz A, Knight S, Dawson S, Jovanović J, Gašević D (2016) About" learning" and" analytics. J Learn Anal 3(2):1–5

Hickey DT, Kelley TA, Shen X (2014) Small to big before massive: scaling up participatory learning analytics. In: Proceedings of the fourth international conference on learning analytics & knowledge, ACM, pp 93–97

Hitlin S, Piliavin J (2004) Values: reviving a dormant concept. Ann Rev Sociol 30:359–393

Hofstede G (2011) Dimensionalizing cultures: the Hofstede model in context. Online Readings Psychol Cult 2(1):8

Hu X, Zhang Y, Chu SKW, Ke X (2016) Toward personalizing an e-quiz bank for primary school students: an exploration with association rule mining and clustering. In: Proceedings of the sixth international conference on learning analytics & knowledge, ACM, pp 25–29

Joksimović S, Manataki A, Gašević D, Dawson S, Kovanović V, de Kereki IF (2016) Translating network position into performance: importance of centrality in different network configurations. In: Proceedings of the sixth international conference on learning analytics & knowledge, ACM, pp 314–323

Kärkkäinen T, Saarela M (2015) Robust principal component analysis of data with missing values. In: Lecture notes in artificial intelligence (9166). Springer International Publishing, pp 140–154

Kjærnsli M, Lie S (2004) PISA and scientific literacy: similarities and differences between the nordic countries. Scand J Educ Res 48(3):271–286

Kriegbaum K, Jansen M, Spinath B (2015) Motivation: a predictor of PISA's mathematical competence beyond intelligence and prior test achievement. Learn Individ Differ 43:140–148

Kruskal W, Wallis W (1952) Use of ranks in one-criterion variance analysis. J Am Stat Assoc 47(260):583–621

Kumpulainen K, Lankinen T (2012) Striving for educational equity and excellence. In: Miracle of education. Springer, pp 69–81

Laney D (2001) 3D data management: controlling data volume, velocity and variety. Technical report, META Group

Li N, Cohen WW, Koedinger KR (2013) Discovering student models with a clustering algorithm using problem content. In: Proceedings of the 6th international conference on educational data mining, pp 98–105

Linnakylä P, Välijärvi J, Arffman I (2011) Finnish basic education—when equity and excellence meet. Equity and excellence in education: towards maximal learning opportunities for all students. Routledge, New York, pp 190–214

Long P, Siemens G (2011) Penetrating the fog: analytics in learning and education. Educ Rev 46(5):30–40

López MI, Luna JM, Romero C, Ventura S (2012) Classification via clustering for predicting final marks based on student participation in forums. In: Proceedings of the 5th international conference on educational data mining, pp 148–151

Marsman M (2014) Plausible values in statistical inference. Universiteit Twente

Merceron A, Blikstein P, Siemens G (2016) Learning analytics: from big data to meaningful data. J Learn Anal 2(3):4–8

Metsämuuronen J, Svedlin R, Ilic J (2012) Change in pupils' and students' attitudes toward school as a function of age—a Finnish perspective. J Educ Dev Psychol 2(2):134–151

Monseur C, Adams R (2008) Plausible values: how to deal with their limitations. J Appl Meas 10(3):320–334

Morgan H (2014) Review of research: the education system in finland: a success story other countries can emulate. Child Educ 90(6):453–457

Mukala P, Buijs J, Leemans M, van der Aalst W (2015) Learning analytics on Coursera event data: a process mining approach

Musik A (2016) Philologenverband bezeichnet pisa-studie als geldverschwendung. http://www.deutschlandfunk.de/bildungsforschung-in-der-kritikphilologenverband.680.de.html?dram:articleid=347675

National Center for Education Statistics (2016) Program for international student assessment. https://nces.ed.gov/surveys/pisa/

OECD (2011) Finland: slow and steady reform for consistently high results. In: Successful reformers in education: lessons from PISA for the United States, OECD, pp 117–135

OECD (2012) PISA 2009 technical report. OECD Publishing

OECD (2013a) PISA 2012 results: ready to learn—students' engagement, drive and self-beliefs (volume III). OECD Publishing, PISA

OECD (2013b) PISA 2012 results: what students know and can do (volume I) student performance in mathematics, reading and science: student performance in mathematics, reading and science. v. 1, OECD Publishing

OECD (2014a) PISA 2012 results in focus: what 15-year-olds know and what they can do with what they know. OECD Publishing. Paris, France

OECD (2014b) PISA 2012 technical report

Olsen RV (2005a) Achievement tests from an item perspective: an exploration of single item data from the PISA and TIMSS studies, and how such data can inform us about students' knowledge and thinking in science. PhD thesis, University of Oslo

Olsen RV (2005b) An exploration of cluster structure in scientific literacy in PISA: evidence for a Nordic dimension? Nord Stud Sci Educ 1(1):81–94

Pardo A, Teasley S (2014) Learning analytics research, theory and practice: widening the discipline. J Learn Anal 1(3):4–6

Peña-Ayala A (2014) Educational data mining: a survey and a data mining-based analysis of recent works. Expert Syst Appl 41(4):1432–1462

Picciano AG (2012) The evolution of big data and learning analytics in american higher education. J Asynchronous Learn Netw 16(3):9–20

Piety PJ, Hickey DT, Bishop M (2014) Educational data sciences—framing emergent practices for analytics of learning, organizations, and systems. In: Proceedings of the fourth international conference on learning analytics & knowledge, ACM, pp 193–202

Rasch G (1960) Studies in mathematical psychology: I. Probabilistic models for some intelligence and attainment tests

Rasmussen J, Bayer M (2014) Comparative study of teaching content in teacher education programmes in Canada, Denmark, Finland and Singapore. J Curriculum Stud 46(6):798–818

Rogers T (2015) Critical realism and learning analytics research: epistemological implications of an ontological foundation. In: Proceedings of the fifth international conference on learning analytics & knowledge, ACM, pp 223–230

Reich J, Tingley DH, Leder-Luis J, Roberts ME, Stewart B (2014) Computer-assisted reading and discovery for student generated text in massive open online courses. J Learn Anal 2(1): 156–184

Rutkowski L, Gonzalez E, Joncas M, von Davier M (2010) International large-scale assessment data issues in secondary analysis and reporting. Educ Res 39(2):142–151

Saarela M, Kärkkäinen T (2014) Discovering gender-specific knowledge from finnish basic education using PISA scale indices. In: Proceedings of the 7th international conference on educational data mining, pp 60–68

Saarela M, Kärkkäinen T (2015a) Analysing student performance using sparse data of core bachelor courses. J Educ Data Min 7(1):3–32

Saarela M, Kärkkäinen T (2015b) Do country stereotypes exist in PISA? A clustering approach for large, sparse, and weighted data. In: Proceedings of the 8th international conference on educational data mining (EDM 2015), pp 156–163

Saarela M, Kärkkäinen T (2015c) Weighted clustering of sparse educational data. In: Proceedings of the European symposium on artificial neural networks, computational intelligence and machine learning, pp 337–342

Saarela M, Kärkkäinen T, Lahtonen T, Rossi T (2016a) Expert-based versus citation-based ranking of scholarly and scientific publication channels. J Informetrics 10(3):693–718

Saarela M, Yener B, Zaki MJ, Kärkkäinen T (2016b) Predicting math performance from raw large-scale educational assessments data: a machine learning approach. In: MLDEAS workshop of the 33rd international conference on machine learning, pp 1–8

Sahlberg P (2011) Finnish lessons. Teachers College Press

Santos JL, Klerkx J, Duval E, Gago D, Rodriiguez L (2014) Success, activity and drop-outs in MOOCs an exploratory study on the UNED COMA courses. In: Proceedings of the fourth international conference on learning analytics & knowledge, ACM, pp 98–102

Schatz M, Popovic A, Dervin F (2016) From PISA to national branding: exploring Finnish education. Discourse: Studies in the Cultural Politics of Education pp 1–13

Sedrakyan G, Weerdt JD, Snoeck M (2016) Process-mining enabled feedback: tell me what I did wrong vs. tell me how to do it right. Comput Hum Behav 57:352–376

Segedy JR, Kinnebrew JS, Biswas G (2015) Using coherence analysis to characterize self-regulated learning behaviours in open-ended learning environments. J Learn Anal 2(1): 13–48

Siemens G (2013) Learning analytics: the emergence of a discipline. Am Behav Sci 57:1380–1400

Siemens G, Baker RS (2012) Learning analytics and educational data mining: towards communication and collaboration. In: Proceedings of the 2nd international conference on learning analytics & knowledge, ACM, pp 252–254

Simola H (2005) The Finnish miracle of PISA: historical and sociological remarks on teaching and teacher education. Comp Edu 41(4):455–470

Skryabin M, Zhang J, Liu L, Zhang D (2015) How the ICT development level and usage influence student achievement in reading, mathematics, and science. Comput Educ 85:49–58

Tømte C, Hatlevik O (2011) Gender-differences in Self-efficacy ICT related to various ICT-user profiles in Finland and Norway. How do self-efficacy, gender and ICT-user profiles relate to findings from PISA 2006. Comput Educ 57(1):1416–1424

Trčka N, Pechenizkiy M, van der Aalst W (2010) Process mining from educational data. Chapman & Hall/CRC

Välijärvi J, Kupari P, Linnakylä P, Reinikainen P, Sulkunen S, Törnroos J, Arffman I (2007) The Finnish success in PISA—and some reasons behind it: PISA 2003. Jyväskylän yliopisto, Koulutuksen tutkimuslaitos

Verbert K, Manouselis N, Drachsler H, Duval E (2012) Dataset-driven research to support learning and knowledge analytics. Educ Technol Soc 15(3):133–148

Vogelsang T, Ruppertz L (2015) On the validity of peer grading and a cloud teaching assistant system. In: Proceedings of the fifth international conference on learning analytics & knowledge, ACM, pp 41–50

Waldow F, Takayama K, Sung YK (2014) Rethinking the pattern of external policy referencing: media discourses over the Asian Tigers: PISA success in Australia, Germany and South Korea. Comp Educ 50(3):302–321

Wang Y, Paquette L, Baker R (2014) A longitudinal study on learner career advancement in MOOCs. J Learn Anal 1(3):203–206

Wise AF, Shaffer DW (2015) Why theory matters more than ever in the age of big data. J Learn Anal 2(2):5–13

Wise AF, Cui Y, Vytasek J (2016) Bringing order to chaos in MOOC discussion forums with content-related thread identification. In: Proceedings of the sixth international conference on learning analytics & knowledge, ACM, pp 188–197

Worsley M, Blikstein P (2014) Analyzing engineering design through the lens of computation. J Learn Anal 1(2):151–186

Wu M (2005) The role of plausible values in large-scale surveys. Stud Educ Eval 31(2):114–128

Wu M, Adams R (2002) Plausible values: why they are important. In: 11th international objective measurement workshop, New Orleans

Xing W, Wadholm B, Goggins S (2014) Learning analytics in CSCL with a focus on assessment: an exploratory study of activity theory-informed cluster analysis. In: Proceedings of the fourth international conference on learning analytics & knowledge, ACM, pp 59–67

Yates L (2013) Revisiting curriculum, the numbers game and the inequality problem. J Curriculum Stud 45(1):39–51

Ye C, Biswas G (2014) Early prediction of student dropout and performance in MOOCs using higher granularity temporal information. J Learn Anal 1(3):169–172

Zaki MJ, Meira Jr W (2014) Data mining and analysis: fundamental concepts and algorithms. Cambridge University Press

Zhong ZJ (2011) From access to usage: the divide of self-reported digital skills among adolescents. Comput Educ 56(3):736–746

Chapter 9
A Learning Analytics Approach for Job Scheduling on Cloud Servers

Mohammad Samadi Gharajeh

Abstract Learning analytics improves the teaching and learning procedures by using the educational data. It uses analysis tools to carry out the statistical evaluation of rich data and the pattern recognition within data. This chapter, firstly, describes four learning analytics methods in educational institutions. Secondly, it proposes a learning analytics approach for job scheduling on cloud servers, called LAJOS. This approach applies a learning-based mechanism to prioritise users' jobs on scheduling queues. It uses the three basic attributes "importance level", "waiting time" and "deadline time" of various jobs on cloud servers. The cloud broker acts as a teacher and local schedulers of cloud sites act as students. The broker learns to local schedulers how to prioritise users' jobs according to the values of their attributes. In the *deployment* phase, the effect of the above attributes on the system throughput is studied separately to select the best attribute. In the *service* phase, users' jobs are prioritised by computer systems according to the selected attribute. Simulation results show that the LAJOS approach is more efficient compared to some of the job scheduling methods in terms of schedule length and system throughput.

Keywords Learning analytics · Teaching procedure · Learning management · Cloud computing · Job scheduling

Abbreviation

Symbol	Phrase
CM	Cyclomatic Complexity
CPU	Central Processing Unit
CSCL	Computer-Supported Collaborative Learning
HCI	Human Computer Interaction
HOU	Hellenic Open University

M.S. Gharajeh (✉)
Young Researchers and Elite Club, Tabriz Branch,
Islamic Azad University, Tabriz, Iran
e-mail: m.samadi@iaut.ac.ir; m.samadi@ieee.org; mhm.samadi@gmail.com

© Springer International Publishing AG 2017
A. Peña-Ayala (ed.), *Learning Analytics: Fundaments, Applications, and Trends*, Studies in Systems, Decision and Control 94,
DOI 10.1007/978-3-319-52977-6_9

IaaS	Infrastructure-as-a-Service
ID	Identifier Number
IT	Information Technology
LAJOS	Learning Analytics approach for JOb Scheduling
QoS	Quality of Service
RAM	Random Access Memory
SL	Schedule Length
SMA	Services Management Agent
ST	System Throughput
VMs	Virtual Machines
WSA	Web Service Agent

9.1 Introduction

Learning is one of the most important characteristics of many people so that learners make the interactions with instructors and tutors, with contents, and/or with other people. Many educational institutions spend more efforts to design their teaching classes in order to enhance the performance of learning techniques. Traditional learning methods involve student evaluation, the analysis of grades, and the instructor's perceptions gathered at the end of any course. They have various constraints such as a limited quantity of data at the end of a course, a limited quality of the self-reported and retrospective data, and an interaction delay between the reported events and the implementation of an intervention (Bandiera and Bruno 2006; Bele and Rugelj 2010; Thrun and Pratt 2012; Park and Choi 2014). These constraints led to emerge learning analytics (Ferguson 2012; Haythornthwaite et al. 2013; Baker and Inventado 2014) for improving the teaching and learning processes.

Analytics tools include potential characteristics to apply the statistical evaluation of big data and the pattern identification within the data. These mechanisms are used to anticipate some imprecise conditions and make the knowledge-based decisions. They attempt to improve the learning outcomes. Hence, learning analytics uses some of the analytic tools to enhance the performance of learning and educational styles. It is worth to noting that learning analytics is associated with other learning fields such as academic analytics (Ferreira and Andrade 2014), action analytics (Scheffel et al. 2012), and educational data mining (Romero and Ventura 2007).

This chapter, first, discusses about four learning analytics methods in educational institutions. Afterward, it describes the proposed learning analytics approach for job scheduling on cloud servers. The current job scheduling methods, in the most cases, do not use learning tools. That is, they cannot be applicable under different network metrics (e.g., number of users). This problem causes the schedule length of cloud servers to be increased and the system throughput of the network to be reduced, considerably.

This chapter proposes an efficient job scheduling on cloud servers. The proposed approach uses the three basic attributes "importance level", "waiting time" and "deadline time" by using the learning analytics techniques. The cloud broker learns to local schedulers how to prioritise users' jobs on cloud servers. The main objectives of this approach are to minimise the schedule length and enhance the system throughput of cloud servers. The approach enhances the performance of cloud servers using some of the learning analytics tools. It uses a multi-criteria learning strategy to make proper decisions based on the three attributes mentioned above. The cloud servers are learned by this approach to work even under imprecise conditions.

The existing job scheduling methods, in the most cases, do not use any learning technique to conduct the scheduling purposes even under uncertain situations. The proposed scheduling approach attempts to improve the cloud efficiency in terms of schedule length and system throughput compared to the Improved Priority based (Patel and Bhoi 2014) and Credit based (Thomas et al. 2015) methods. It uses several learning analytics techniques to obtain the above objectives. Note that some of the learning techniques presented in Sect. 9.3 are used in the proposed approach to increase the performance of cloud servers.

The remainder of the chapter is organised as follows. Section 9.2 presents an overall view of the learning analytics. Section 9.3 introduces various models of learning analytics in educational institutions. It explains four learning analytics methods in this area: the relationship between student social networks and sense of community, the learning analytics methodology for student profiling, the genetic and participation based student prediction model, and the learning analytics method based on academic achievements of students and interaction data. Section 9.4 represents some of the job scheduling methods in cloud computing. Section 9.5, firstly, presents the main features of learning analytics in cloud computing and, secondly, describes the proposed approach for job scheduling on cloud servers. The approach is compared to some of the current scheduling methods in Sect. 9.6. Section 9.7 discusses about the efficiency of the proposed approach compared to the other scheduling methods. Finally, the chapter is concluded by Sect. 9.8.

9.2 A Glance on Learning Analytics

Learning analytics indicates the modelling analytics to anticipate learning behaviours, act based on the predictions, and use the predicted results through a learning process. This process is conducted to apply the teaching techniques in systematic environments. Furthermore, learning analytics introduces new tools to develop the learning and teaching skills for individual students and instructors. This section describes two basic concepts of learning analytics: learning analytics methods and learning analytics models.

9.2.1 Learning Analytics Methods

Learning analytics is a powerful field in educational institutions so that some of the analytic tools are applied to improve the learning and education procedures. Various representations of the analytical process can be expanded in different disciplines. According to Baker (2007), Knowledge Continuum is an actionable and conceptual framework to utilise learning analytics via multiple learning tools. As shown in Fig. 9.1, it consists of the four elements: data, information, knowledge, and wisdom. The bottom of continuum indicates the raw data which is composed of characters, symbols, and other meaningless inputs. Information is defined by attaching the meaning concepts to the raw data. It can answer the questions *who, what, when* and *where*. Upon information is evaluated and synthesised, the knowledge is obtained to answer the questions *why* and *who*. Finally, the knowledge is transformed into wisdom to achieve the desired goals. Knowledge Continuum indicates that the data can be processed for transformation process into meaningful things.

9.2.2 Learning Analytics Models

Dron and Anderson (2009) have presented Collective Application Model that defines the executive process of learning analytics. As shown in Fig. 9.2, this model is composed of the five layers categorised into three cyclical phases. The model illustrates the cyclical nature of analytical processes. It contains various features to improve the learning systems via the three successive cycles: gathering, processing, and presentation. The gathering cycle contains the selection and capture operations. The processing cycle includes various aggregation methods to make suitable predictions by using the obtained information. The presentation cycle involves the use, refinement, and knowledge sharing. This cycle is used to improve the performance of learning systems.

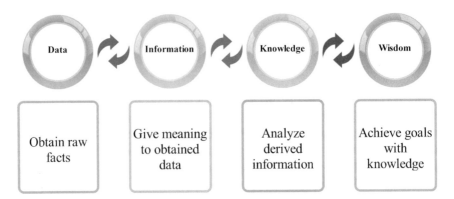

Fig. 9.1 The main elements of knowledge continuum (Baker 2007)

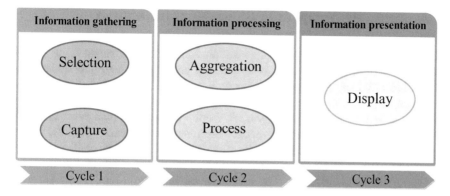

Fig. 9.2 The main components of collective application model (Dron and Anderson 2009)

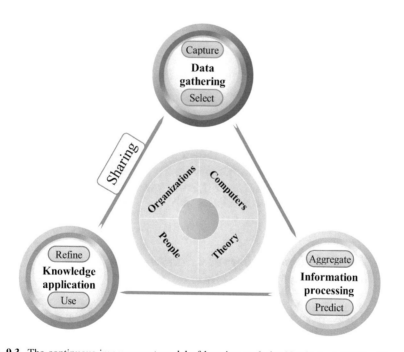

Fig. 9.3 The continuous improvement model of learning analytics Norris et al. (2008a, b)

Figure 9.3 shows a completed and extended model of learning analytics (Norris et al. 2008a, b). This model contains four types of the technology resources to achieve the learning goals through the three cycles mentioned above. It enables various organisations, computers, people, and theory to obtain the needs of teachers, educators, and administrators. This model could manage the high-quality data. In the first cycle, data is gathered by the *capture* and *select* operations. In the second cycle, information is processed by the *aggregate* and *predict* operations. In the third cycle, knowledge is derived by the *refine* and *use* operations.

9.3   Learning Analytics in Educational Institutions

Learning analytics can be used in educational institutions to enhance the student success. These institutions can put educational programmes into public classes to study the activities of all students, not only specific students. These programmes enhance the student success by integrating a student into the institution through various academic procedures. They allow faculty members to send personal emails to students in order to report their performance in a special course. Besides, faculty members will be able to encourage their students to visit various course resources on the campus or office places. These activities lead to the students become more integrated into the institution. The remainder of this section presents some of the learning analytics methods in educational institutions.

9.3.1   The Course Signals Method

Educational institutions can apply learning analytics to enable the real-time integration on student performance using the demographic and academic information. The learning activities offer an intentional environment for the students who persist to the graduation (Tinto 2006). Table 9.1 represents success results of the students participated in Course Signals activities (Arnold and Pistilli 2012). The Course Signals method is a student success system for enabling the faculty members to offer the meaningful feedback to students via the predictive models. The analysis results indicate that the use of Course Signals system leads to the success rate of educational institutions enhances, considerably.

9.3.2   The Relationship Between Student Social Networks and Sense of Community

The social network analysis is considered by researchers in behavioural sciences. It represents the exchange between various resources of the social actors. Dawson

Table 9.1 Success rate of course signals method for the year 2009 (Arnold and Pistilli 2012)

The number of courses	The number of units	Analyzed for one year (%)	Analyzed for two years (%)
Without course signals system	3164	87.67	81.89
At least one instance	2962	90.34	83.22
Exactly one instance	2296	87.72	80.87
Two or more instances	666	99.40	91.44

(2008) presents a learning analytics method to educational institutions to define a relationship between student social networks and the sense of community. This method illustrates the relationship between the students' positions in a classroom social network and the community level. The community level and students' positions into a classroom are analysed by Classroom Community Scale (Rovai 2002b). Furthermore, this relationship can be clarified using the discussion forum content analysis and student interviews (Vonderwell 2003).

This method applies a technical procedure that consists of both quantitative and qualitative measures to answer the researchers' questions. It uses overall information of the students participated in undergraduate and postgraduate courses at a large metropolitan university. A greater insight relationship is provided by the three social network calculations: betweenness, closeness, and degrees. The *betweenness* metric indicates the frequency of an individual occurrence within the shortest path between multiple actors, the *closeness* metric indicates the degree of relationship between an actor and the network, and the *degrees* metric indicates the number of connections between the possesses of each actor on the network. Figure 9.4 illustrates an example to indicate the relationship between the student social networks and the sense of community. It shows a high closeness centrality in a way that *I* is the academic staff member associated with the teaching unit, *II* considers the

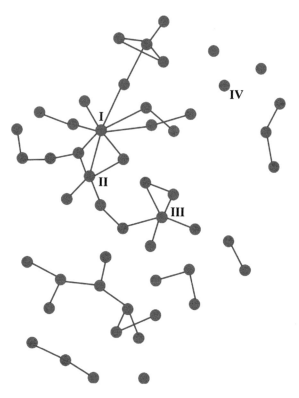

Fig. 9.4 An instance to represent the relationship between student social net-works and the sense of community (Dawson 2008)

Table 9.2 Percentages of classroom community scale of the total interactions (Dawson 2008)

Time (weeks)	Admin	Learning	Social
1	10	28	64
2	2	41	58
3	4	42	57
4	8	33	60
5	15	27	56
6	3	55	45
7	37	35	30
8	8	38	56
9	7	52	46
10	1	50	47
11	0	47	52
12	2	45	54
13	3	43	55

student having a high betweenness centrality within the network, *III* indicates the student having a high degree centrality, and *IV* represents the student disconnected from the main network.

Table 9.2 represents content analysis of the forum in the first six weeks of the teaching period. The evaluation results indicate that learning interactions are increased from 28 to 45% for each communication interaction in the teaching interval time. They demonstrate that the performance of learning analytics is better than the admin interactions. Moreover, the mean percentage achieved by the social interactions is enhanced by nearly 80% more than that achieved by the admin interactions and by nearly 20% more than that achieved by the learning interactions.

9.3.3 The Learning Analytics Methodology for Student Profiling

The participation of students in e-learning forums generates a big volume of data every day. Lotsari et al. (2014) have presented the learning analytics methodology for student profiling that utilises a threefold analysis of big data. The data are revealed by the participation of students in the online forums of several universities. It applies two techniques to conduct two major actions: text mining techniques and social network analysis techniques. The text mining techniques, efficiently, evaluate contents of the enormous messages posted in the forums. The social network analysis techniques evaluate a network of the students who interact via the online forums. Both techniques provide a combined knowledge for educators to give the practical and valuable information.

The text mining techniques extract the text of all the messages posted in the forums. Afterward, they convert the text to the corpus that does not contain any

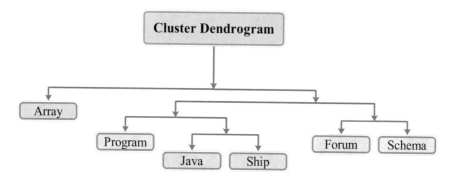

Fig. 9.5 Various clusters of terms representing frequently co-occurring terms in an online forum

punctuations, numbers, and hyperlinks. Finally, a document-term matrix is built by a transformation process of the corpus. This matrix involves multiple rows and columns in a way that each row represents a term, each column indicates a document, and each entry contains various occurrences of the term into the document. The matrix contains the words that are saved in the dictionary. The graph mining techniques are applied to locate any link between the words and the groups of words. As illustrated in Fig. 9.5, each term is placed at its own cluster so that all the terms are associated with the same cluster.

The social network analysis techniques consider possible interactions amongst the students on online forums and existing correlations among the terms of online forums. As illustrated in Fig. 9.6, a network of students is formed to indicate any interaction amongst the students in the same class. Each node indicates a student and each edge indicates the correlation amongst two students. The label size of each vertex is defined in the graph according to the degree of participation. Moreover, the width of each edge is specified based on the weight. It is worth to noting that the thicker edges indicate the higher degrees between the various correlations. The graph represents how a student is influential within the social network. The labels *id1*, *id2* and *id3* represent the course/module instructors. The students who have a high level of the participation are located at the centre of the network.

The authors have used real information of the online forums related to Hellenic Open University (HOU) to represent the performance of the presented method. They have utilised the analytics tools to analyse the inner structure and contents of the messages posted in the forums. Table 9.3 represents the experimental results of the six homework grades based on the *participation* factor. It consists of two clusters in a way that the first cluster contains the students having the highest homework grades and the lowest forum participation. Furthermore, the second cluster contains the students having the lowest homework grades and the highest forum participation. These results indicate that the most students are participated in the first cluster.

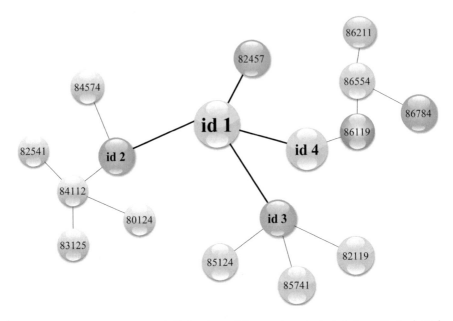

Fig. 9.6 A network of students built by the social network analysis techniques (Lotsari et al. 2014)

Table 9.3 The evaluation results of the learning analytics methodology for the pair of emerging clusters (Lotsari et al. 2014)

Attribute	More active (42)	Less active (22)
Homework 1	9.15	8.36
Homework 2	8.56	5.74
Homework 3	9.59	6.64
Homework 4	8.43	3.34
Homework 5	8.26	6.20
Homework 6	7.72	5.27
Participation	2.05	5.05
Start	0.57	1.41

9.3.4 The Genetic and Participation Based Student Prediction Model

One of the main requirements for evaluating the big data is to form the student performance prediction models. Xing et al. (2015) have presented the genetic and participation based student prediction model that uses a prediction model based on the data of a collaborative geometry-problem solving environment. Two learning analytics methods, educational data mining and HCI theory, are synthesised to discover the development of prediction models. The authors have applied the activity theory to quantify the students' participation in the CSCL (Computer-supported

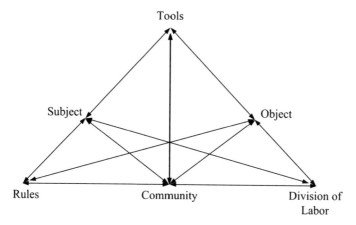

Fig. 9.7 A schematic of the activity theory (Xing et al. 2015)

Collaborative Learning) course according to the online participation theory (Hrastinski 2009). Moreover, they have defined the six learning variables: Subject, Rules, Tools, Division of Labor, Community, and Object.

The student prediction model considers the activity theory of Fig. 9.7. Activity theory uses a social, psychological, multidisciplinary logic to make naturalistic decisions. It offers a holistic framework to indicate the different practical activities while having a link between individual behaviour and social behaviour (Barab et al. 2002). Moreover, the prediction model applies the activity system (Leont'ev 1974; Engeström 1999) to utilise the six interaction components in decision making process. Object involves some of the learning tasks (e.g., solving a problem). Subject consists of the learning activities, which contain individual students in the learning areas. Tools are composed of computers, online tools, systems, and environments to achieve the learning activity. Community is the direct and indirect communications to enable an individual subject. Rules are the implicit and explicit rules to restrict some of the students' activities. Finally, Division of Labor indicates the fundamental contributions of any individual in the system.

Figure 9.8 illustrates flowchart of the genetic programming considered by the presented prediction model. In Step 1, a population of N models is randomly generated to determine the possible solutions. In Step 2, each model is used in the current population on the training data. Besides, the fitness of each model is determined in the current population. In Step 3, the parent models and genetic operators generate the offspring models until the certain population size is reached, completely. In the steps *4* and *5*, the *N* old models are replaced by the new generated *N* models. The steps *2–5* are repeated until the maximum generations are reached by the model. Finally, the rule with the best fitness level is selected as the best result of the algorithm in Step 6.

Fig. 9.8 The workflow of the genetic programming algorithm (Xing et al. 2015)

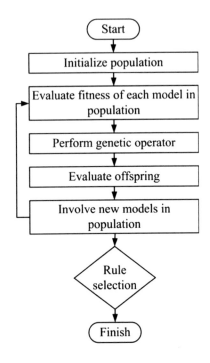

Table 9.4 Final student performance results of the student prediction model (Xing et al. 2015)

Type	Prediction model	Prediction result (%)
Fitness	Overall prediction	80.2
	At-risk prediction	89.5
Sensitivity	Overall prediction	80.3
	At-risk prediction	85.0
Specificity	Overall prediction	80.3
	At-risk prediction	94.4

Table 9.4 represents the performance results of the student prediction model. The *Fitness* type indicates an overall reflection of the model. The main goal of the education prediction models is to identify the *At-Risk* students. Therefore, the *At-Risk* prediction is selected as the important parameter to analyse this method. The performance results indicate that the fitness values of the *Overall* prediction have the noticeable summary among all of the prediction models.

The evaluation results of the *Sensitivity* type for both the *Overall* and *At-Risk* prediction models are efficient compared to the other results. Furthermore, the performance results of the *Specificity* type for the *At-Risk* prediction model are better than the *Fitness* and *Sensitivity* types.

9.3.5 The Learning Analytics Method Based on Academic Achievements of Students and Interaction Data

Vahdat et al. (2015) have presented a learning analytics method to study the learning behaviour of students. They use some of the technical learning tools to obtain their objectives. This method establishes a relationship between the academic achievements of students and the evaluation data of an educational simulator. The authors conduct the work based on the interaction data gathered from the six laboratory sessions where the first-year students of Computer Engineering Department at the University of Genoa were working with a digital electronics simulator. Furthermore, they apply the mining algorithm to analyse the learning procedures of students.

This method executes the data analytics process via two parts: creating the process models and comparing the models to each other. It uses fuzzy miner algorithm to achieve the process models from the interaction logs. A given knowledge is used to compare the process models of students. The work uses the *Disco* tool to form the process models. Figure 9.9 illustrates a student process model that is aggregated by the fuzzy miner algorithm via one of the course sessions. Each *A* refers to an activity and each *light to dark blue* transformation indicates the activities having the low to high frequency. The complexity of student processes is measured according to the learning data achievements in order to evaluate the performance of process models (McCabe 1976).

Table 9.5 represents the performance results of student clusters in the average Cyclomatic Complexity (CM). It indicates the difference among results when *n* equals 0 versus 2 for the second level of granularity. The experimental results of

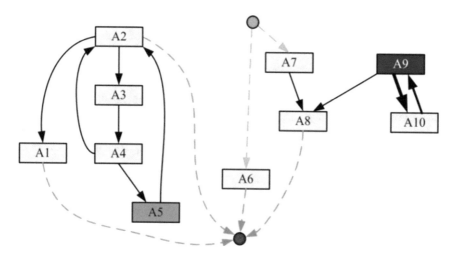

Fig. 9.9 An instance of a student process model achieved by Disco tool (Vahdat et al. 2015)

Table 9.5 Comparison of student clusters for the second level of granularity in the learning method based on academic achievements and interaction data (Vahdat et al. 2015)

Session	CM (n = 0) Low-graded	CM (n = 0) High-graded	CM (n = 2) Low-graded	CM (n = 2) High-graded
1	88	90	88	96
2	102	109	98	112
3	57	69	51	69
4	75	75	72	79
5	72	80	69	81
6	104	119	108	119

the second and third levels are better than the results of the first level. The reason is that the out-of-task events in the first level cause to a disturbance in the results. The results represent that the measured complexity has a positive correlation in accordance with the final grades of students.

9.4 Background on Job Scheduling in Cloud Computing

The Random algorithm uses a random process to accomplish the users' jobs by an appropriate virtual machine. It does not analyse the traffic status of machines in which they have low or heavy load. The process complexity of Random algorithm is very low because it does not need any overhead. The Round Robin algorithm completes users' jobs by some of the available machines in a round order where each job is executed in a justified state. It attempts to complete the users' jobs in a round form. The Opportunistic Load Balancing algorithm executes users' jobs on the machine that contains the lowest load balance compared to the other ones. The main goal of this algorithm is to reduce the traffic load of each virtual machine (Mohialdeen 2013).

Job priority is one of the most important issues for job scheduling in cloud computing. A job scheduling method should apply a job priority strategy to enhance the performance of cloud systems. Patel and Bhoi (2014) have presented an improved and priority based algorithm to carry out job scheduling in cloud computing. The *makespan* and *consistency* features obtained by this algorithm are better than those obtained by other scheduling algorithms. The algorithm uses an iterative process to prioritise the jobs and resources, properly.

Thomas et al. (2015) have presented a credit based job scheduling algorithm that considers the three scenarios to schedule users' jobs on cloud systems. The first scenario works based on the length of users' tasks, the second scenario works based on the task priority, and the third scenario works based on both the length of users' tasks and task priority. The *makespan* feature achieved by this algorithm is more efficient than that achieved by other algorithms. Furthermore, the *makespan* feature improves after a certain number of tasks.

Abdullahi et al. (2016) have introduced a Discrete Symbiotic Organism Search (DSOS) algorithm to conduct an optimal task scheduling on cloud resources. Symbiotic Organism Search (SOS) presents a novel developed metaheuristic optimisation technique to solve some of the numerical optimisation problems. SOS discusses about the symbiotic relationships (i.e., mutualism, commensalism, and parasitism) presented by organisms into an ecosystem. DSOS uses the Particle Swarm Optimization (PSO), which is one of the most popular heuristic optimisation techniques, for task scheduling problems.

Yang et al. (2016) have presented an integer programming model to formulate the existing problems and propose multiple solution methods in order to organise the acquisition and scheduling plans. The model applies ten well-known heuristics of parallel-machine scheduling to fit into the studied problem for offering the initial solutions. Furthermore, it uses tabu search and genetic algorithm for reflecting the problem nature to improve upon the initial solutions. The authors have utilised a series of computational experiments to analyse the efficiency of the presented model.

9.5   Learning Analytics in Cloud Computing

Cloud computing is a one of the technical model to enable a ubiquitous and on-demand network access to a pool of computing resources (e.g., data storages, applications, and services). It can be provided by the minimal management tools or service providers, easily. Cloud-based systems allow organisations to offer the Information Technology (IT) services for their business activities. Cloud networks, nowadays, are used in educational environments to facilitate the teaching procedures. Cloud learning is a technology-based system so that each cloud server uses some of the learning analytics techniques. It enables the reuse capability of learning resources via a distributed process. This process is offered using some of the educational tools. Cloud learning assists the teaching procedures through an unobtrusive way and, also, improves their scalability features (Li 2009; Antonopoulos and Gillam 2010; Ghanbari and Othman 2012; Gharajeh 2015; Sánchez et al. 2016).

9.5.1   Main Analytics Components in Educational Systems

Learning analytics can be used in cloud systems to obtain some of the service requirements (e.g., job scheduling). Sánchez et al. (2016) have presented a learning analytics method for cloud computing on smart educational environments. This method enables the invocation of educational Web services to indicate the smart objects using the ambient intelligence. This goal is obtained using a transparent process to improve the users' tasks.

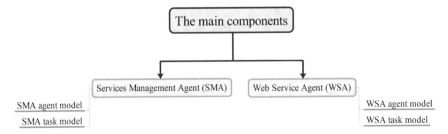

Fig. 9.10 The main components of the learning analytics method for cloud computing in smart educational environments

As shown in Fig. 9.10, this method is composed of two components: Services Management Agent (SMA) and Web Service Agent (WSA). Each component consists of an agent model and a task model. The *SMA* agent model has a goal-based intelligent agent, who learns how to localise the Web services according to some previous experiences. The *SMA* task model uses three tasks: receiving the data of Web services to localise, selecting an appropriate WSA to achieve the requested requirements, and removing a Web service. The *WSA* agent model is a typical reflex model to invoke the Web service. The *WSA* task model uses four tasks: receiving the feed data of Web services, inferring the method called by the Web service, calling the Web service method, and transmitting the execution results to the requested agent.

9.5.2 Job Scheduling in Cloud Computing

Cloud computing is established by computer networks to offer various network resources (e.g., storage disks) to users. In cloud computing, similar to any computer network, there are a large number of users' jobs that should be responded by various cloud-based systems. These systems should accomplish the requested jobs based on their basic features (e.g., required hardware specifications and response time). This subsection describes the cloud computing and job scheduling profile.

Cloud Computing. Cloud computing is a technology-based model that enables ubiquitous, on-demand network access to various configurable computing resources (e.g., file storages). On-demand self-service, broad network access, resource pooling, rapid elasticity, and measured service are the main characteristics of cloud computing. The *on-demand self-service* feature indicates that cloud systems can be used as a non-demand service without need any manual intervention. The *broad network access* feature represents that the network connection is high-efficient and available for the particular services. The *resource pooling* feature indicates that the service provider on cloud servers provides the requested resources of the users using the modern technologies (e.g., virtualisation). The *rapid elasticity* feature represents that the resources necessary can be achieved without any intervention process when no longer needed. Finally, the *measured service* feature represents that a cloud service is measurable in terms of the resources used (e.g., "pay as you go" and "pay-per-use") (Wang et al. 2015; Gharajeh 2015; Ratten 2016).

Job Scheduling Profile. Job scheduling on cloud systems prioritises the users' jobs on various cloud resources. This process is conducted using the defined rules of resource requirements under a certain circumstances. At this time, there are not any uniform standard for job scheduling on cloud servers. Job scheduling and resource management are the basic techniques in cloud computing so that they play a key role in the resource management systems. A cloud scheduler should order a large number of users' jobs in a way that the quality of all the desired services is improved as well as the fairness of the jobs are maintained, simultaneously. Figure 9.11 illustrates the main scheduling metrics of cloud computing. These performance criteria are indicated to evaluate the efficiency of scheduling algorithms (Mohialdeen 2013; Kalra and Singh 2015).

The *execution time* metric indicates the time that is required to execute a user's job. The *response time* metric represents the time that a cloud server will be free to execute a user's job. The *trust* metric means that a cloud system should be able to

Fig. 9.11 Job scheduling metrics for cloud computing

execute all of the jobs requested by cloud users. The *makespan* metric represents the finish time of the latest job. The *utilization* metric specifies how well an available data on a cloud network is used by cloud resources. The *flowtime* metric is the certain time that a flow unit spends to execute a requested job. The *cost* metric represents the financial cost to accomplish users' jobs. The *successful execution* metric is the success rate of cloud servers that can be calculated based on the number of all jobs and the number of successful jobs. The *reliability* metric indicates the number of successful jobs on cloud servers. The *fairness* metric means that a cloud server should execute users' jobs according to their arrival time. The *load balancing* metric represents that all users' jobs should be distributed over multiple servers to balance the traffic load on the whole network. Finally, the *responsible complexity* metric indicates the networking complexity among cloud systems (Burkimsher et al. 2013; Gharajeh 2015; Liu et al. 2015).

9.5.3   Scheduling Framework

A job scheduling framework in cloud computing should be designed by considering various features (Jena 2015; Kao and Chen 2016). An efficient scheduling framework includes some of the following features:

- Energy efficiency and load balancing of the data centres on cloud servers
- The Quality of Service (QoS) features (e.g., schedule length and throughput)
- Security requirements (e.g., trust and privacy)
- A fairness resource allocation of users' jobs
- Especially, a learning-based mechanism to conduct job scheduling efficiently

Because it is not possible to consider all of the above features by an individual scheduling framework, some of them should be used to provide an efficient scheduling approach. Patel and Bhoi (2014) have presented the Improved Priority based job scheduling on cloud environments. The main goal of this algorithm is to improve the *makespan* and *consistency* features. Thomas et al. (2015) have presented the Credit based scheduling algorithm on cloud servers. It works based on the priority and length features of tasks. The task with the shortest length will be placed at the beginning of the scheduling queue as well as the task with the highest length will be placed at the last of the scheduling queue. The current job scheduling methods, in the most cases, involve the following drawbacks:

- They only consider one or two scheduling metrics instead of considering most of the crucial metrics.
- They do not use the learning analytics tools.

The above drawbacks cause the job scheduling on cloud servers not to be efficient. Therefore, it is essential to propose a novel job scheduling approach that uses most of the crucial metrics through a learning analytics mechanism.

9.5.4   The Network Model

Figure 9.12 shows a schematic of the considered network model. The whole network is controlled by a cloud broker. The broker manages all of the processes on the network. Various users connect to the broker to hire the required hardware resources (e.g., storage disks, memory, and CPU). It is worth to noting that the proposed scheduling approach uses an Infrastructure-as-a-Service (IaaS) cloud. The whole network is partitioned into multiple cloud sites in a way that the sites are placed at different geographical locations.

Each site consists of a large number of computer systems (i.e., physical machines) that include various hardware specifications. Moreover, it is managed by a local cloud scheduler so that each scheduler has a job scheduling queue. Each computer system involves a given number of virtual machines (VMs) and a data centre. When a user requests a cloud service from the broker, the broker looks for

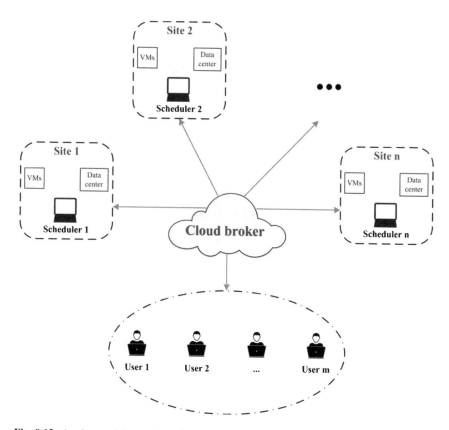

Fig. 9.12 A scheme of the considered network model

an appropriate computer system to execute the requested service. The proposed job scheduling approach prioritises the users' jobs on the scheduling queues to enhance the network efficiency.

9.5.5 The Proposed Learning Analytics Approach for Job Scheduling on Cloud Servers

Job scheduling is one of the essential activities in cloud computing. A job scheduling approach is proposed in this subsection to prioritise the users' jobs on scheduling queues. It attempts to enhance the efficiency of cloud servers. This approach is a learning analytics method for job scheduling on cloud servers, called LAJOS, which uses some of the intelligent procedures with the aid of learning techniques. Cloud servers will be learned by the LAJOS approach based on various features of the requested jobs. By using this approach, the computer systems located at the cloud sites will be able to act like a human for prioritising users' jobs on cloud servers.

Main Characteristics of the Proposed LAJOS Approach. This approach contributes with the learning analytics methods presented in Sect. 9.3 and the learning-based mechanisms proposed in this subsection. The proposed mechanism consists of two phases: deployment and service. In the *deployment* phase, it is analysed that which of the three attributes "importance level", "waiting time" and "deadline time" has a high impact on the schedule length and system throughput. Note that the values of importance level and deadline time are determined by users. In the *service* phase, users' jobs are prioritised on scheduling queues of computer systems based on the selected attribute. The *deployment* phase is conducted on all available cloud sites in the three period times *T1*, *T2* and *T3* through four stages.

In Stage 1, each computer system connects to its local scheduler as well as local schedulers and users connect to the cloud broker, directly. These connections are the same as the connections of the methods presented in Dawson (2008) and Lotsari et al. (2014). This process is carried out in which the cloud broker is the most important node, the local schedulers are important nodes, and users are ordinary nodes. Furthermore, the connections between the broker and some of the local schedulers are more strength than the other connections because they contain the high-performance cloud resources. In Stage 2, the three attributes "importance level", "waiting time" and "deadline time" are applied to prioritise users' jobs on scheduling queues. Each attribute is analysed in a certain period time. In Stage 3, the schedule length and system throughput of all cloud sites are calculated by the local schedulers for each attribute in the period times *T1*, *T2* and *T3*, separately. Afterward, they are reported to the cloud broker to estimate overall status of each cloud site. In Stage 4, the broker selects the attribute with the lowest schedule length and the highest system throughput. This process is conducted by fuzzy miner algorithm (Günther and Van Der Aalst 2007) the same as one of the learning analytics methods presented in Vahdat et al. (2015). Because the *deployment* phase is sequentially organised in the three period times as well as learning analytics techniques are used at each period time, the proposed approach follows the sequential analysis presented in Ozturk et al. (2014). The main objectives of the proposed scheduling approach are as follows:

- Minimising the schedule length: if the schedule length of cloud sites is minimised, users' jobs will be executed quickly and efficiently.
- Increasing the system throughput: if the throughput of cloud sites is increased considerably, the network performance will be high. The system throughput is calculated based on the number of all users' jobs and the number of successful jobs.

Representations of the LAJOS Approach Based on Learning Techniques. Figure 9.13 illustrates the activity model applied by the proposed approach. It works similar to the learning procedures presented in Xing et al. (2015) and Ma et al. (2015). In the *deployment* phase, the cloud broker transmits three commands to all of the cloud schedulers. The first command indicates that computer systems should prioritise users' jobs based on the first attribute. After the period time *T1*, computer systems report their schedule length and system throughput to the cloud schedulers. Besides, cloud schedulers report schedule length of T1 (SL1) and system throughput of T1 (ST1) to the broker. After the period time *T2*, all of the schedulers report

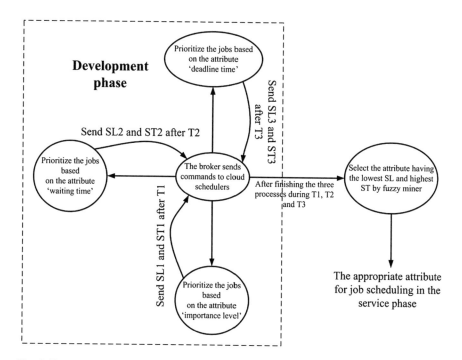

Fig. 9.13 A schematic of the activity model for the proposed LAJOS approach

schedule length of T2 (SL2) and system throughput of T2 (ST2) to the cloud broker. Finally, after the period time *T3*, all of the cloud schedulers report schedule length of T3 (SL3) and system throughput of T3 (ST3) to the broker.

After the above processes are completed by the network elements, the attribute with the lowest schedule length and the highest system throughput is selected by the cloud broker with the aid of the fuzzy miner algorithm. It is worth to noting that the best attribute is selected based on the average values of schedule length and system throughput for the whole network.

Calculating the Elements of LAJOS Approach. Schedule length of computer systems equals the number of jobs on the scheduling queue. In contrast, system throughput of the systems is calculated by the system itself using Eq. (9.1).

$$ST_i = \frac{N_S}{N_{RJ}} \tag{9.1}$$

where i indicates the identifier number of the system, N_S indicates the number of successful jobs, and N_{RJ} indicates the number of requested jobs.

After all of the computer systems report their schedule length and system throughput to the local schedulers, the mean values of these parameters are independently calculated for each cloud site by local schedulers. The average schedule length of a cloud site is calculated by the scheduler based on the parameters represented in Eq. (9.2).

$$SL_S = \frac{\sum_{i=1}^{n} SL_i}{n} \tag{9.2}$$

where i is the identifier number of the computer system, SL_i is the schedule length of system i, and n is the number of computer systems within the cloud site. Besides, the average system throughput of each cloud site is determined by the cloud scheduler based on average values of the reported parameters as indicated in Eq. (9.3).

$$ST_S = \frac{\sum_{i=1}^{n} ST_i}{n} \tag{9.3}$$

where i indicates the identifier number of the system, ST_i indicates the system throughput of system i, and n indicates the number of computer systems.

After the cloud broker obtains the average schedule length and average system throughput of all local schedulers, the average values of schedule length and system throughput for the whole network will be calculated by the broker. This process is carried out based on the overall information of network. The average schedule length of the network is calculated by using Eq. (9.4).

$$SL_N = \frac{\sum_{i=1}^{n} SL_i}{n} \qquad (9.4)$$

where i is the identifier number of each cloud site, SL_i is the schedule length of cloud site i, and n is the number of cloud sites. Moreover, the average system throughput of the network is determined by Eq. (9.5).

$$ST_N = \frac{\sum_{i=1}^{n} ST_i}{n} \qquad (9.5)$$

where i indicates the identifier number of the cloud site, ST_i indicates the system throughput of site i, and n indicates the number of cloud sites. Note that the average schedule length and average system throughput of the whole network will be used in the *deployment* phase.

In the proposed approach, the cloud broker acts like a teacher and all cloud schedulers act like students. The cloud broker learns to the local schedulers how to prioritise the requested jobs in the *service* phase with regarding the experimental results of the *deployment* phase. Table 9.6 represents various hardware characteristics of the cloud resources that can be requested by users. Each type is similar to the question asked by a teacher from his/her students into a classroom. Because a student answers to the questions based on some of the educational factors (e.g., correct and fast), a scheduler and a computer system will execute users' jobs based on the requested cloud resources (e.g., RAM and file storage).

Learning Analytics Components in the LAJOS Approach. As mentioned before, data, information, knowledge, and wisdom are the main components of the learning analytics considered by Knowledge Continuum (Baker 2007). Figure 9.14 illustrates how these components are applied by the proposed LAJOS approach.

The *data* component includes the raw data of cloud servers such as users' jobs, scheduling attributes, and the physical characteristics of computer systems. The *information* component contains the data packets transmitted from local schedulers to the broker (i.e., schedule length and system throughput). The *knowledge* component analyses the information of all cloud sites to select the best attribute for job scheduling on cloud sites. Finally, the *wisdom* component gives the objectives of the proposed approach using the available knowledge. The proposed LAJOS approach, dynamically and independently, makes proper decisions based on the overall situations of all cloud sites, instead of considering some of the limited situations. This process leads the efficiency of cloud servers to be enhanced, considerably.

Table 9.6 Various types of the cloud resources

No.	Type
#1	Amount of required RAM (GB)
#2	Volume of required disk storage (TB)
#3	Amount of required CPU speed (GHz)
#4	Number of required CPU cores

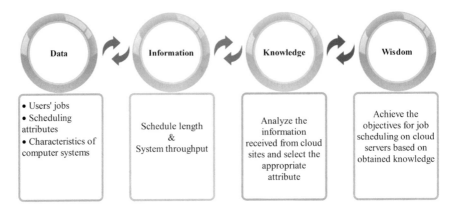

Fig. 9.14 An overall view of the learning analytics components in the LAJOS approach

Table 9.7 Representations of the LAJOS approach based on content analysis codification schema

Dimension	Definition
Spirit	The feeling of being connections in cloud server
Trust	The feeling that cloud schedulers can be trusted regarding the selected attribute
Interaction	The interaction between computer systems, schedulers, and the broker
Learning	The notion that knowledge is constructed by the cloud broker

Content Analysis Codification Schema in the LAJOS Approach. Rovai (2002a) has presented a learning schema to study the four dimensions of classroom community: spirit, trust, interaction, and learning. Table 9.7 describes that the proposed approach how to acts the same as this schema. The *spirit* factor represents the feeling of the connections between the cloud broker, cloud schedulers, and computer systems. Hence, each network element knows how to establish a connection with other elements. The *trust* factor indicates the feeling of computer systems to trust the cloud schedulers which apply the selected attribute for job scheduling on servers. The *interaction* factor represents the interaction between computer systems, schedulers, and the cloud broker. It includes some of the network parameters such as users' jobs, schedule length, and system throughput. The *learning* factor indicates the knowledge gathered by the cloud broker.

Case Study. Table 9.8 represents an example to indicate how the best attribute is selected in the *deployment* phase from among the three attributes "importance level", "waiting time" and "deadline time". This process is carried out on two cloud sites and nine computer systems in the three period times *T1*, *T2* and *T3*. In the *deployment* phase, it is considered that the requested jobs are randomly executed by the cloud sites. N_{RJ} is the number of requested jobs and N_S is the number of successful jobs.

The schedule length *SL* equals the number of jobs on the scheduling queues and the system throughput *ST* is calculated by Eq. (9.1). The schedule length of each

Table 9.8 An example to select the best attribute in the deployment phase

Period time	System ID	Cloud site	N_{RJ}	N_S	SL	ST	SL_S	ST_S	SL_N	ST_N
T1	1	Site 1	145	78	67	0.54	46	0.53	50.5	0.63
	2	Site 1	51	26	25	0.51				
	3	Site 2	205	150	55	0.73	55	0.73		
T2	4	Site 1	15	14	1	0.93	1	0.93	18.25	0.76
	5	Site 2	60	55	5	0.92	35.5	0.59		
	6	Site 2	89	23	66	0.26				
T3	7	Site 1	154	120	34	0.78	20	0.86	48.5	0.59
	8	Site 1	80	74	6	0.93				
	9	Site 2	112	35	77	0.31	77	0.31		

cloud site SL_S is determined by Eq. (9.2) and the system throughput of each site ST_S is determined by Eq. (9.3), separately. Afterwards, the average schedule length of the network SL_N is calculated by Eq. (9.4) and the average system throughput of the network ST_N is calculated by Eq. (9.5). Finally, the fuzzy miner algorithm estimates that the network efficiency in the period time T2 is better than the period times T1 and T3. Therefore, the attribute "waiting time" is selected as the best attribute to execute users' jobs in the *service* phase.

9.6 Evaluation Results

This section represents the evaluation results of the proposed LAJOS approach. The approach is compared to the Improved Priority based (Patel and Bhoi 2014) and Credit based (Thomas et al. 2015) methods in order to demonstrate the efficiency of the proposed mechanisms. The comparison results are carried out in terms of schedule length and system throughput. Besides, the effects of various input parameters (i.e., job generation rate and number of users) on the above terms are investigated separately.

9.6.1 Simulation Setup

The simulation process is carried out in 5 h, the number of cloud sites is 10, the number of computer systems within each cloud site is 100, and the scheduling queue of each computer system can hold 50 jobs. Job generation rate of the whole network is varied from 0.1 to 1 job/s and the number of users connected to the cloud broker is 500.

Table 9.9 Simulation parameters

Parameter	Value
Simulation time	5 h
Number of cloud sites	10
Number of computer systems within each site	100
Scheduling queue size (job)	50
Job generation rate	0.1–1 job/s
Number of users	500
Period time of the *deployment* phase	10 min

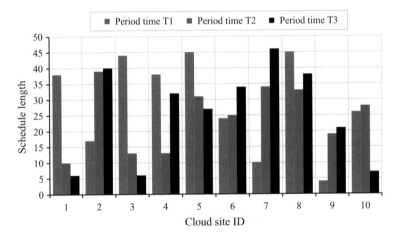

Fig. 9.15 The schedule length obtained by the proposed approach for each cloud site

Table 9.9 represents the simulation parameters and their default values. The resource discovery process is conducted based on the discovery method presented in Chung et al. (2013) and VMs are determined based on the virtual machine placement method presented in Kavvadia et al. (2015). It is worth to noting that the simulation process is planned based on the network model illustrated by Fig. 9.12.

9.6.2 Simulation Results

Figure 9.15 shows the schedule length of the LAJOS approach for each cloud site in the *deployment* phase. The simulation results are distinguished from each other based on the period times *T1*, *T2* and *T3*. The simulation results illustrate the schedule length of cloud sites for each period time, separately. The average values of the results will be calculated to select the period time which has the lowest average schedule length.

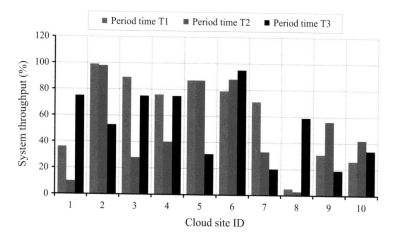

Fig. 9.16 The system throughput achieved by the proposed approach for each cloud site

Figure 9.16 depicts the system throughput of the LAJOS approach for each cloud site in the *deployment* phase. The simulation results indicate the system throughput of the cloud sites in the period times *T1*, *T2* and *T3*. The mean values of these results will be considered to select the period time which has the highest average system throughput.

9.6.3 Comparison Results

Figure 9.17 shows the effect of the job generation rate on the schedule length obtained by the LAJOS approach compared to the Improved Priority based and Credit based scheduling methods. The comparison results demonstrate that the schedule length obtained by the proposed approach is less than those obtained by the other methods. These results are carried out under various job generation rates. If the job generation rate increases, the number of requested jobs and the network traffic enhance consequently. Therefore, the schedule lengths of all the simulated methods increase in the simulation process.

Figure 9.18 shows the effect of the job generation rate on the system throughput carried out by the LAJOS approach compared to the Improved Priority based and Credit based scheduling methods. The comparison results illustrate that the system throughput achieved by the proposed approach is more than those achieved by the other scheduling methods. This progress is kept under various changes on the job generation rate. As depicted in the comparison results, the system throughputs of all the simulated methods reduce when the job generation rate increases. The reason is that the heavy traffic causes the number of successful jobs and system throughput to be decreased, considerably.

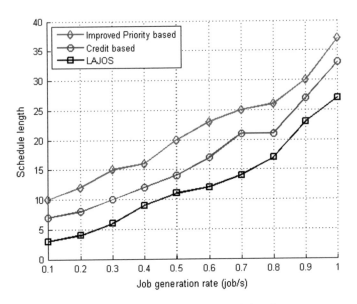

Fig. 9.17 The effect of the job generation rate on the schedule length

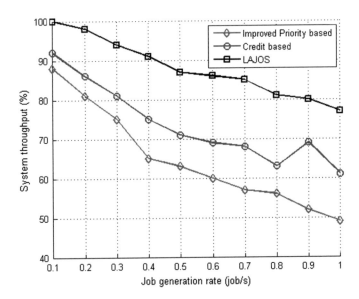

Fig. 9.18 The effect of the job generation rate on the system throughput

Figure 9.19 shows the effect of the number of users on the schedule length for each scheduling method. The results are carried out in a way that the job generation rate is considered as 0.5 job/s. The comparison results indicate the performance evaluation of the proposed approach compared to the other simulated methods.

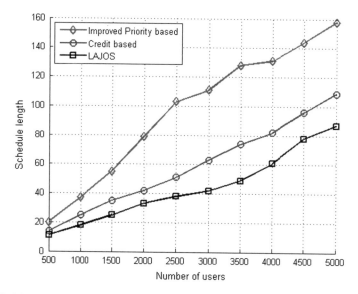

Fig. 9.19 The effect of the number of users on the schedule length

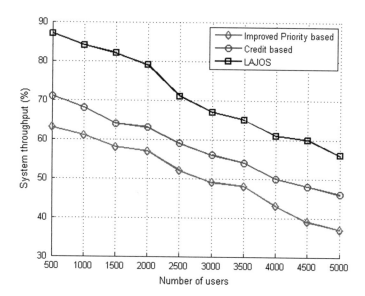

Fig. 9.20 The effect of the number of users on the system throughput

They represent that the schedule length obtained by the proposed approach is less than those obtained by the other methods. Note that the schedule lengths of all the simulated methods enhances when the number of users increases.

Figure 9.20 shows the impact rate of the number of users on the system throughput for each simulated scheduling method. It is carried out in a way that the job generation rate equals 0.5 job/s. The comparison results indicate the effect of the number of users on the system throughput. It describes that the number of users has a high effect on the system throughputs of all the simulated methods. These results demonstrate that the system throughput achieved by the LAJOS approach is more than those achieved by the Improved Priority based and Credit based methods under various simulation parameters.

9.7   Discussion

The concepts of the learning analytics techniques are defined based on various research fields such as educational data mining, recommender systems, academic analytics, and adaptive learning (Fernández-Delgado et al. 2014). Zhang et al. (2007) have implemented a CMS log analysis tool, called Moodog, to trace the students' online learning activities. This tool offers the interaction instructors between the students and the online course materials. Moreover, it enables students to evaluate their own progress compared to the other students of the class.

Picciano (2012) have focused on evolving the big data analytics in American higher education. This work describes that the online learning is a proper solution to ensure that the students can complete their degrees. Furthermore, the data-driven decision making can be applied to assist colleges in improving the students' progress using the big data and learning analytics fields. Chatti et al. (2012) have presented a reference model for learning analytics. They have used the four dimensions of learning procedures: data and environments (what?), stakeholders (who?), objectives (why?), and methods (how?). Fernández-Gallego et al. (2013) have presented a learning analytics framework for the 3D educational virtual worlds. This work describes the learning flows and, also, evaluates the efficiency of presented framework by using the process mining techniques. It discusses about the applications which reveals the learning flows of students in virtual worlds.

Job scheduling is one of the key requirements in cloud computing. A cloud server should use an appropriate mechanism to prioritise users' jobs, efficiently (Selvarani and Sadhasivam 2010; Dutta and Joshi 2011; Li et al. 2012; Maguluri and Srikant 2012; Kaur and Kinger 2014). The schedule length and system throughput are two important parameters on cloud servers to evaluate the job scheduling methods. Figures 9.15 and 9.16 illustrated simulation results of the proposed scheduling approach to execute users' jobs on scheduling queues. Furthermore, the comparison results of Figs. 9.17, 9.18, 9.19, and 9.20 demonstrated that the LAJOS approach is more efficient compared to the other scheduling methods.

There are various simulation parameters in cloud computing that effect on the efficiency of the presented methods. The effect of two simulation parameters (i.e., job generation rate and number of users) is independently investigated on the output parameters (i.e., schedule length and system throughput). The comparison results indicated that the schedule length obtained by the proposed approach is less than those obtained by the Improved Priority based and Credit based scheduling methods. Furthermore, the system throughput achieved by the LAJOS approach is more than those achieved by the other simulated scheduling methods.

The proposed LAJOS approach is compared to the Improved Priority based and Credit based scheduling methods in terms of schedule length and system throughput. The comparison results indicated that the schedule length obtained by the proposed approach could be decreased by nearly 45% less than that obtained by the Improved Priority based method and by nearly 30% less than that obtained by the Credit based method. Moreover, the system throughput achieved by the LAJOS approach could be increased by nearly 30% more than that achieved by the Improved Priority based method and by nearly 20% more than that achieved by the Credit based method.

9.8 Conclusions

Job scheduling is one of the main requirements in cloud computing. An efficient scheduling mechanism enhances the performance of cloud servers, considerably. This chapter, firstly, introduced the four learning analytics methods in educational institutions: the relationship between student social networks and sense of community, the learning analytics methodology for student profiling, the genetic and participation based student prediction model, and the learning analytics method based on academic achievements of students and interaction data. It describes the main characteristics and objectives of these methods according to their simulation or experimental results. Secondly, the chapter proposed a learning analytics approach for job scheduling on cloud servers, called LAJOS.

The LAJOS approach uses the three attributes "importance level", "waiting time" and "deadline time" to prioritise users' jobs on scheduling queues. It consists of the *deployment* and *service* phases in a way that the initial decision made in the *deployment* phase will be used in the *service* phase. In the *deployment* phase, the attribute which increases the performance of cloud servers are selected based on the schedule length and system throughput in the three period times $T1$, $T2$ and $T3$. In the *service* phase, users' jobs will be prioritised based on the selected attribute. Since the LAJOS approach applies the intelligent features of learning analytics methods and makes the appropriate decisions with regarding the hardware characteristics of cloud systems, it enhances the efficiency of cloud servers considerably. Because cloud computing and grid computing contain several common features, the proposed approach can be used in grid systems too.

The learning techniques (e.g., machine learning and deep learning) can eliminate the existing problems of computer networks. In the future work, some of these techniques will be used in the proposed job scheduling approach to enhance the performance of cloud servers more than now.

References

Abdullahi M, Ngadi MA, Abdulhamid SM (2016) Symbiotic organism search optimization based task scheduling in cloud computing environment. Future Gener Comp Sy 56:640–650

Antonopoulos N, Gillam L (2010) Cloud computing: principles, systems and applications. Springer Science & Business Media

Arnold KE, Pistilli MD (2012) Course signals at Purdue: using learning analytics to increase student success. Paper presented at the 2nd international conference on learning analytics and knowledge, ACM, New York, NY, USA, pp 267–270

Baker BM (2007) A conceptual framework for making knowledge actionable through capital formation. Dissertation, University of Maryland University College

Baker RS, Inventado PS (2014) Educational data mining and learning analytics. In: Larusson JA, White B (eds) Learning analytics. Springer, pp 61–75

Bandiera M, Bruno C (2006) Active/cooperative learning in schools. J Biol Educ 40:130–134

Barab SA, Barnett M, Yamagata-Lynch L et al (2002) Using activity theory to understand the systemic tensions characterizing a technology-rich introductory astronomy course. Mind Cult Act 9:76–107

Bele JL, Rugelj J (2010) Comparing learning results of web based and traditional learning students. Lect Notes Comput Sci, pp 375–380

Burkimsher A, Bate I, Indrusiak LS (2013) A survey of scheduling metrics and an improved ordering policy for list schedulers operating on workloads with dependencies and a wide variation in execution times. Future Gener Comp Sy 29:2009–2025

Chatti MA, Dyckhoff AL, Schroeder U et al (2012) A reference model for learning analytics. Int J Technol Enhanc Learn 4:318–331

Chung W-C, Hsu C-J, Lai K-C et al (2013) Chung. Direction-aware resource discovery in large-scale distributed computing environments. J Supercomput 66:229–248

Dawson S (2008) A study of the relationship between student social networks and sense of community. Educ Technol Soc 11:224–238

Dron J, Anderson T (2009) On the design of collective applications. Paper presented at the IEEE international conference on computational science and engineering (CSE'09), Vancouver, BC, 29–31 Aug 2009, pp 368–374

Dutta D, Joshi RC (2011) A genetic: algorithm approach to cost-based multi-QoS job scheduling in cloud computing environment. Paper presented at the international conference & workshop on emerging trends in technology (ICWET'11), New York, NY, USA, 2011, pp 422–427

Engeström Y (1999) Activity theory and individual and social transformation. In: Engeström Y, Miettinen R, Punamäki R-L (eds) Perspectives on activity theory. Cambridge University Press, pp 19–38

Ferguson R (2012) Learning analytics: drivers, developments and challenges. Int J Techn Enhanc Learn 4:304–317

Fernández-Delgado M, Mucientes M, Vázquez-Barreiros B et al (2014) Learning analytics for the prediction of the educational objectives achievement. Paper presented at the IEEE frontiers in education conference (FIE), Madrid, Spain, 22–25 Oct 2014, pp 1–4

Fernández-Gallego B, Lama M, Vidal JC et al (2013) Learning analytics framework for educational virtual worlds. Procedia Comput Sci 25:443–447

Ferreira SA, Andrade A (2014) Academic analytics: mapping the genome of the University. IEEE Rev Iberoam Technol Aprendizaje 9:98–105

Ghanbari S, Othman M (2012) A priority based job scheduling algorithm in cloud computing. Procedia Eng 50:778–785

Gharajeh MS (2015) The significant concepts of cloud computing: technology, architecture, applications, and security. CreateSpace Independent Publishing Platform

Günther CW, Van Der Aalst WMP (2007) Fuzzy mining–adaptive process simplification based on multi-perspective metrics. In: Alonso G, Dadam P, Rosemann M (eds) Business process management. Springer, pp 328–343

Haythornthwaite C, De Laat M, Dawson S et al (2013) Introduction to learning analytics and networked learning minitrack. Paper presented at the IEEE 46th Hawaii international conference on system sciences (HICSS), Wailea, Maui, HI, 7–10 Jan 2013, p 3077

Hrastinski S (2009) A theory of online learning as online participation. Comput Educ 52:78–82

Jena RK (2015) Multi objective task scheduling in cloud environment using nested PSO framework. Procedia Comput Sci 57:1219–1227

Kalra M, Singh S (2015) A review of metaheuristic scheduling techniques in cloud computing. Egypt Inform J 16:275–295

Kao Y-C, Chen Y-S (2016) Data-locality-aware mapreduce real-time scheduling framework. J Syst Softw 112:65–77

Kaur R, Kinger S (2014) Analysis of job scheduling algorithms in cloud computing. Int J Comp Trends Technol 9:379–386

Kavvadia E, Sagiadinos S, Oikonomou K et al (2015) Elastic virtual machine placement in cloud computing network environments. Comput Netw 93:435–447

Leont'ev AN (1974) The problem of activity in psychology. Sov Psychol 13:4–33

Li L (2009) An optimistic differentiated service job scheduling system for cloud computing service users and providers. Paper presented at the IEEE third international conference on multimedia and ubiquitous engineering (MUE'09), Qingdao, 4–6 June 2009, pp 295–299

Li J, Qiu M, Ming Z et al (2012) Online optimization for scheduling preemptable tasks on IaaS cloud systems. J Parallel Distr Com 72:666–677

Liu X, Zha Y, Yin Q et al (2015) Scheduling parallel jobs with tentative runs and consolidation in the cloud. J Syst Softw 104:141–151

Lotsari E, Verykios VS, Panagiotakopoulos C et al (2014) A learning analytics methodology for student profiling. In: Likas A, Blekas K, Kalles D (eds) Artificial intelligence: methods and applications. Springer, pp 300–312

Ma J, Han X, Yang J et al (2015) Examining the necessary condition for engagement in an online learning environment based on learning analytics approach: The role of the instructor. Internet High Educ 24:26–34

Maguluri ST, Srikant R, Ying L (2012) Stochastic models of load balancing and scheduling in cloud computing clusters. Paper presented at the IEEE INFOCOM, Orlando, Florida, USA, 25–30 Mar 2012, pp 702–710

McCabe TJ (1976) A complexity measure. IEEE Trans Softw Eng 4:308–320

Mohialdeen IA (2013) Comparative study of scheduling algorithms in cloud computing environment. J Comput Sci Technol 9:252–263

Norris D, Baer L, Leonard J et al (2008a) Action analytics: measuring and improving performance that matters in higher education. Educause Rev 43:42–67

Norris D, Baer L, Leonard J et al (2008b) Framing action analytics and putting them to work. Educause Rev 43:1–10

Ozturk HT, Deryakulu D, Ozcinar H et al (2014) Advancing learning analytics in online learning environments through the method of sequential analysis. Paper presented at the IEEE international conference on multimedia computing and systems (ICMCS), Marrakech, 14–16 Apr 2014, pp 512–516

Park EL, Choi BK (2014) Transformation of classroom spaces: traditional versus active learning classroom in colleges. High Educ 68:749–771

Patel SJ, Bhoi UR (2014) Improved priority based job scheduling algorithm in cloud computing using iterative method. Paper presented at the IEEE fourth international conference on advances in computing and communications (ICACC), Cochin, 27–29 Aug 2014, pp 199–202

Picciano AG (2012) The evolution of big data and learning analytics in american higher education. J Learn Asynchronous Netw 16:9–20

Ratten V (2016) Continuance use intention of cloud computing: innovativeness and creativity perspectives. J Bus Res 69:1737–1740

Romero C, Ventura S (2007) Educational data mining: a survey from 1995 to 2005. Expert Syst Appl 33:135–146

Rovai AP (2002a) Building sense of community at a distance. Int Rev Res Open Distr Learn 3

Rovai AP (2002b) Development of an instrument to measure classroom community. Internet High Educ 5:197–211

Sánchez M, Aguilar J, Cordero J et al (2016) Cloud computing in smart educational environments: application in learning analytics as service. In: Rocha A, Correia AM, Adeli H et al (eds) New advances in information systems and technologies. Springer, pp 993–1002

Scheffel M, Niemann K, Leony D et al (2012) Key action extraction for learning analytics. In: Ravenscroft A, Lindstaedt S, Kloos CD et al (eds) 21st century learning for 21st century skills. Springer, pp 320–333

Selvarani S, Sadhasivam GS (2010) Improved cost-based algorithm for task scheduling in cloud computing. Paper presented at the IEEE international conference on computational intelligence and computing research (ICCIC), Tamil Nadu, India, 28–29 Dec 2010, pp 1–5

Thomas A, Krishnalal G, Raj VPJ (2015) Credit based scheduling algorithm in cloud computing environment. Procedia Comput Sci 46:913–920

Thrun S, Pratt L (2012) Learning to learn. Springer Science & Business Media

Tinto V (2006) Research and practice of student retention: what next? J Coll Stud Ret Res Theory Pract 8:1–19

Vahdat M, Oneto L, Anguita D et al (2015) A learning analytics approach to correlate the academic achievements of students with interaction data from an educational simulator. In: Conole G, Klobučar T, Rensing C et al (eds) Design for teaching and learning in a networked world. Springer, pp 352–366

Vonderwell S (2003) An examination of asynchronous communication experiences and perspectives of students in an online course: A case study. Internet High Educ 6:77–90

Wang B, Qi Z, Ma R et al (2015) A survey on data center networking for cloud computing. Comput Netw 91:528–547

Xing W, Guo R, Petakovic E et al (2015) Participation-based student final performance prediction model through interpretable Genetic Programming: Integrating learning analytics, educational data mining and theory. Comput Hum Behav 47:168–181

Yang C-N, Lin BMT, Hwang FJ et al (2016) Acquisition planning and scheduling of computing resources. Comput Oper Res 76:167–182

Zhang H, Almeroth K, Knight A et al (2007) Moodog: Tracking students' online learning activities. Paper presented at world conference on educational multimedia, hypermedia & telecommunications (ED-MEDIA), 25 June 2007, pp 4415–4422

Author Index

Printed in the United States
By Bookmasters